T0091773

HOW THE BODY WORKS

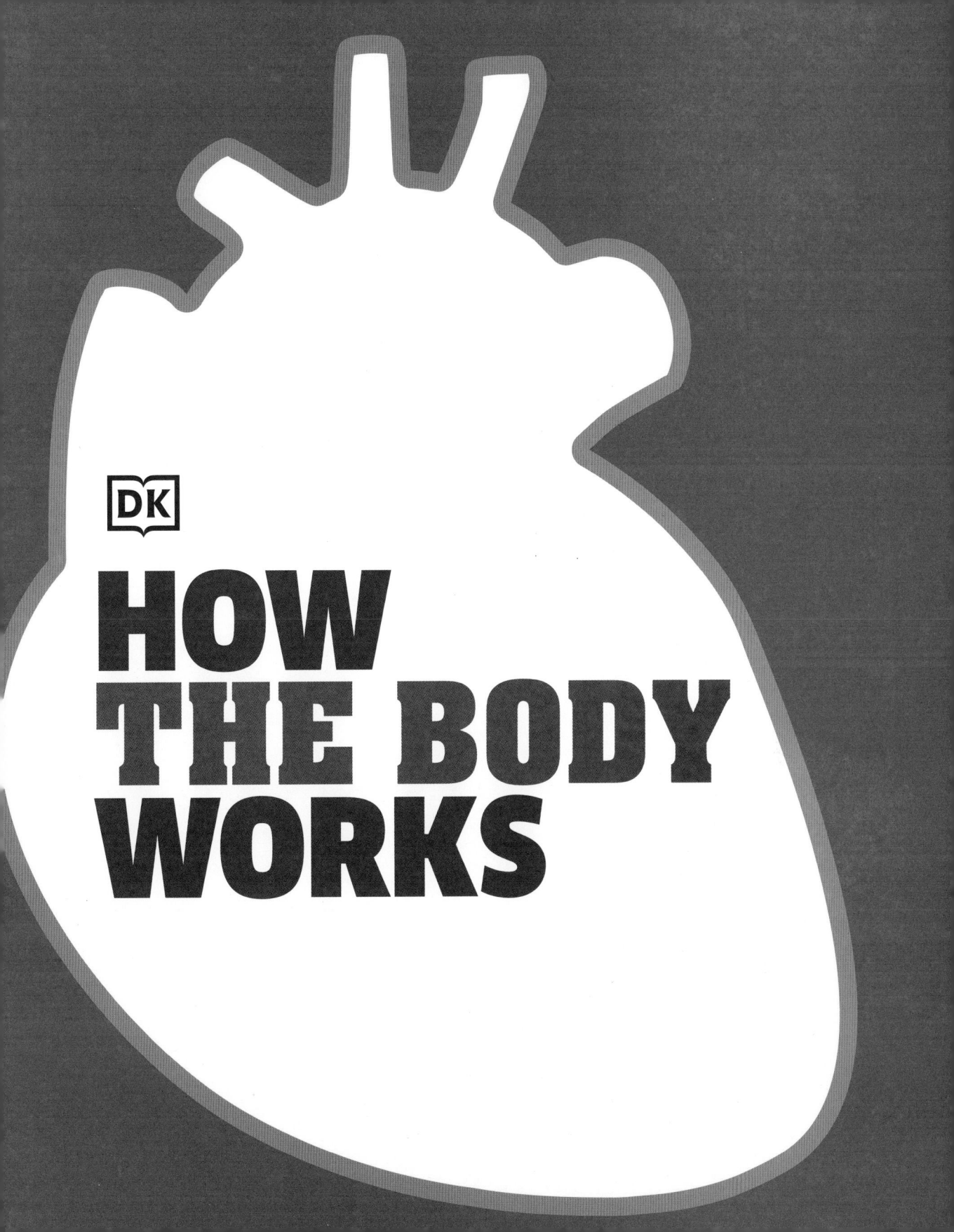
DK
HOW
THE BODY
WORKS

Penguin Random House

Editorial consultant
Dr Sarah Brewer

Project Art Editor
Francis Wong

Designers
Paul Drislane, Charlotte Johnson, Shahid Mahmood

Illustrators
Mark Clifton, Phil Gamble, Mike Garland, Mik Gates, Alex Lloyd, Mark Walker

Managing Art Editor
Michael Duffy

Jacket Designer
Mark Cavanagh

Jacket Editor
Claire Gell

Jacket Design Development Manager
Sophia MTT

Art Director
Karen Self

Contributors
Ginny Smith, Nicola Temple

Senior Editor
Rob Houston

Editors
Wendy Horobin, Andy Szudek, Miezan van Zyl

Assistant Editor
Francesco Piscitelli

US Editor
Jill Hamilton

Managing Editor
Angeles Gavira Guerrero

Producer, Pre-production
Nikoleta Parasaki

Producer
Nancy-Jane Maun

Publisher
Liz Wheeler

Publishing Director
Jonathan Metcalf

First American Edition, 2016
Published in the United States by DK Publishing
1745 Broadway, 20th Floor, New York, NY 10019

Copyright © 2016 Dorling Kindersley Limited
DK, a Division of Penguin Random House LLC
23 24 25 26 13 12 11 10
058–274815–May/2016

All rights reserved. Without limiting the rights under the copyright reserved above, no part of this publication may be reproduced, stored in or introduced into a retrieval system, or transmitted, in any form, or by any means (electronic, mechanical, photocopying, recording, or otherwise), without the prior written permission of the copyright owner.
Published in Great Britain by Dorling Kindersley Limited

A catalog record for this book is available from the Library of Congress.
ISBN 978-1-4654-2993-3

DK books are available at special discounts when purchased in bulk for sales promotions, premiums, fund-raising, or educational use.
For details, contact: DK Publishing Special Markets, 1745 Broadway, 20th Floor, New York, NY 10019
SpecialSales@dk.com

Printed and bound in China

For the curious
www.dk.com

MIX
Paper | Supporting responsible forestry
FSC™ C018179
FSC www.fsc.org

This book was made with Forest Stewardship Council™ certified paper – one small step in DK's commitment to a sustainable future.
For more information go to www.dk.com/our-green-pledge

CONTENTS

UNDER THE MICROSCOPE

Who's in charge? **10**
Organ to cell **12**
How cells work **14**
What is DNA? **16**
How cells multiply **18**
How genes work **20**
How genes make different cells **22**
Stem cells **24**
When DNA goes wrong **26**

HOLDING IT TOGETHER

Skin deep 30
Outer defenses 32
The extremities 34
Pillars of support 36
Growing bones 38
Flexibility 40
Biting and chewing 42
The grinder 44
Skin damage 46
Breaking and mending 48
Wearing thin 50

ON THE MOVE

Pulling power 54
How do muscles pull? 56
Working, stretching, pulling, braking 58
Sensory input, action output 60
The control center 62
Communication hub 64
Sparking into life 66
Act or relax? 68
Bumps, sprains, and tears 70

SENSITIVE TYPES

Feeling the pressure 74
How do you feel? 76
Pain's pathway 78
How the eye works 80
Forming an image 82
Vision in the brain 84
Vision problems 86
How the ear works 88
How the brain hears 90
Balancing act 92
Hearing problems 94
Catching a scent 96
On the tip of the tongue 98
Body position sense 100
Integrated senses 102
Using your voice 104
Reading faces 106
What you don't say 108

THE HEART OF THE MATTER

Filling your lungs 112

From air to blood 114

Why do we breathe? 116

Coughs and sneezes 118

The many tasks of our blood 120

How the heart beats 122

How blood travels 124

Broken blood vessels 126

Heart problems 128

Exercising and its limits 130

Fitter and stronger 132

Maximizing your fitness 134

IN AND OUT

Feeding the body 138

How does eating work? 140

A mouth to feed 142

Gut reaction 144

Up, down, and out 146

Bacterial breakdown 148

Cleaning the blood 150

Water balance 152

How the liver works 154

What the liver does 156

Energy balance 158

The sugar trap 160

Feast of fast? 162

Digestive problems 164

FIT AND HEALTHY

Body battleground 168

Friend or foe? 170

Germs are us 172

Damage limitation 174

Infectious diseases 176

Looking for trouble 178

Assassination squad 180

Cold and flu 182

Vaccine action 184

Immune problems 186

CHEMICAL BALANCE

Hormone factories 190

How hormones work 192

Inner balance 194

Hormonal changes 196

Daily rhythms 198

Diabetes 200

THE CIRCLE OF LIFE

Sexual reproduction 204

Monthly cycle 206

Tiny beginnings 208

The generation game 210

Growing life 212

Mother's new body 214

The miracle of birth 216

Primed for life 218

Growing up 220

Hormonal teenagers 222

Getting older 224

The end of life 226

MIND MATTERS

Learning skills 230

Making memories 232

Falling asleep 234

Entering your dreams 236

All emotional 238

Fight or flight 240

Emotional problems 242

Feeling attraction 244

Extraordinary minds 246

INDEX 248

ACKNOWLEDGMENTS 256

UNDER THE MICROSCOPE

Who's in charge?

To perform any task, the body's many parts work together in groups of organs and tissues called systems. Each system is in charge of a function, such as breathing or digestion. Most of the time, the brain and spinal cord are the main coordinators, but the body's systems are always communicating and giving each other instructions.

ARE THERE ANY BODY SYSTEMS WE CAN LIVE WITHOUT?

All our body systems are vital. Unlike some organs—such as the appendix—if an entire system fails it usually results in death.

A matter of organization

Systems are communities of body parts with a single function. However, some body parts have more than one job. The pancreas, for example, is part of the digestive system because it pipes digestive juices into the gut. It also acts as part of the endocrine system, releasing hormones into the bloodstream.

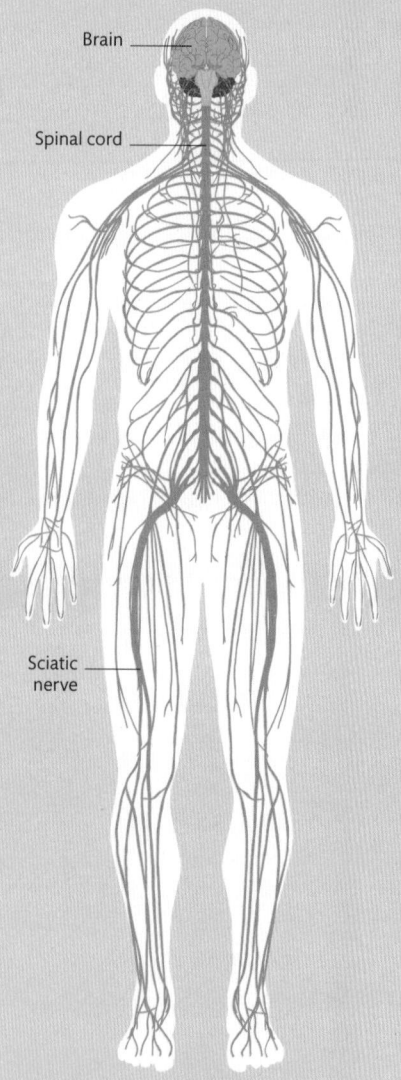

Central nervous system
The brain and spinal cord process and act upon information received from all over the body through an extensive network of nerves.

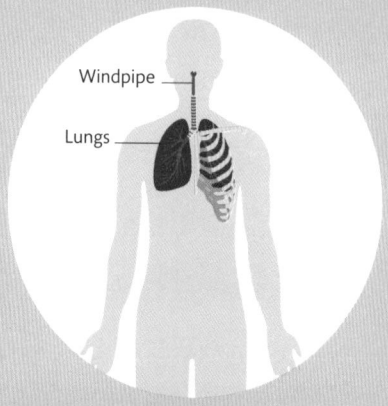

Respiratory system
The lungs bring air into contact with blood vessels so that oxygen and carbon dioxide can be exchanged.

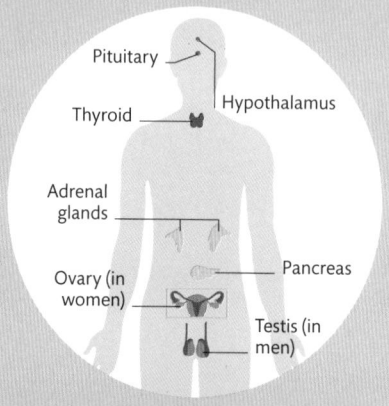

Endocrine system
This system of glands secretes hormones, which are the body's chemical messengers, sending information to other body systems.

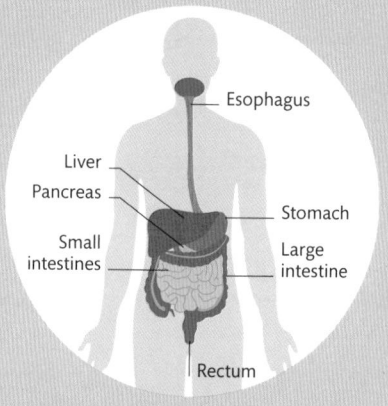

Digestive system
The stomach and intestines are the major parts of this system, which turns food into nutrients needed by the body.

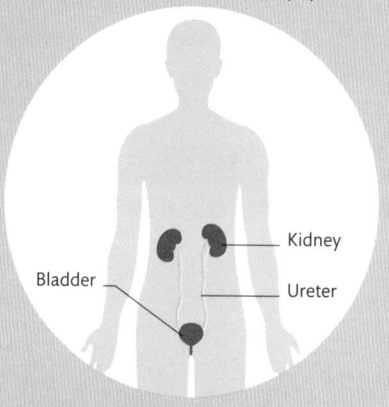

Urinary system
The kidneys filter blood to remove unwanted substances, which are stored in the bladder and expelled as urine.

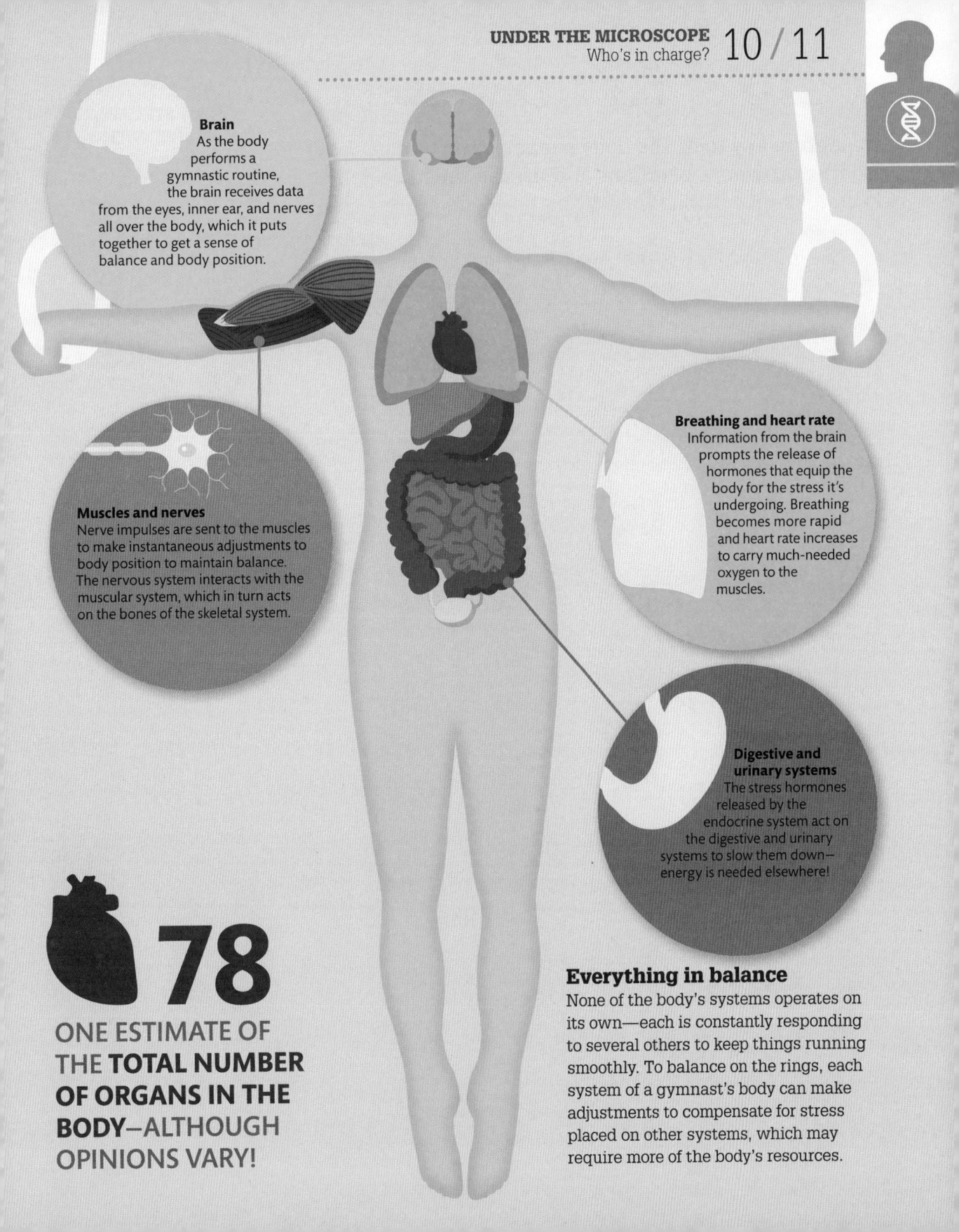

Brain
As the body performs a gymnastic routine, the brain receives data from the eyes, inner ear, and nerves all over the body, which it puts together to get a sense of balance and body position.

Muscles and nerves
Nerve impulses are sent to the muscles to make instantaneous adjustments to body position to maintain balance. The nervous system interacts with the muscular system, which in turn acts on the bones of the skeletal system.

Breathing and heart rate
Information from the brain prompts the release of hormones that equip the body for the stress it's undergoing. Breathing becomes more rapid and heart rate increases to carry much-needed oxygen to the muscles.

Digestive and urinary systems
The stress hormones released by the endocrine system act on the digestive and urinary systems to slow them down—energy is needed elsewhere!

78

ONE ESTIMATE OF THE **TOTAL NUMBER OF ORGANS IN THE BODY**—ALTHOUGH OPINIONS VARY!

Everything in balance

None of the body's systems operates on its own—each is constantly responding to several others to keep things running smoothly. To balance on the rings, each system of a gymnast's body can make adjustments to compensate for stress placed on other systems, which may require more of the body's resources.

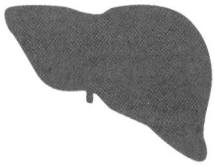

ONE IN 10,000 PEOPLE HAS ALL THE INTERNAL ORGANS **ON THE WRONG SIDE OF THE BODY**

Organs

The organs within the body are typically self-contained and perform a specific function. The tissues that make up that organ help it function in a particular way. The stomach, for example, is largely made of muscle tissue that can expand and contract to accommodate the intake of food.

Stomach structure
Muscle is the main tissue of the stomach, but it is also lined with glandular tissue, which secretes digestive juices, and epithelial tissue, which forms a protective barrier on both the inner and outer surfaces.

Organ to cell

Each organ in the body is distinct and recognizable to the naked eye. Cut through an organ, however, and layers of different tissues are revealed. Within each tissue are different types of cells. They all work together to carry out the functions of the organ.

OESOPHAGUS

Stomach has three layers of smooth muscle

STOMACH

Entrance to intestines

Inner wall is lined with cells that secrete mucus or acid

Outer layer is covered with epithelial cells

WHICH IS THE LARGEST ORGAN?

The liver is the largest of the internal organs but the skin is actually the biggest organ of the body. It weighs roughly 6 lb (2.7 kg).

Tissues and cells

Tissues are made up of a group of connected cells. Some tissues come in different types, such as the smooth muscle that forms the walls of the stomach and skeletal muscle, which is attached to the bones and makes them move. As well as cells, the tissue might contain other structures, such as collagen fibers in connective tissue. A cell is a self-contained living unit—the most basic structure of all living organisms.

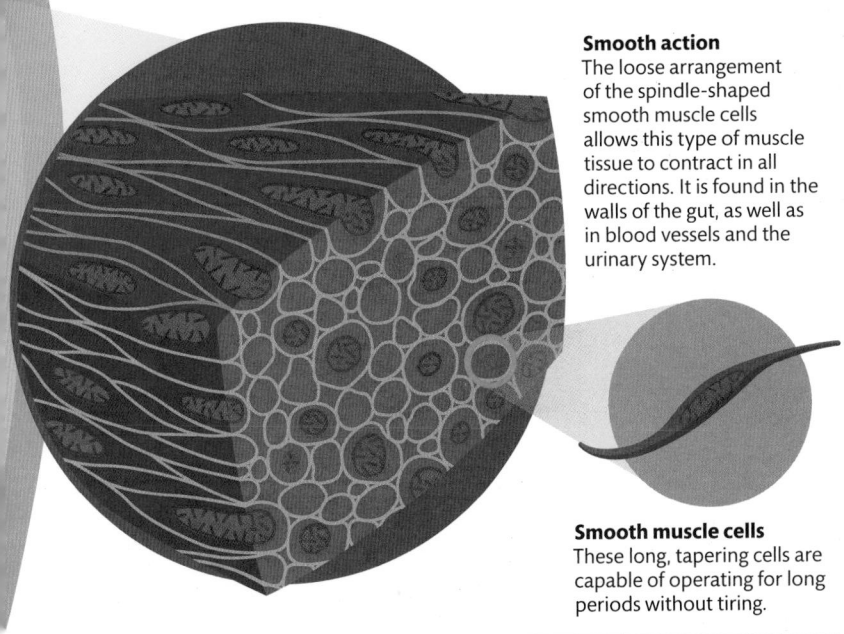

Smooth action
The loose arrangement of the spindle-shaped smooth muscle cells allows this type of muscle tissue to contract in all directions. It is found in the walls of the gut, as well as in blood vessels and the urinary system.

Smooth muscle cells
These long, tapering cells are capable of operating for long periods without tiring.

Tissue types

There are four basic types of tissue found in the human body. These are subdivided into different subtypes, for example, blood and bone are both connective tissues. Each type has different properties—such as strength, flexibility, or movement—that makes it suited to a specific task.

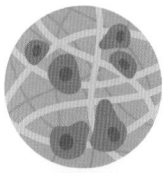

Connective tissue
Connects, supports, binds, and separates other tissues and organs.

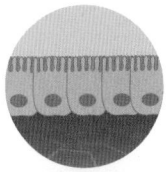

Epithelial tissue
Closely packed cells in one or more layers that form barriers.

Muscle tissue
Long, thin cells that relax and contract to create movement.

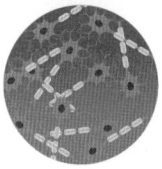

Nervous tissue
Cells that work together to transmit electrical impulses.

Types of cells

There are around 200 different types of cells in the human body. They look very different under a microscope, but most have common features, such as a nucleus, cell membrane, and organelles.

Red blood cells
Lack a nucleus and carry as much oxygen as possible.

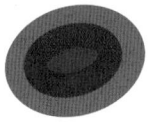

Nerve cells
Carry electrical signals between the brain and all parts of the body.

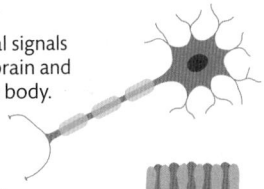

Epithelial cells
Line the surfaces and cavities of the body to form a tight barrier.

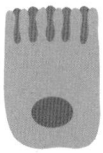

Adipose cells
Store molecules of fat that help insulate the body and can be turned into energy.

Skeletal muscle cells
Arranged into fibrous bundles that contract to move bones.

Reproductive cells
The female egg and male sperm combine to form a new embryo.

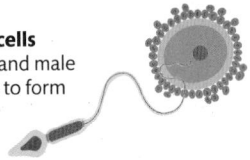

Photoreceptor cells
Line the back of the eye and respond to light falling on them.

Hair cells
Pick up sound vibrations being transmitted through the fluid of the inner ear.

How cells work

Your body is made up of approximately 10 trillion cells, and each one is a self-contained, living unit. Each cell uses energy, multiplies, eliminates waste, and communicates. Cells are the basic units of all living things.

Cell function

Most cells have a nucleus—a structure in their center that contains genetic data, or DNA. They rely on this data to build various molecules that are essential to life. All of the resources they need to do this are contained within the cell. Structures called organelles carry out specialized functions, similar to the organs of the body. Organelles are held in the cytoplasm, the space between the nucleus and the cell membrane. Molecules are brought into the cell and others are shipped out, just like in an efficient factory.

1 Receiving instructions
Everything that happens in a cell is controlled by instructions in the nucleus. These instructions are exported on long molecules called messenger ribonucleic acid (mRNA)—these molecules travel out of the nucleus and into the cytoplasm.

2 Manufacture
The mRNA travels to an organelle attached to the nucleus called the rough endoplasmic reticulum. There, it attaches to ribosomes that stud the organelle, and the instructions are made into a chain of amino acids that becomes a protein molecule.

3 Packaging
The proteins travel in vesicles—little cellular bubbles—that float through the cytoplasm to the Golgi body. This organelle acts much like the mail room of the cell—packaging the proteins and putting labels on them, which determine where they are sent next.

4 Shipping
The Golgi body places the proteins into different types of vesicles depending on their labeled destination. These vesicles bud off, and those destined for outside the cell fuse with the cell membrane and release the proteins outside of the cell.

Inside a cell
Numerous organelles comprise the internal structure of cells—the specific types vary from cell to cell.

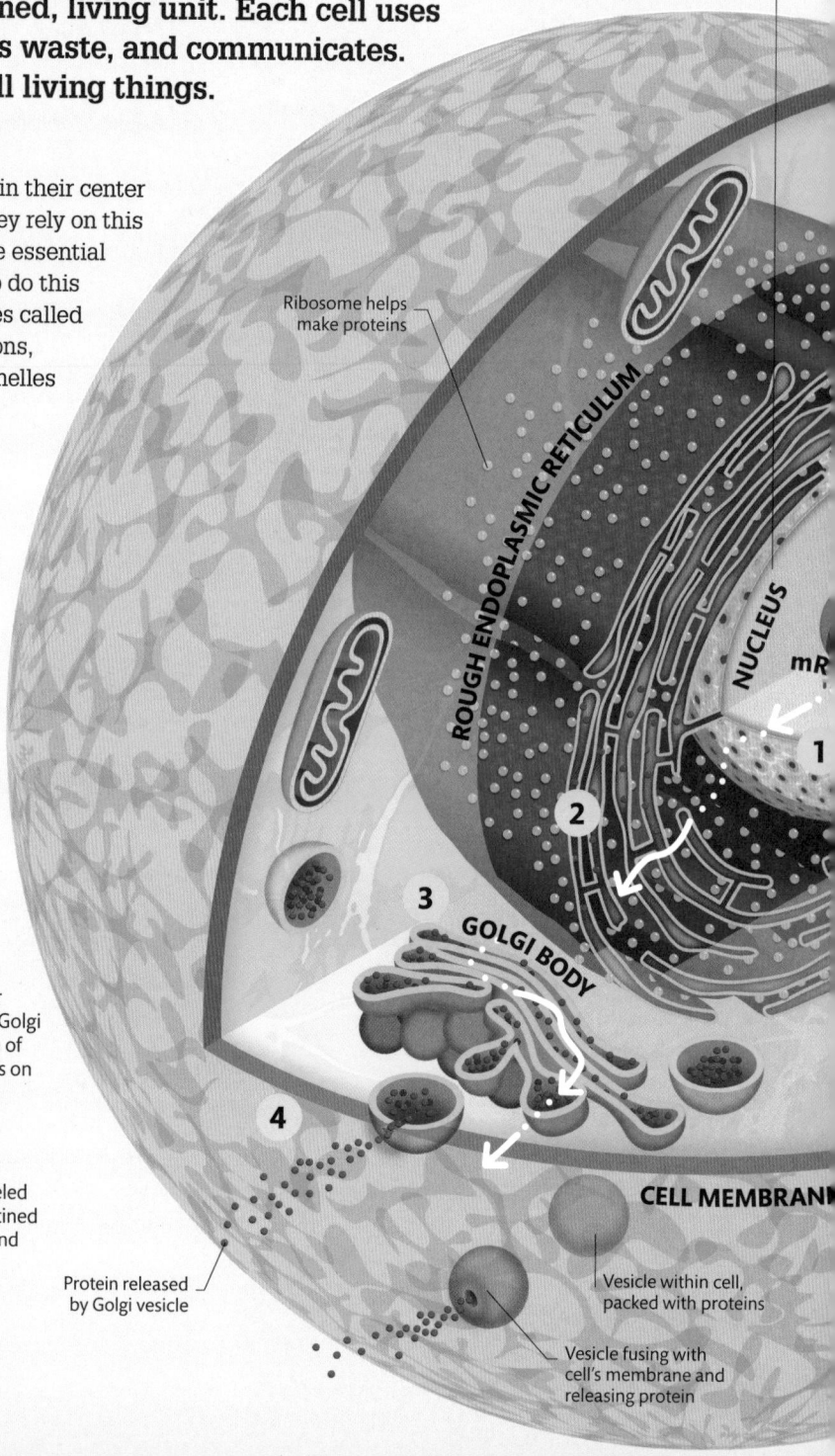

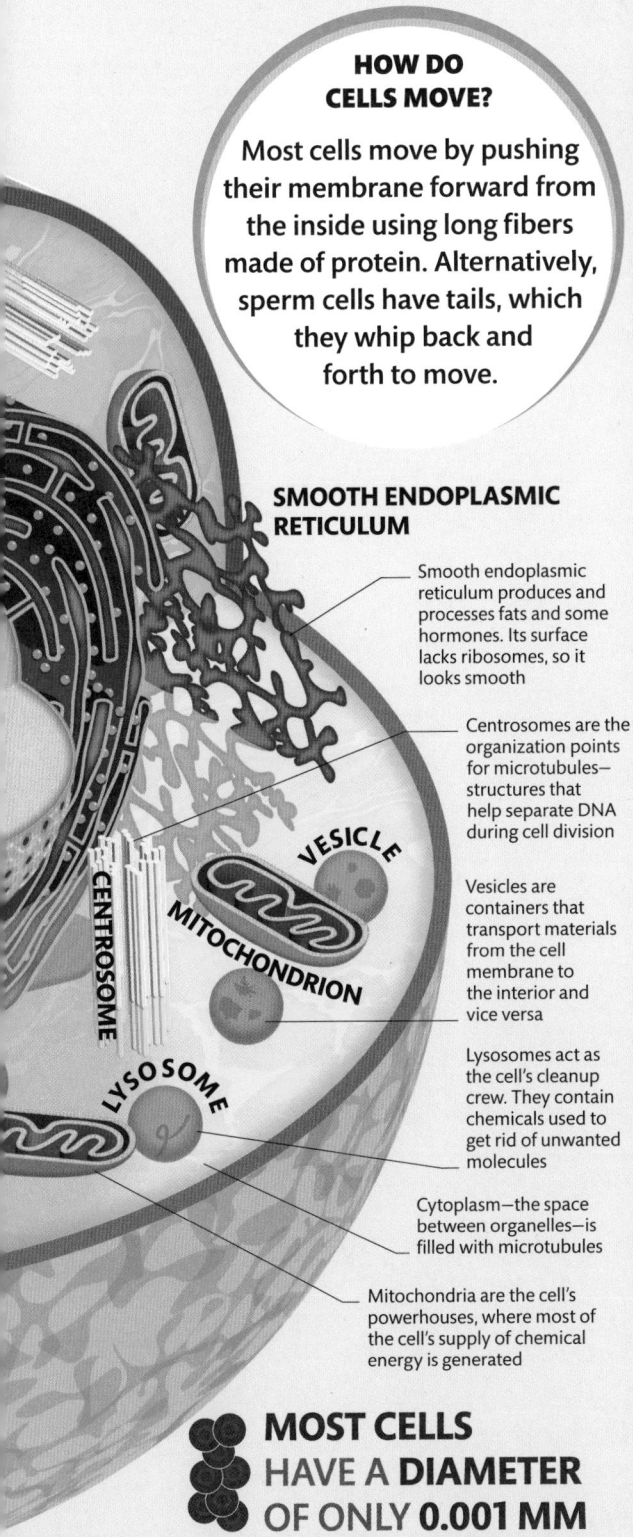

HOW DO CELLS MOVE?

Most cells move by pushing their membrane forward from the inside using long fibers made of protein. Alternatively, sperm cells have tails, which they whip back and forth to move.

MOST CELLS HAVE A DIAMETER OF ONLY 0.001 MM

Cell death

When cells have reached the natural end of their life cycle they undergo apoptosis—a deliberate series of events that causes the cell to dismantle itself, shrink, and fragment. Cells can also die prematurely due to infections or toxins. This causes necrosis, a process in which the cell's internal structure detaches from its membrane, causing the membrane to burst and the cell to die.

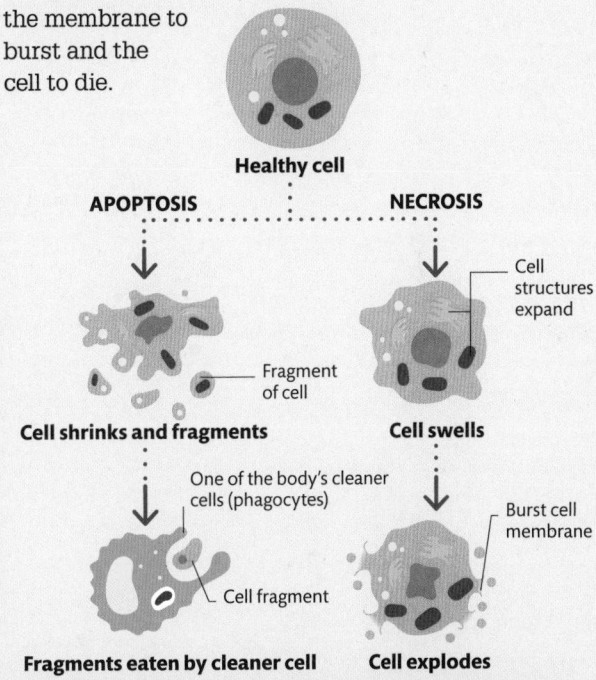

CELL COMMUNICATION

Cells communicate with one another and respond to their environment using signaling molecules produced by distant cells, nearby cells, or even the same cell. Signaling molecules bind with receptors, which are themselves molecules, on the cell's membrane. The binding event triggers changes in the cell, such as activating a gene.

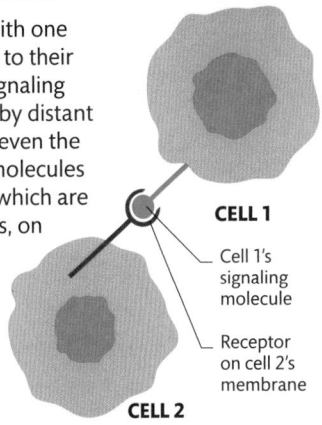

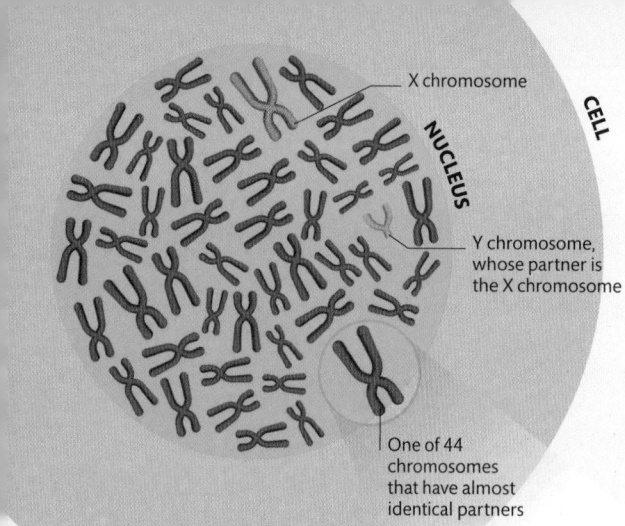

Control centre
DNA is stored in the nucleus of every cell (except for red blood cells, which lose their DNA as they mature). In each cell nucleus, there are 6 ft (2 m) of DNA tightly coiled into 23 pairs of chromosomes—making a total of 46. We inherit one chromosome of each pair from our father, the other from our mother.

Boy or girl?
While 22 of our chromosome pairs are duplicates—with only a slightly different version of each gene on each chromosome—the final, 23rd pairing is different. In most people, it determines our sex. Females usually have two X chromosomes, while males have an X and a Y. Only a few of the X chromosome genes are repeated on the shorter Y chromosome, which mostly carries genes that produce masculine features.

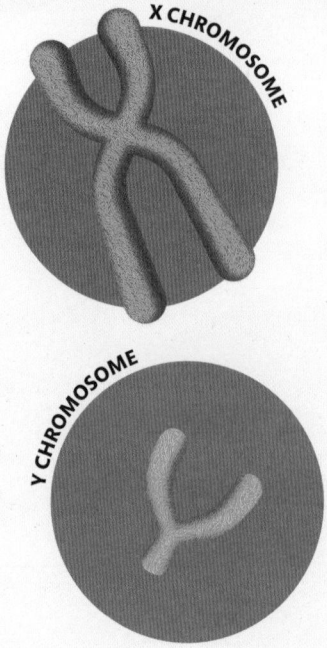

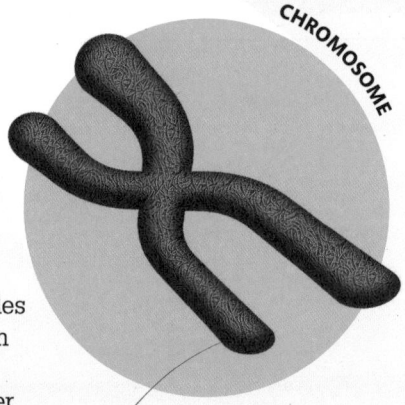

Human library

DNA is a long molecule that provides all the information necessary for an organism to develop, survive, and reproduce. It is like a twisted ladder with rungs made of a pair of chemical bases. These bases form long sequences called genes that are coded instructions for building proteins. When a cell needs to duplicate its DNA or make a new protein, the two halves of the ladder unzip so that a copy of the gene can be made. Humans have more than 3 billion bases in their DNA and nearly 20,000 genes.

Body builders
The genes that build our bodies may range from a few hundred bases to more than 2 million bases in length—longer than the small section shown here. Each gene produces a single protein. These proteins are the building blocks of the body, forming cells, tissues, and organs. They also regulate all the body's processes.

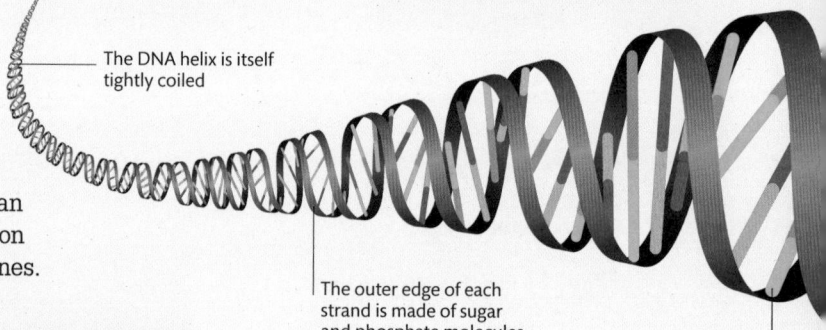

What is DNA?

DNA (deoxyribonucleic acid) is a chain molecule that exists in nearly all living things. The chain is made up of a sequence of molecular components, known as bases. Incredibly, the sequence acts as coded instructions for making an entire living organism. We inherit our DNA from our parents.

The colored bars show the four bases—adenine, thymine, guanine, and cytosine—which are arranged in a particular, meaningful sequence

Express yourself

The majority of genes are the same in everybody because they code for molecules that are essential for life. However, around 1 percent have slight variations—known as alleles—that give us our unique physical characteristics. While many of these are harmless traits, such as hair or eye color, they may also result in more problematic conditions, such as hemophilia or cystic fibrosis. Because alleles come in pairs, one may override the effect of the other so that the trait remains hidden.

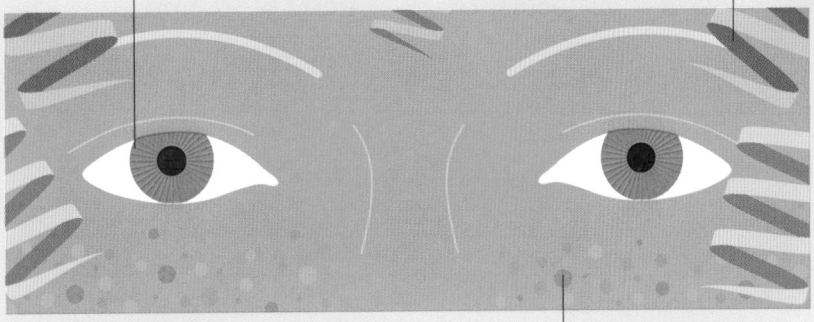

Unpredictable outcomes
Many of our physical features are under the control of more than one gene. This may result in unexpected combinations.

Unraveling DNA

Chromosomes help package DNA to fit into the nucleus. The DNA is wrapped around spool-like proteins that run through the center of each chromosome. The helix is made of two strands of sugar phosphate linked together by a pair of bases. The bases always form the same pairs, but the sequences of bases along the strand are specific to the proteins they will eventually produce.

DO HUMANS HAVE THE MOST GENES?

Humans have a relatively low number of genes. We have more than a chicken (16,000) but fewer than an onion (100,000) or an amoeba (200,000). This is because we lose unwanted genes faster from our DNA than they do.

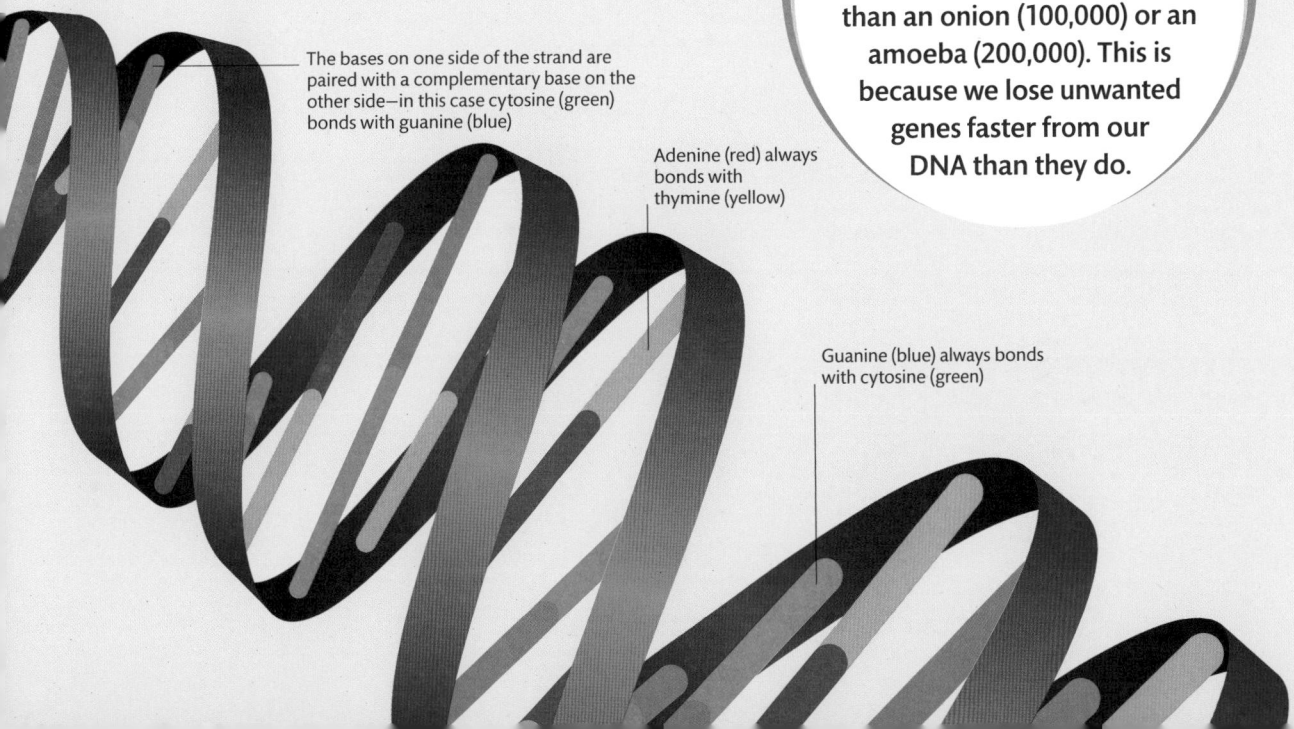

How cells multiply

We all start life as a single cell, so to develop specific tissues and organs and enable our body to grow, our cells need to multiply. Even as adults, cells need to be replaced because they get damaged or complete their life cycle. There are two processes by which this happens—mitosis and meiosis.

OUT OF CONTROL

Many cancers occur when a mutant cell begins to multiply rapidly. This is because the cell can override the usual checks during mitosis, enabling it to replicate itself more quickly than surrounding cells and take up more of the available oxygen and nutrients.

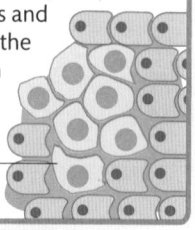

Wear and tear
Mitosis happens whenever new cells are needed. Some cells, such as neurons, are rarely replaced, but others, such as those lining the gut or tastebuds, undergo mitosis every few days.

1 Resting
The parent cell gets ready for mitosis by checking its DNA for damage and making any repairs needed.

Cell
Nucleus
Four of cell's 46 chromosomes

2 Preparation
Each chromosome in the parent cell makes an exact copy of itself prior to entering mitosis. The copies join at a region called the centromere.

Centromere

3 Lining up
Each of the doubled chromosomes attaches to special fibers, which help line them up in the middle of the cell.

Fiber

4 Separation
The chromosomes split at their attachment point (centromere) and each half is pulled to an opposite end of the cell.

Centromere

5 Splitting
A nuclear membrane forms around each group of chromosomes and the cell membrane starts to pull apart to form two cells.

6 Offspring
Two daughter cells are formed, each containing a nucleus with an exact copy of the DNA from the parent cell.

Mitosis

Every cell enters a phase in its life cycle called mitosis. During mitosis, the cell's DNA is duplicated and then divides equally to form two identical nuclei, each containing the exact same DNA as the original parent cell. The cell then divides up its cytoplasm and organelles to form two daughter cells, each containing a single nucleus. There are a number of checkpoints throughout the DNA replication and division processes to repair any damaged DNA, which could lead to permanent mutations and disease.

1 Preparation
Each of the cell's chromosomes duplicates and the copies join together at the centromere.

Cell
Nucleus
Chromosome
Centromere

2 Pairing and crossover
Chromosomes with similar lengths and centromere locations line up with one another and undergo gene swapping.

3 First separation
The chromosomes line up and, just like in mitosis, are pulled to opposite ends of the cell along special fibers.

Fiber

4 Two offspring
The cell divides, and two cells containing half the chromosomes are formed. Each is genetically distinct from each other and from the parent cell.

5 Second separation
The chromosomes line up along the midline of each cell and are pulled apart so that each new cell receives half of the chomosome pair.

6 Four offspring
Four cells are produced, each with half the number of chromosomes of the original parent cell, and each genetically unique.

Gene swapping
Meiosis features a unique process that shuffles the DNA passing into the daughter cells. DNA is exchanged between the chromosomes, which creates new combinations of DNA. Some new combinations may be beneficial.

Meiosis

Egg and sperm cells are produced through a specialized type of cell division known as meiosis. The aim is to reduce the number of chromosomes from the parent cell by half so that when an egg and sperm fuse during fertilization, the new cell has a full complement of 46 chromosomes. Meiosis produces four daughter cells that are each genetically different from the parent cell. It is the process of gene swapping during meiosis that introduces the genetic diversity that helps make each of us unique individuals.

DOWN SYNDROME

Sometimes mistakes can happen during meiosis. Down syndrome is caused by an extra copy of chromosome 21 in some or all of the body's cells. This usually happens when the chromosome doesn't separate properly during the meiosis of an egg or sperm cell—a condition known as trisomy 21. Having an extra chromosome means that some genes are overexpressed by the cell, which can cause problems in how it functions.

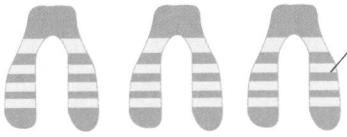

The extra 310 genes can result in overproduction of some proteins.

THREE COPIES OF CHROMOSOME 21

How genes work

If our DNA is the body's recipe book, then a gene within that DNA is equivalent to a single recipe in the book; it is the instructions for building a single chemical or protein. It's estimated that humans have around 20,000 genes that code for different proteins.

Genetic blueprint

To translate a gene into a protein, the DNA is first copied (transcribed) in the nucleus of a cell by enzymes, forming a strand of messenger RNA (mRNA). The cell will only copy those genes that it needs, not the entire DNA sequence. The mRNA then travels outside the nucleus where it can be translated into a chain of amino acids, which will build the protein.

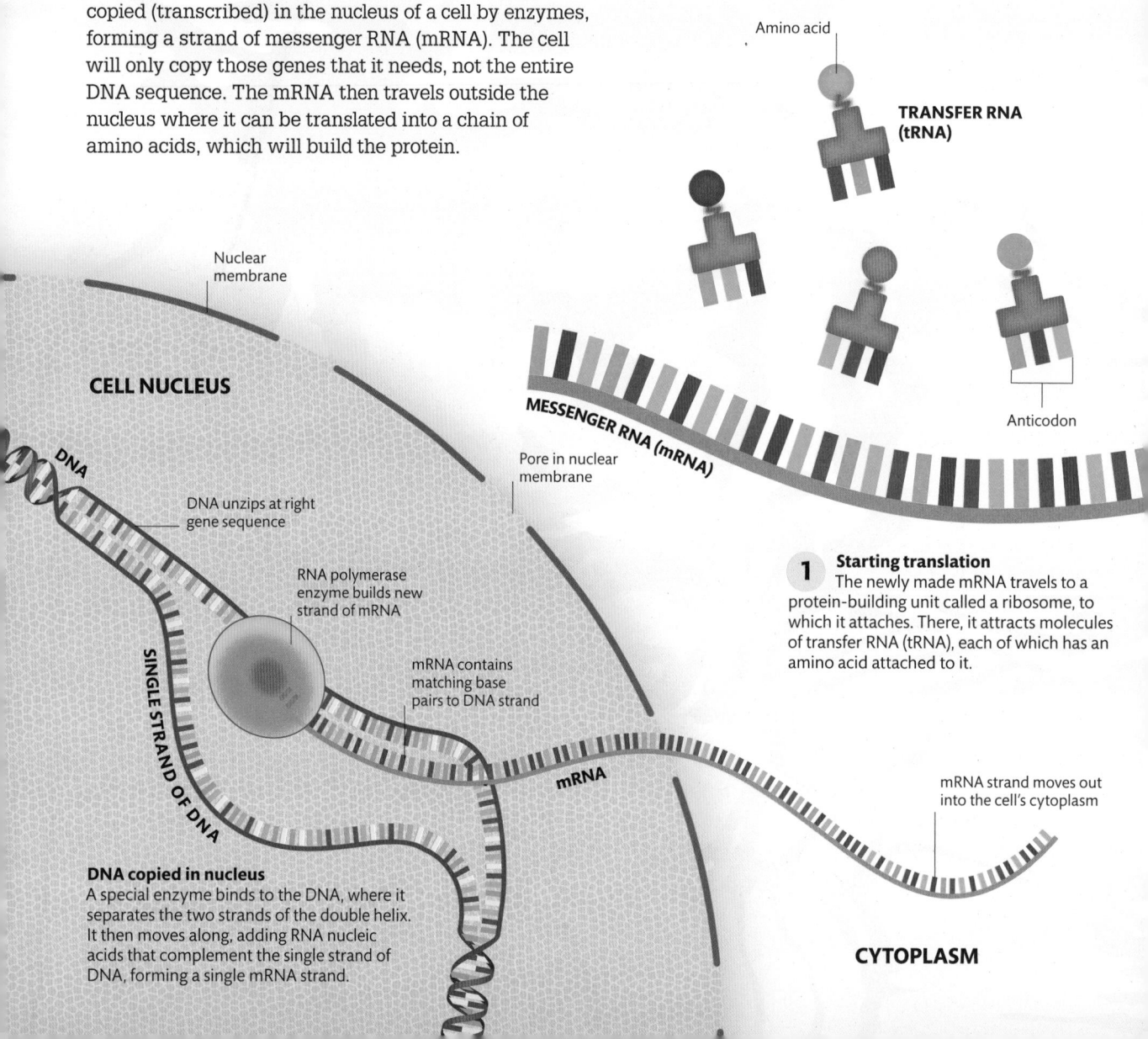

1 Starting translation
The newly made mRNA travels to a protein-building unit called a ribosome, to which it attaches. There, it attracts molecules of transfer RNA (tRNA), each of which has an amino acid attached to it.

DNA copied in nucleus
A special enzyme binds to the DNA, where it separates the two strands of the double helix. It then moves along, adding RNA nucleic acids that complement the single strand of DNA, forming a single mRNA strand.

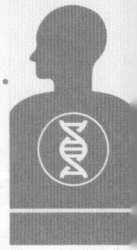

Making proteins

Every three bases in the mRNA is known as a codon and each codon specifies a particular amino acid. There are 21 different amino acids and a single protein may be made up of a chain of hundreds of these amino acids.

4 Amino acids folded into proteins
When the ribosome reaches a stop codon at the end of the mRNA strand, the long chain of amino acids is complete. The sequence of the amino acids determines how the chain folds up into a protein.

CHAIN FOLDED INTO PROTEIN

Chain of amino acids builds as ribosome moves along mRNA strand

2 Ribosome attaches amino acids
As the ribosome moves along the mRNA strand, the tRNA molecules attach to the mRNA in a specific order. This order is determined by the matching up of codons—a sequence of three nucleic acid bases on the mRNA strand—and their complementary three bases—called anticodons—on the tRNA molecule.

RIBOSOME

3 Building a chain
The amino acid detaches from the tRNA molecule and is joined to the previous amino acid with a peptide bond, forming a chain.

tRNA, once it has dropped off its animo acid, floats off into cytoplasm

Codon

LOST IN TRANSLATION

Gene mutations can cause changes in the amino acid sequence. A single mutation in the 402nd base of the gene that codes for the hair protein keratin causes the amino acid lysine to be put in place of glutamate. This changes the shape of the keratin, making the hair look beaded.

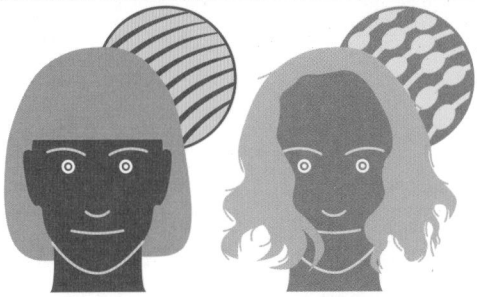

STRAIGHT HAIR BEADED HAIR

WHAT HAPPENS TO mRNA AFTER TRANSLATION?

A strand of mRNA may be translated into a protein many times before it eventually degrades within the cell.

How genes make different cells

DNA contains all of the blueprints for life, but cells pick and choose only the plans (genes) they need. These genes are used by the cell to build the proteins and molecules that not only define what the cell looks like, but what it does within the body.

HOW DO CELLS KNOW WHAT TO DO?

The chemical environment around the cell or signals from other cells tell it that it is part of a particular tissue or organ, or in a certain stage of development.

Gene expression

Each cell uses, or "expresses," only a fraction of its genes. As it becomes more specialized, more genes are switched off. This process is highly regulated and happens in a specific order, usually when the DNA is being transcribed to RNA (see pp.20–21).

1 Regulation
Transcription of a required gene is controlled by a series of genes that sit in front of it. These include regulator, promoter, and operator genes. The gene won't be transcribed until conditions are right.

2 Repressor protein
If a repressor protein is blocking the gene, transcription can't take place. The gene can only be turned on when a change in the environment removes the repressor protein.

3 Activation
When an activator protein binds to the regulator protein and there are no repressor proteins blocking the gene, transcription can start.

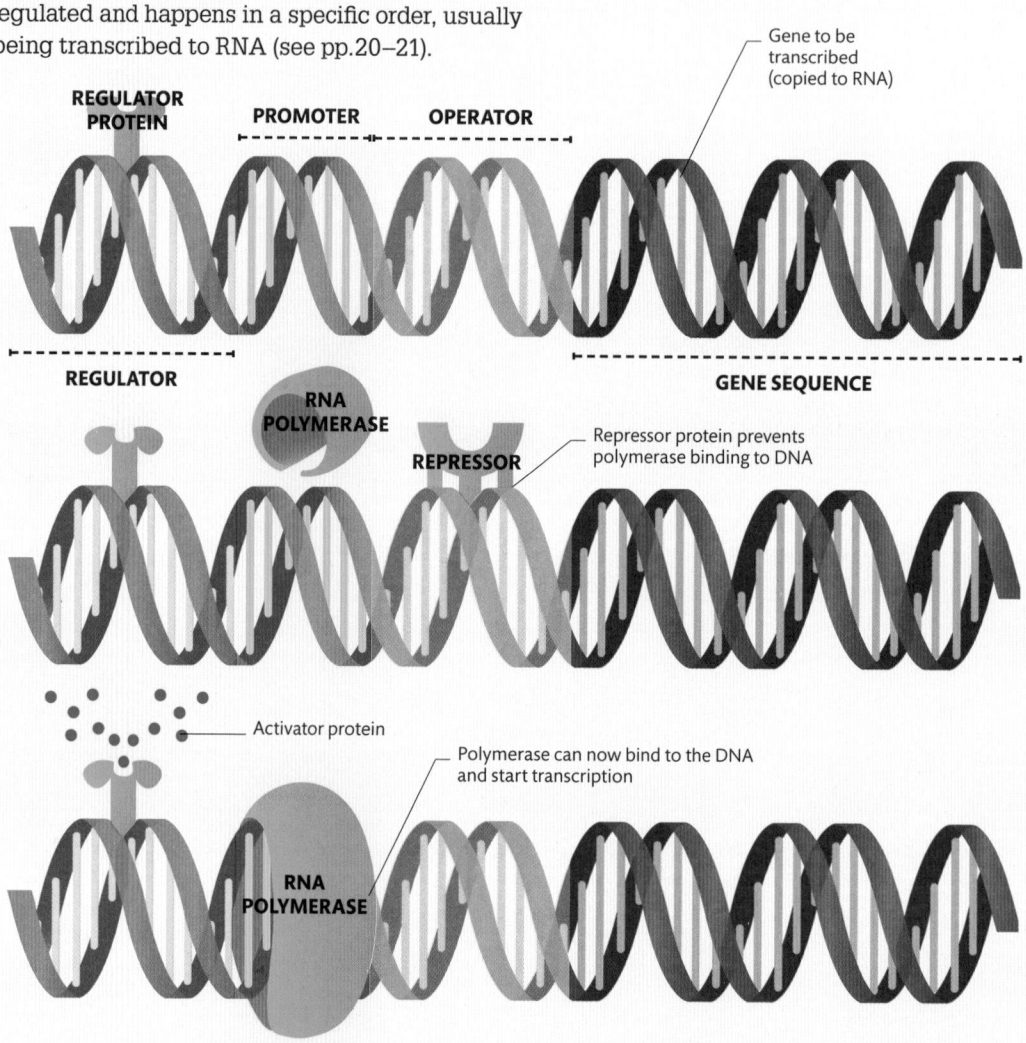

On or off?

Embryonic cells start out as stem cells—cells with the ability to turn into different cell types. Stem cells initially have the same set of genes switched on and they simply keep growing and dividing to produce more cells. As an embryo develops, its cells need to specialize and organize into tissues and eventually organs. So when signaled, the cells start shutting off some genes and switching on others to turn into a specific type of cell.

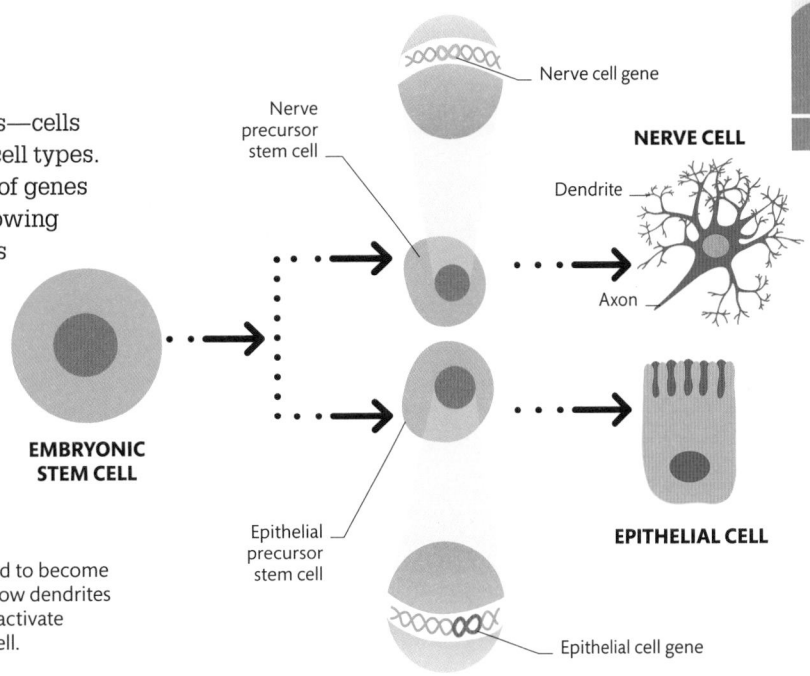

Making a difference
As an embryo is developing, a stem cell destined to become a nerve cell will turn on the genes needed to grow dendrites and an axon, whereas another stem cell might activate different genes to become an epithelial (skin) cell.

Housekeeping proteins

Some proteins, such as DNA repair proteins or enzymes needed for metabolism, are called housekeeping proteins, because they are essential to the basic functioning of all cells. Many are enzymes, while others add structure to cells or help transport substances in and out of cells. The genes for these proteins are always turned on.

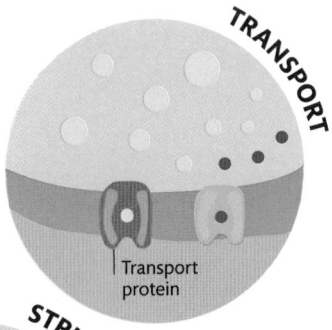

On the move
Special proteins are needed to move materials around the body or help them cross cell membranes.

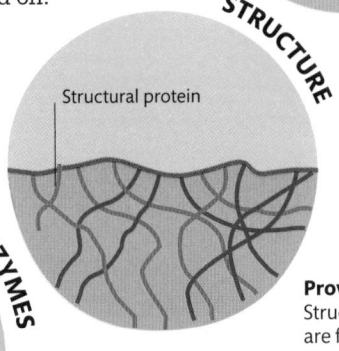

Providing support
Structural proteins are found in all cells. They give the cell its shape and hold the organelles in place.

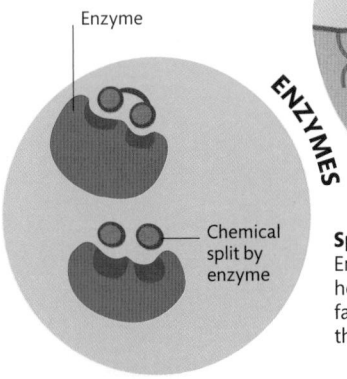

Speeding things up
Enzymes are proteins that help chemical reactions go faster, such as those used in the breakdown of food.

BOY OR GIRL?

At 6 weeks, an embryo has all the internal organs needed to be either male or female. If it is genetically a male embryo, a gene on the Y chromosome will turn on at this stage and produce the hormones that develop the male reproductive organs and cause the female organs to degenerate. The reason why men have seemingly pointless nipples is that these are formed in the first 6 weeks, but their further development depends on whether they are in a male or female hormonal environment.

Adult stem cells

Adult stem cells have been found in the brain, bone marrow, blood vessels, skeletal muscles, skin, teeth, heart, gut, liver, ovaries, and testes. These cells can sit inactive for a long time until they are called into action to replace cells or repair damage, when they begin to divide and specialize. Researchers can manipulate these cells to become specific cell types that can then be used to grow new tissues and organs.

WHERE DO ADULT STEM CELLS COME FROM?

This is currently being investigated, but one theory is that some embryonic stem cells remain in various tissues after development.

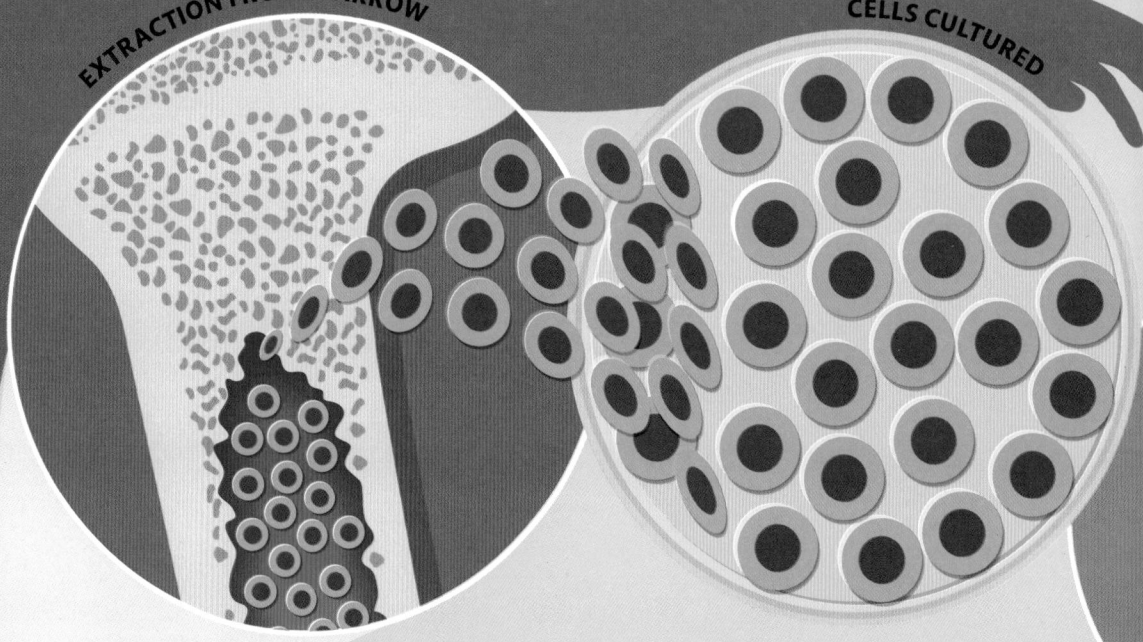

1 Harvest
Stem cell therapy may help repair damaged heart tissue following a heart attack. A small sample of the patient's bone marrow is taken because stem cells are more concentrated there.

2 Culture
The sample is filtered to remove non-stem cell material and then taken to a lab that will identify the stem cells. The lab cultures these cells, getting them to multiply and specialize.

Stem cells

Stem cells are unique because they can specialize into many different types of cells. Stem cells are the foundation for the body's repair mechanisms, which makes them potentially useful in helping repair damage in the body.

ADULT OR EMBRYONIC CELLS?

Embryonic stem cells can develop into any cell type, but research on them is controversial, because embryos—created using donor eggs and sperm—are grown specifically for the purpose of harvesting the cells. Adult stem cells are less flexible, forming only different types of blood cells, for instance, but new treatments can now be used to turn them into a wider range of cells.

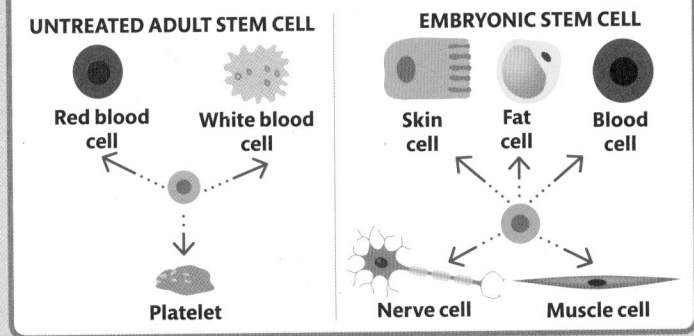

Engineering tissues

Researchers have found that the physical structure of the supporting matrix (scaffold) used to grow stem cells is critical to the way they grow and specialize.

1 Taking shape
To repair the eye's cornea, stem cells are extracted from a healthy tissue (the cornea of the unaffected eye) and grown on a dome-shaped mesh.

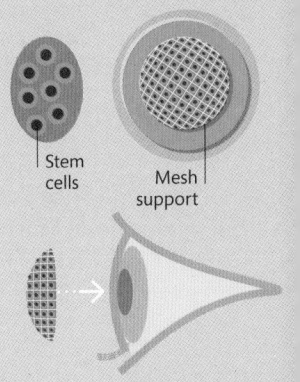

2 Transplant
The damaged cells on the cornea of the eye are removed and replaced with the mesh structure. After several weeks, the mesh dissolves leaving the grafted cells, which have restored the patient's sight.

Potential uses of stem cells
Stem cell research has improved our understanding of embryonic development and the natural repair mechanisms in the body. The most active area of research is their use in growing replacement organs and reconnecting the spinal cord so that paralyzed people can walk again.

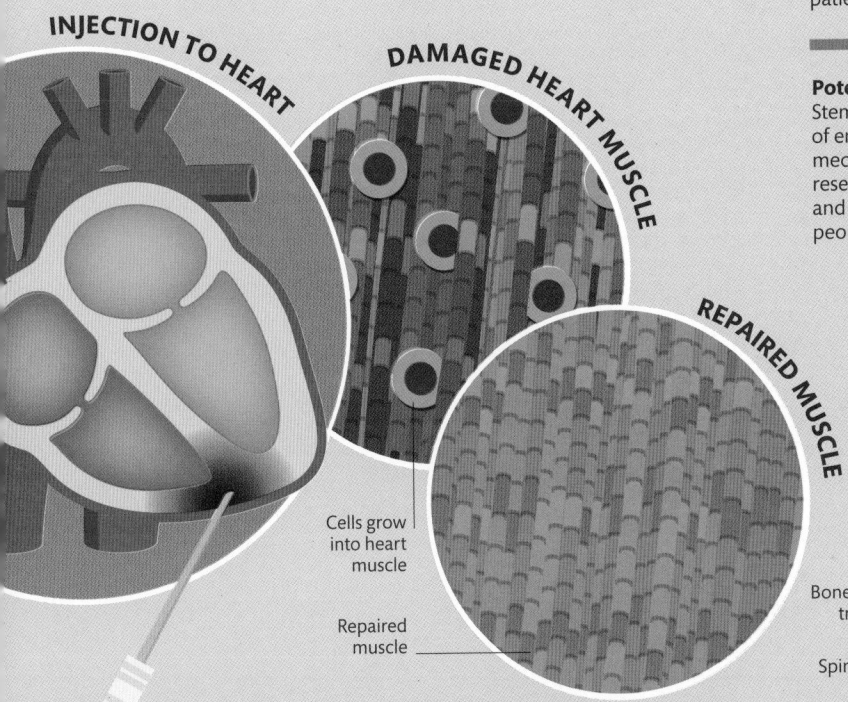

3 Inject
The cells are injected into the damaged heart muscle where they attach to the damaged fibers and begin to grow into new tissue.

4 Repair
After several weeks, the damaged heart muscle is rejuvenated. This process also reduces scarring that would restrict the heart's movement.

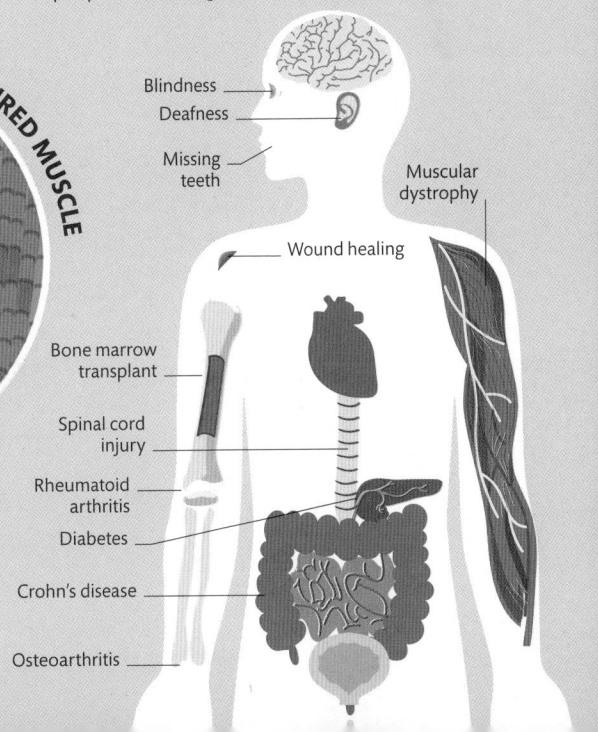

Environmental assault

Each of our cells is inundated daily by chemicals and energy that can cause damage to our DNA. Solar radiation (UV), environmental toxins, and even the chemicals produced through our own cellular processes can cause changes to our DNA that affect how it works, including how it can be copied or how it produces proteins. If this damage becomes a permanent change in the DNA, it is called a mutation.

20,000

THE NUMBER OF **DAMAGED BASES REMOVED** AND **REPLACED** IN EVERY CELL **EVERY DAY**

CAN THE DAMAGE ALWAYS BE REPAIRED?

Our ability to repair DNA diminishes as we get older. Damage starts to accumulate and this is thought to be one of the main reasons behind aging.

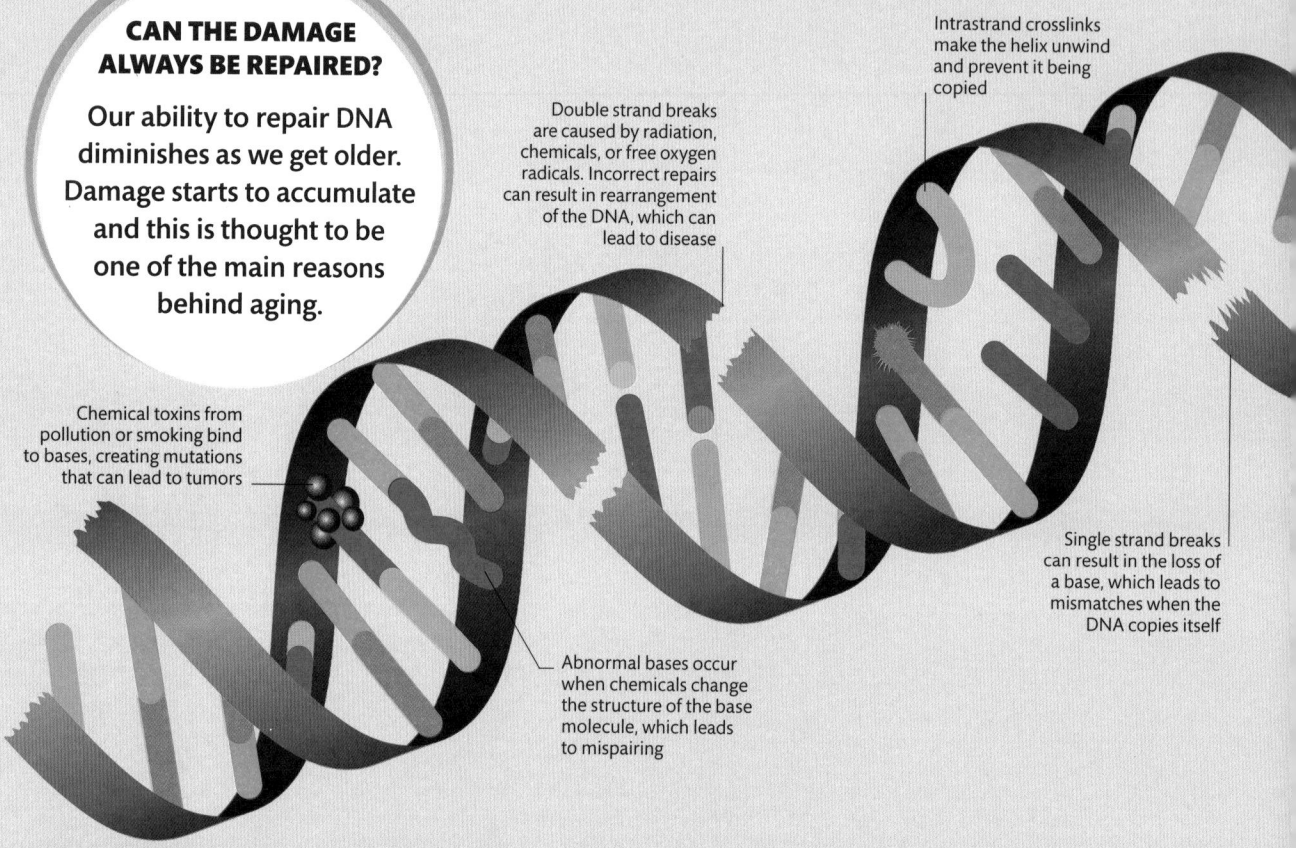

When DNA goes wrong

Every day, the DNA in cells is damaged—whether by natural processes or environmental factors. This damage can affect DNA copying or how specific genes function and if it can't be repaired, or is repaired incorrectly, it can lead to disease.

UNDER ATTACK
This DNA strand is shown under many kinds of stress. However, some types of DNA damage can be used to advantage. Many chemotherapy drugs are designed to cause damage to the DNA in cancerous cells. Cisplatin, for example, forms crosslinks in the DNA, which triggers cell death. Unfortunately it also causes damage in normal healthy cells.

Interstrand crosslinks between the same bases halt DNA copying because they prevent the strands from unzipping

Base mismatches occur when an extra base has been added or one has been skipped in the replication process

The insertion or deletion of bases means that when the code is being read during copying, the wrong proteins will be produced

Gene therapy

When DNA damage causes a mutation, it can stop a gene from working properly and result in disease. While drugs might help treat the symptoms of the disease, they can't solve the underlying genetic problem. Gene therapy is an experimental method that's exploring ways to fix the defective gene.

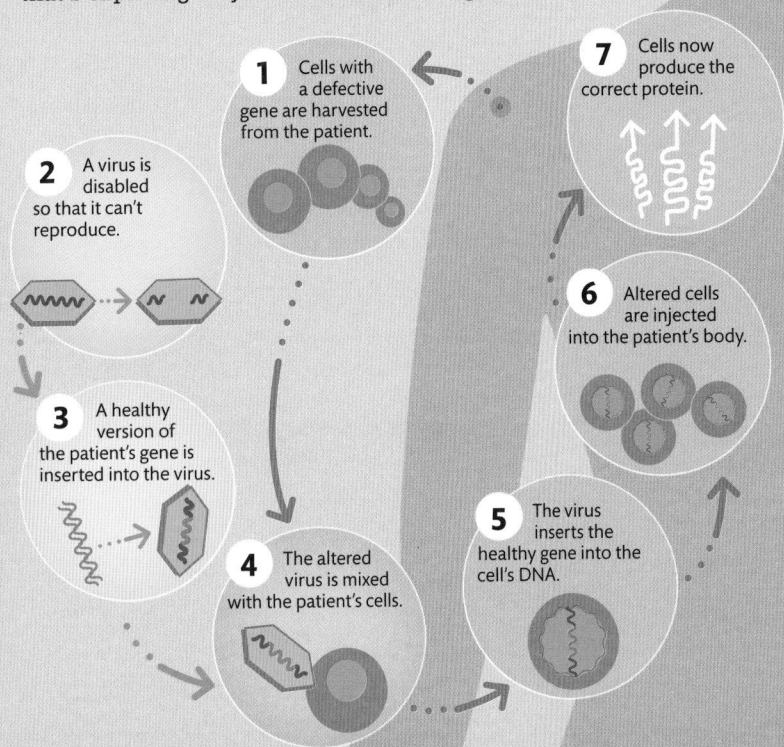

REPAIRING DNA

Cells have builtin safety systems that help identify and repair damage to their DNA. These systems are constantly active and if they are unable to fix the damage quickly, they will stop the cell cycle temporarily so they can take some extra time to work on it. If it's not repairable, they will trigger the death of the cell by apoptosis (see p.15).

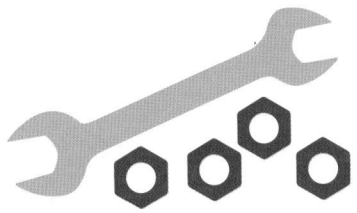

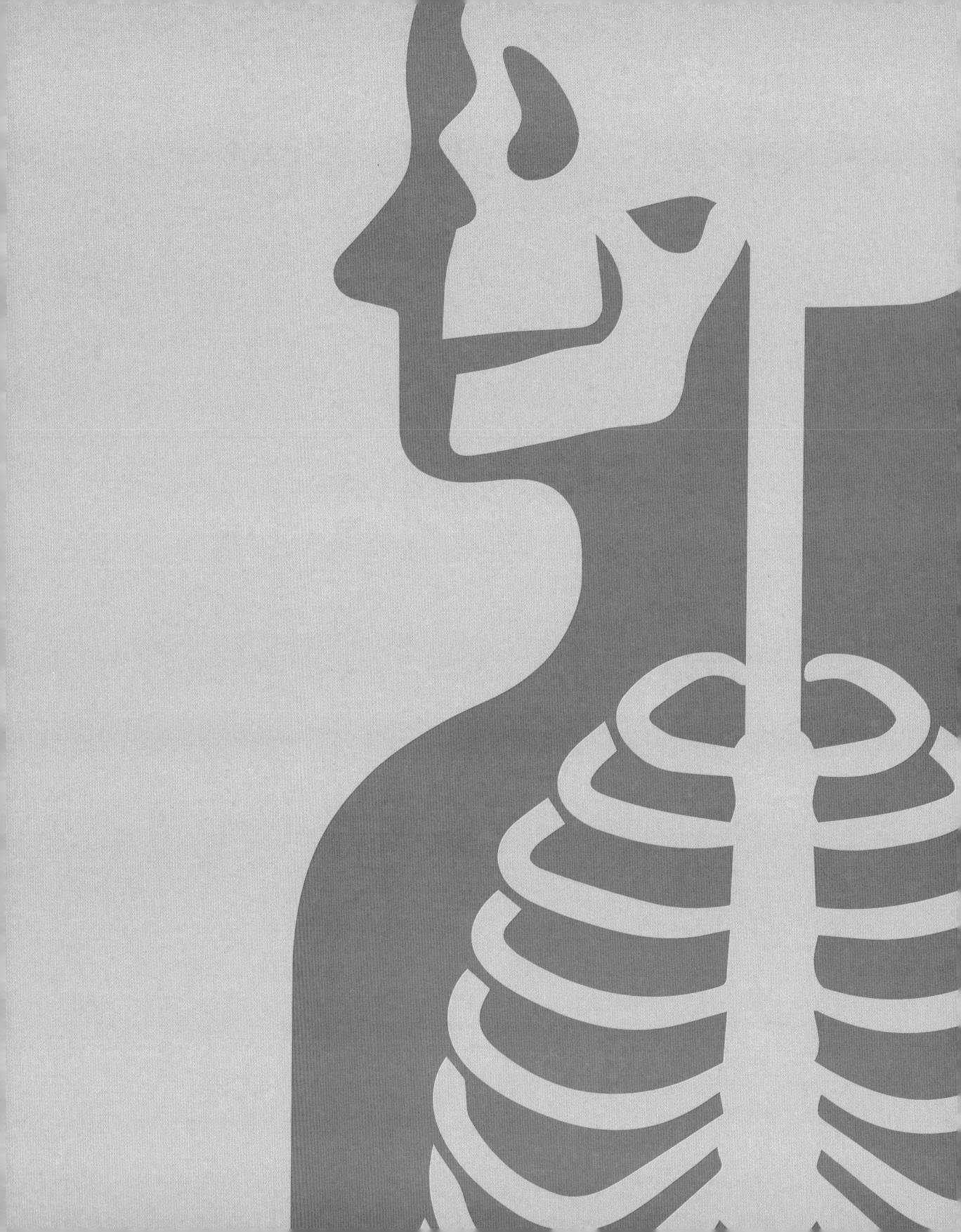

HOLDING IT TOGETHER

Skin deep

The skin is the largest organ of the human body. It protects us from physical damage, dehydration, overhydration, and infection, but also regulates body temperature, makes vitamin D, and has an extraordinary array of special nerve endings (see pp.74–75).

Keeping cool and staying warm

Humans have adapted to survive in the heat of the tropics, the cold of the arctic, and the temperate climates in between. Although we have lost most of our body hair and rely on clothes to keep us warm, even fine body hair plays a role in controlling body temperature. In hot weather, it is vital to drink plenty of water to replace the sweat that helps keep us cool.

Hot-weather skin

Each day, the skin's 3 million sweat glands secrete 2.1 pints (1 liter) of sweat, or up to 21 pints (10 liters) daily in extreme conditions. Evaporation of sweat takes the heat energy away from the body. Ring-shaped muscles around the blood vessels also help by diverting blood to the skin, which lets heat escape from deep in the body.

Cold-weather skin

In cold weather, the skin goes into heat-retention mode. Tiny muscles stand our body hairs upright, trapping warmth close to the skin. Meanwhile, the capillary-network muscles stop warm blood from flowing into the skin's surface layers.

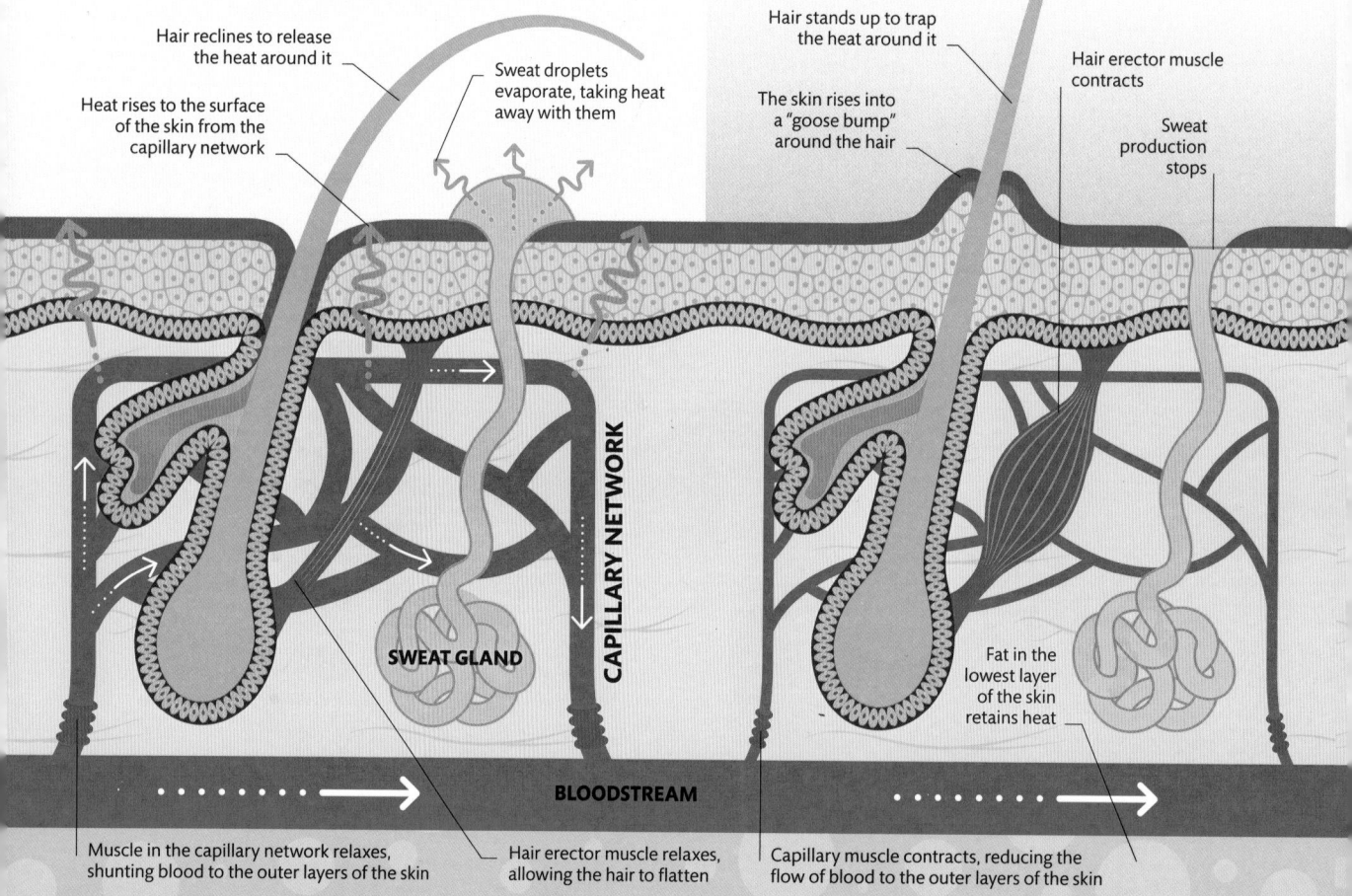

Defensive barriers

The skin is made up of three layers, each of which plays a vital role in our survival. The upper layer, called the epidermis, is an ever-regenerating defense system (see pp.32–33) and has its roots in the middle layer, called the dermis. The inside layer is the hypodermis—a cushion of fat that keeps us warm, protects our bones, and keeps us supplied with energy (see pp.158–59).

THE SKIN OF AN AVERAGE ADULT MEASURES 20 SQ FT (1.9 SQ M) IN AREA

Microbe

Sebum

Antibacterial oil
Glands secrete an oil called sebum into hair follicles to condition the hair and waterproof the skin. Sebum also suppresses the growth of bacteria and fungi.

Ultraviolet light

Ultraviolet light protection
The skin uses ultraviolet light to synthesize vitamin D—but too much ultraviolet light can cause skin cancer. A skin pigment called melanin helps maintain a balance between the two (see pp.32–33).

DO GOOSE BUMPS REALLY HELP?

Goose bumps do help us retain heat in cold weather. However, they were much more effective millions of years ago, when we were covered in thick hair. The thicker the hair, the more heat is trapped when the hair stands on end.

Sebaceous gland secretes sebum

Constantly regenerating epidermal cells

NICOTINE PATCH

EPIDERMIS

HAIR SHAFT

DERMIS

Nicotine reaches the bloodstream

One of the skin's many types of nerve endings (see pp.74–75)

Letting things pass
Although skin is a barrier, it is selectively permeable, letting through drugs, such as nicotine and morphine, from patches applied to the skin's surface. Various creams, such as sunblock, moisturizer, and antiseptic cream, can also cross the barrier.

HAIR BULB

The epidermis stretches all the way under the hair bulb

HYPODERMIS

Outer defenses

The skin is the frontier between us and the outside world—a boundary at which enemies are fought and friends let in. Key features of its defenses are a self-renewing outer layer and a pigment that shields us from ultraviolet light.

ARE FINGERPRINTS REALLY UNIQUE?

The curls, loops, and swirls of each finger are unique, and each grows back the same way after injury—a handy fact for police detective work.

The self-renewing layer

The epidermis is a conveyor belt of cells, which are constantly forming at its base—the basal layer—and traveling upward to the surface. As they move, they lose their nucleus, flatten, and fill with a tough protein called keratin, and so form a protective, outer layer. This layer is constantly being worn away and replaced by new, upthrusting cells. Each cell dies by the time it reaches the surface. The dead cells fall off and contribute to the dust in our houses.

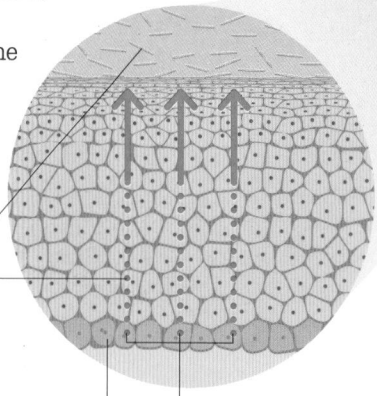

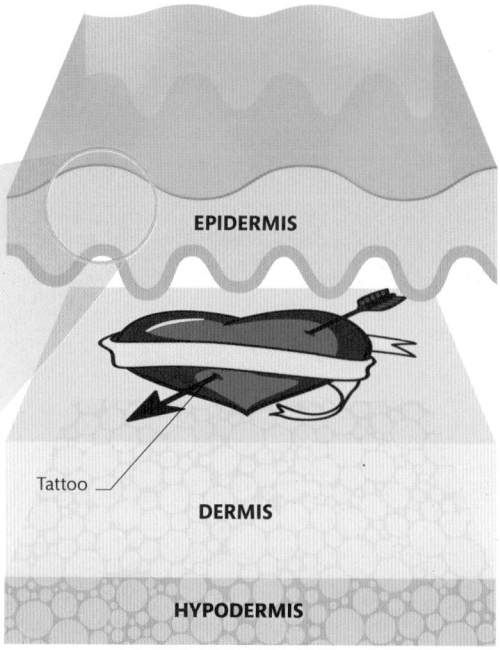

Transparent defense
Because the epidermis sheds its cells, tattoos have to be inscribed beneath it, on the dermis. The epidermis is transparent, so tattoos can be seen through it.

Scaffolding

Beneath the epidermis lies the dermis, a thick layer that gives the skin its strength and flexibility. It contains the skin's nerve endings, sweat glands, oil glands, hair roots, and blood vessels. It is made primarily of collagen and elastin fibers, which form a kind of scaffolding that enables the skin to stretch and contract in response to pressure.

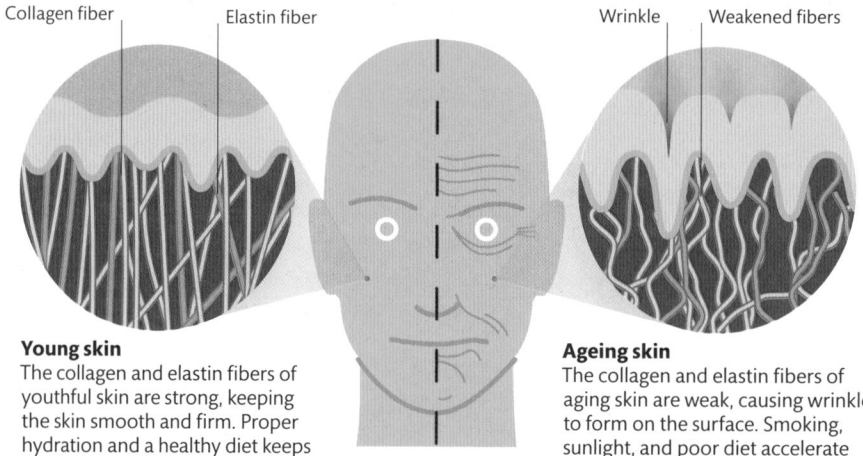

Young skin
The collagen and elastin fibers of youthful skin are strong, keeping the skin smooth and firm. Proper hydration and a healthy diet keeps the skin youthful.

Ageing skin
The collagen and elastin fibers of aging skin are weak, causing wrinkles to form on the surface. Smoking, sunlight, and poor diet accelerate the aging process.

Skin color

One of the skin's many functions is to make vitamin D, which it does by harnessing ultraviolet (UV) light from the Sun. However, UV light is also very dangerous (it can cause skin cancer), so we also need protection against it. As protection, the skin produces melanin—a pigment that serves as a Sun shield, and so determines skin color.

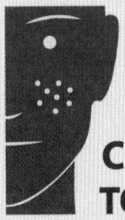

FRECKLES ARE CAUSED BY MELANOCYTES CLUMPING TOGETHER

Dark skin

At the equator, the sun's rays strike the Earth almost vertically, and with great intensity. This means that people born near the equator have a great need of UV protection. To provide this, the skin produces large amounts melanin—which results in dark skin.

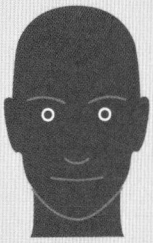

2 Dendrites
Melanocytes have fingerlike extensions called dendrites. Each of these touches around 35 neighboring cells.

1 Melanocytes
Melanin is produced by special cells called melanocytes. These are embedded in the base of the epidermis.

Intense rays of UV light

5 UV shield
Melanosomes break apart, spreading melanin across the skin. This forms a shield against UV rays.

4 Absorption
Melanosomes are absorbed by neighboring skin cells.

3 Melanosomes
Melanin moves along the dendrites in packets called melanosomes.

Dendrite

Melanocyte

Melanosome

Basal layer

Pale skin

North and south of the equator, the sun's rays hit the Earth at increasingly shallow angles. The shallower the angle, the less intense the light, and less need for UV protection. In response, the skin produces smaller amounts of melanin—which results in pale skin.

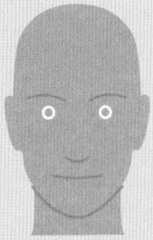

1 Melanocytes
In pale skin, the melanocytes are less active, and have fewer dendrites.

Mild rays of UV light

3 Weaker shield
The weaker melanin shield is sufficient against weaker UV rays.

2 Paler melanosomes
Melanosomes are paler and taken up by fewer surrounding cells.

Dendrite

Melanocyte

Melanosome

The extremities

Hair and nails are both made of a tough, fibrous protein called keratin. Nails strengthen and protect the tips of your fingers and toes, while hair reduces heat loss from the body to help keep you warm.

Hair color, thickness, and curliness

Each hair has a spongy core (medulla) and a middle layer (cortex) of flexible protein chains that give it wave and bounce. An outer layer (cuticle) of scales reflects light so hair looks shiny, but if these are damaged, hair looks dull. The color, curliness, thickness, and length of your hair are determined by the size and shape of your follicles (in which they grow), and the pigments they produce.

Thick, straight, and red
A mixture of pale and dark melanin produces hair that is gold, auburn, or red. Large, round follicles produce thick hair. Thickness also depends on the number of active follicles present. Redheads tend to have relatively few follicles.

Fine, straight, and blonde
Cells at the base of each follicle feed melanin pigments through to the root. Blonde hair contains a pale melanin pigment that is only present in the middle of the shaft (medulla). Small, round follicles produce straight, fine hair.

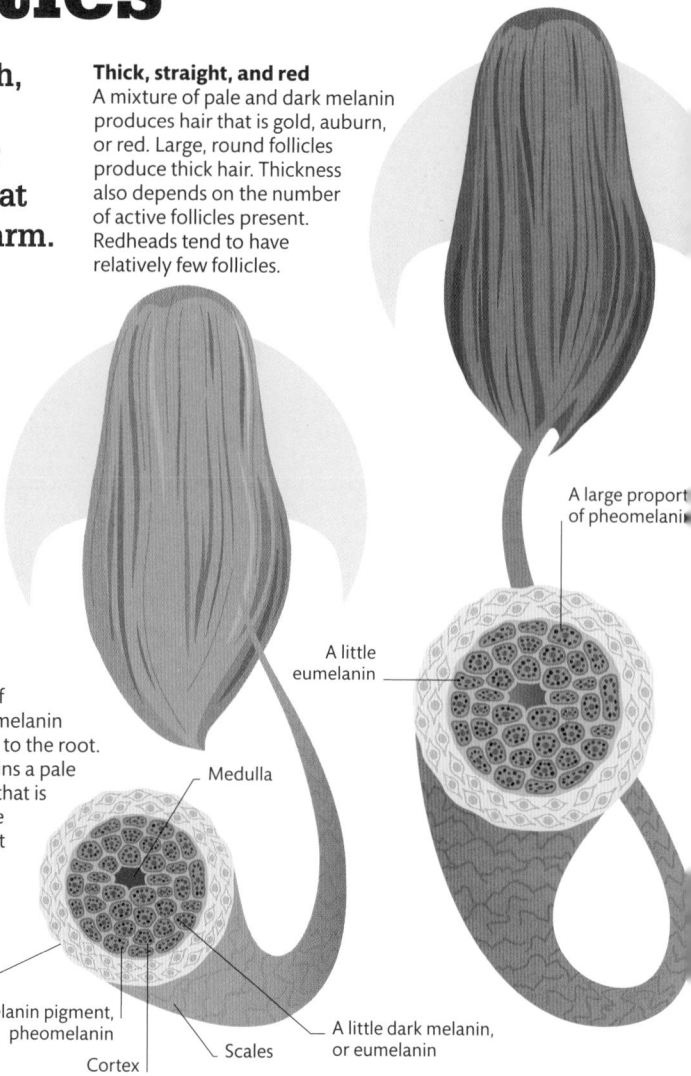

WHY DOES HAIR LENGTH VARY?

Scalp hair can grow for years, but hair found elsewhere on the body only grows for weeks or months. That's why body hair is usually short—it falls out before it can grow very long.

Hair growth

Each hair follicle goes through around 25 cycles of hair growth during its lifespan. Each cycle has a growth stage when it lengthens, followed by a resting phase in which the hair remains the same length, starts to loosen, and falls out. After the resting phase, the follicle reactivates and starts to produce a new hair.

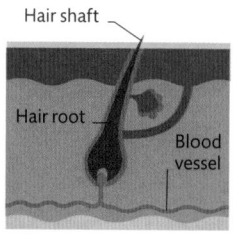

1 Early growth
The follicle activates, producing new cells within the hair root. These die and are pushed upward to form the shaft.

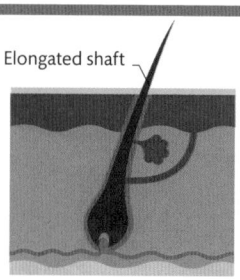

2 Late growth
The shaft elongates over a period of 2–6 years. A longer growth period (more common in women) produces longer hair.

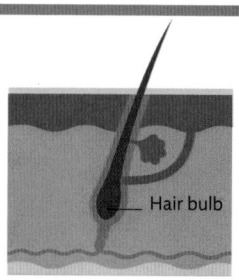

3 Resting
The follicle shrinks and the hair stops growing as the bulb pulls away from the root. This takes 3–6 weeks.

Thick, black, and curly
Dark hair contains black melanin pigment in both the cortex and the medulla, producing more depth of color. Wavy hair grows from oval-shaped follicles. As follicles become flatter, hair curliness increases.

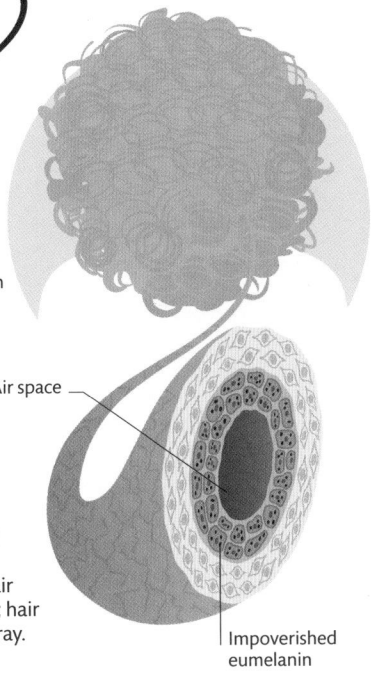

Curly and gray
Hair turns gray due to reduced activity of an enzyme that produces melanin pigment. Hair without melanin is snow white; hair with a little pigment appears gray.

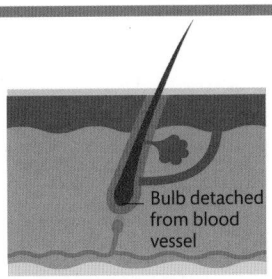

4 Detatchment
The loose hair is shed naturally, or dislodged by brushing or combing. Sometimes it's pushed out by a new hair growing.

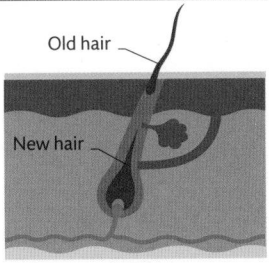

5 New growth
The follicle starts its next cycle. With age, fewer follicles reactivate, so hair becomes thinner, recedes, and bald areas may appear.

Nails

Nails are transparent plates of keratin. They act as splints to stabilize the soft flesh of your fingertips, and improve your grip on small objects. Nails also contribute to the overall sensitivity of your fingertips. However, because they project from the body, nails are easily damaged.

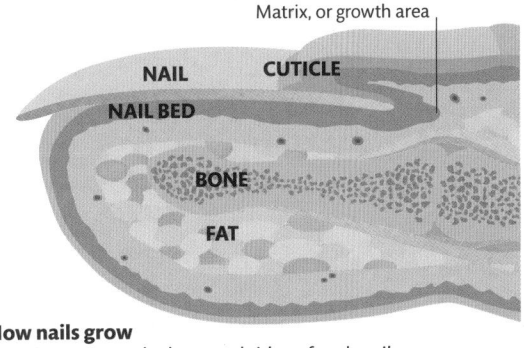

How nails grow
Growing areas at the base and sides of each nail are protected by folds of skin called cuticles. Cells in the nail beds are among the most active in the body. They are constantly dividing, and nails grow up to ½ in (5 mm) per month.

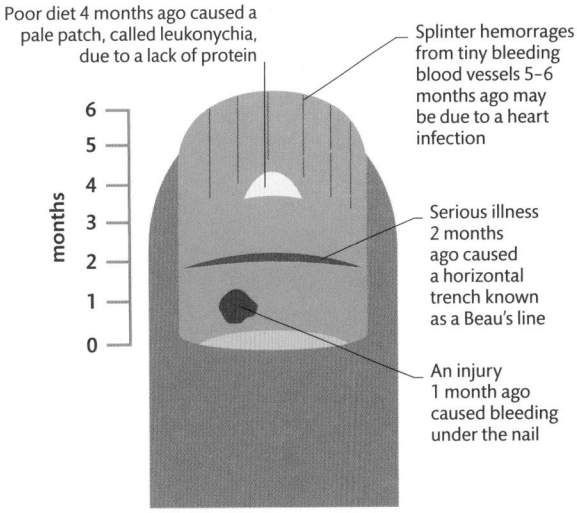

Diary of a nail
Because nails are nonessential structures, blood and nutrients are diverted away from the nail beds in times of deficiency. Nails are therefore a good indicator of your general health and diet. A doctor glances quickly at a patient's hands because the nails can indicate a number of illnesses.

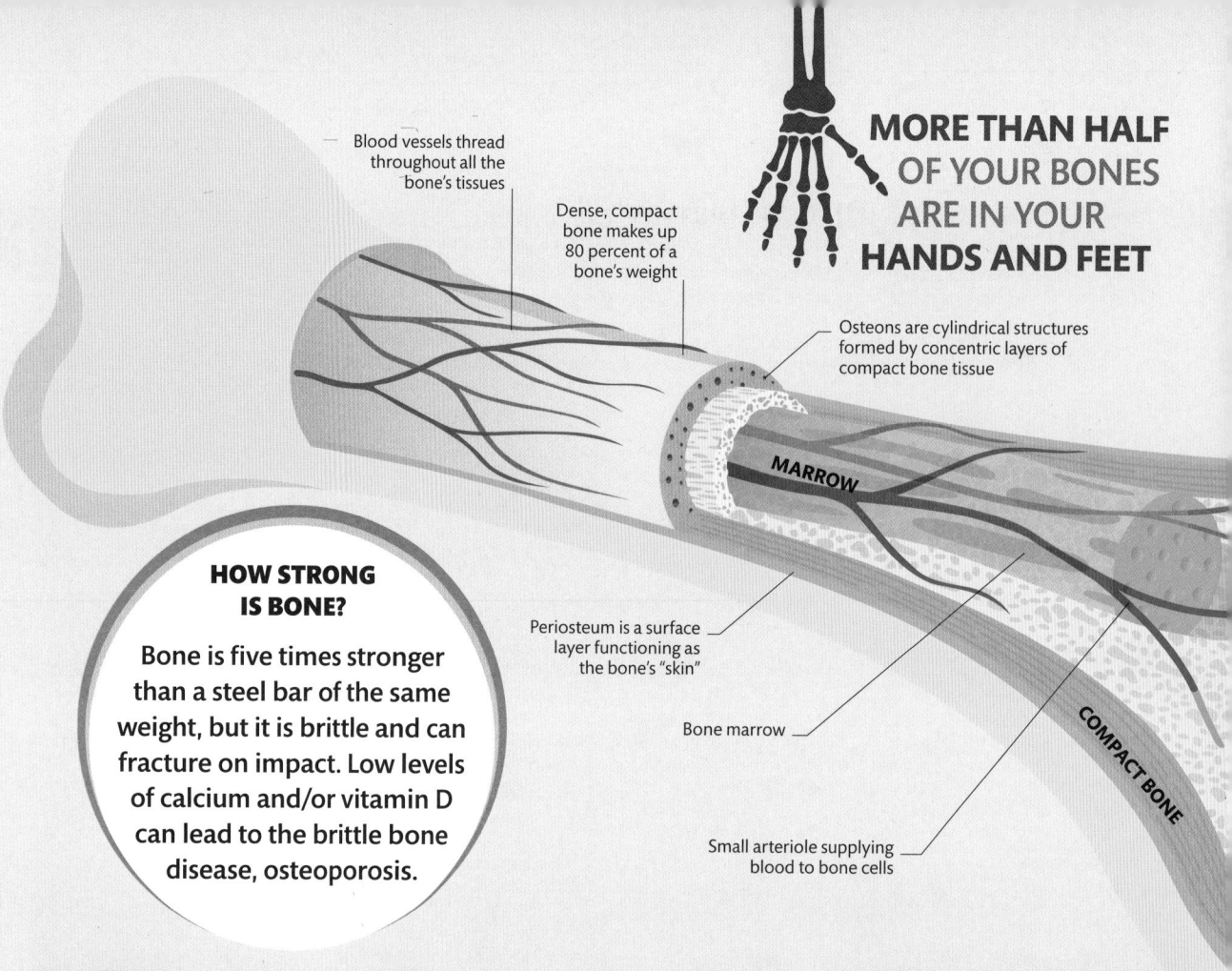

HOW STRONG IS BONE?

Bone is five times stronger than a steel bar of the same weight, but it is brittle and can fracture on impact. Low levels of calcium and/or vitamin D can lead to the brittle bone disease, osteoporosis.

Pillars of support

Your skeleton is rather like a coat hanger on which your flesh is draped. As well as giving your body support and shape, your bones provide protection and, through their interaction with muscles, allow your body to move and adopt different poses.

Living tissue

Bone is a living tissue made up of collagen protein fibers filled with minerals—calcium and phosphate—which give them rigidity. Bones contain 99 percent of all the calcium in your body. Bone cells constantly replace old, worn-out bone with new bone tissue. Blood vessels supply these cells with oxygen and nutrients. A surface layer of skinlike periosteum covers a shell of compact bone, which provides strength. Beneath this is a spongelike network of struts that reduces the overall weight. Bone marrow in certain bones, including the ribs, breast bone, shoulder blades, and pelvis, has a special job—it produces new blood cells.

THE SMALLEST BONES

The stapes in the middle ear is the smallest named bone. You also have small sesamoid bones (named after the sesame seeds they resemble) in long tendons at sites of pressure to prevent the tendons from wearing away.

STIRRUP (OR STAPES) EAR BONE

How the skeleton fits together

The skeleton can be divided into two main parts. The axial skeleton consists of the skull, vertebral column (spine), and ribcage and protects the internal organs and the central nervous system. The appendicular skeleton includes the upper and lower limbs, plus the shoulder and pelvic girdles that attach them to the axial skeleton. It anchors the muscles muscles that bring about conscious movement.

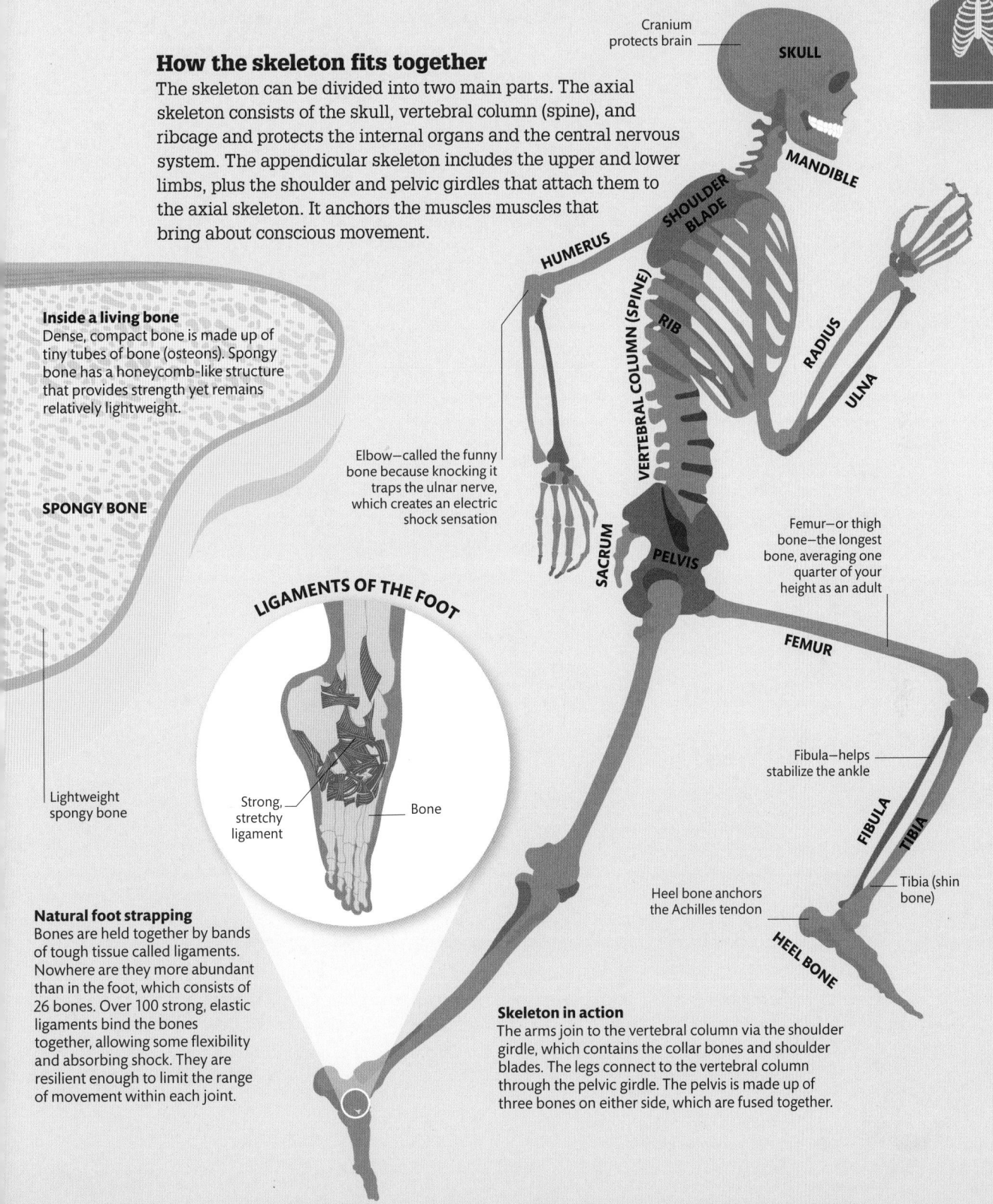

Inside a living bone
Dense, compact bone is made up of tiny tubes of bone (osteons). Spongy bone has a honeycomb-like structure that provides strength yet remains relatively lightweight.

Natural foot strapping
Bones are held together by bands of tough tissue called ligaments. Nowhere are they more abundant than in the foot, which consists of 26 bones. Over 100 strong, elastic ligaments bind the bones together, allowing some flexibility and absorbing shock. They are resilient enough to limit the range of movement within each joint.

Skeleton in action
The arms join to the vertebral column via the shoulder girdle, which contains the collar bones and shoulder blades. The legs connect to the vertebral column through the pelvic girdle. The pelvis is made up of three bones on either side, which are fused together.

Growing bones

A healthy baby measures 18–22 in (46–56 cm) in length at birth. Growth is rapid during infancy as the long bones elongate. Bone growth slows during childhood, but then speeds up again at puberty. Bones stop growing at around 18 years of age, when final adult height is reached.

NEWBORN BABY WEIGHT

An average newborn baby weighs 5½–9½ lb (2.5–4.3 kg). Babies normally lose weight in the first days after birth, but by 10 days, most have regained their birth weight and start to put on around 1 oz (28 g) per day.

How bones grow

Growth in height occurs at special growth plates at the ends of the long bones. Bone growth is controlled by growth hormone, with an additional growth spurt occurring in response to sex hormones at puberty (see pp.223–23). The cartilage growth plates fuse by adulthood, after which no further increases in height are possible.

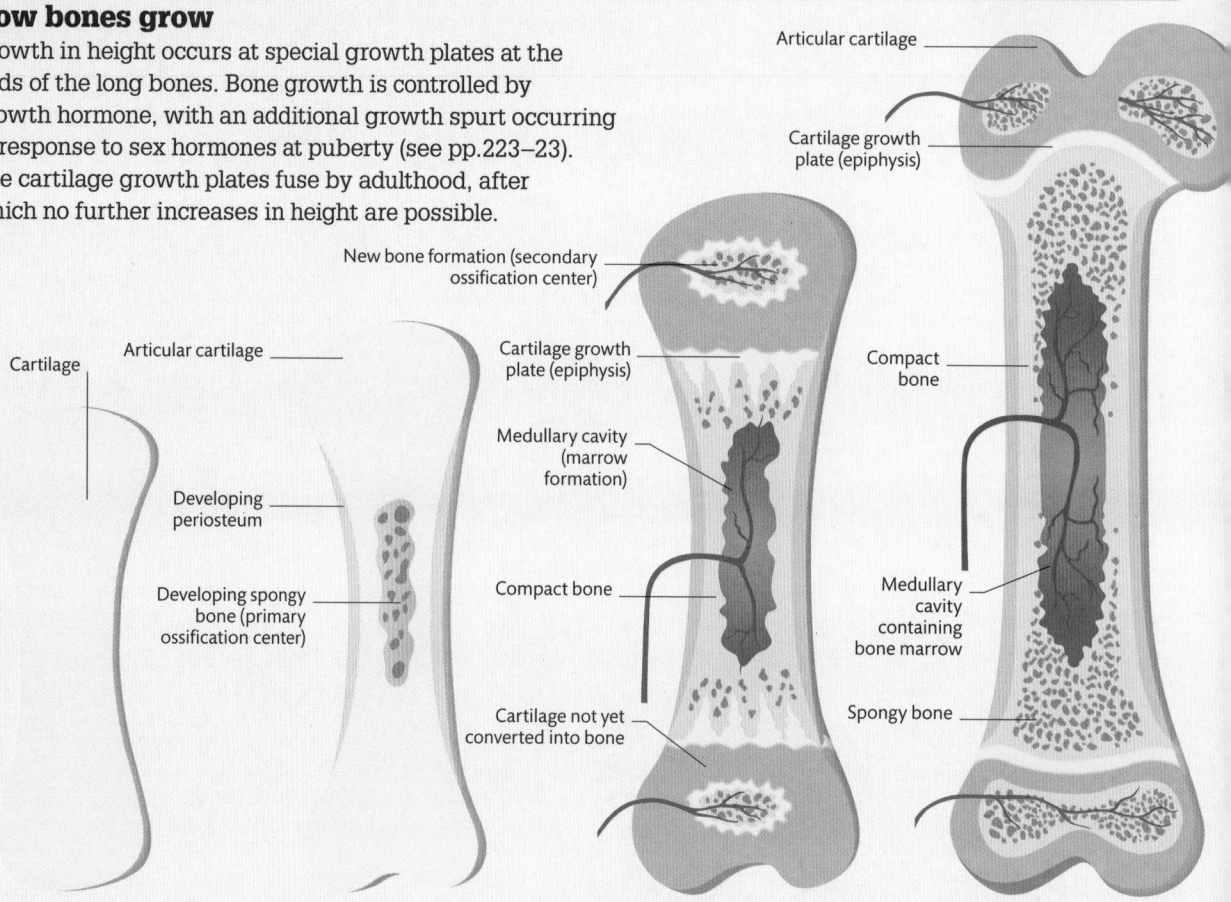

1 Embryo
Bones initially form from soft cartilage that acts as a scaffold on which minerals are laid down. Hardened bone starts forming when the fetus reaches 2–3 months of development in the womb.

2 Newborn baby
At birth, bones still consist mostly of cartilage, but there are active sites of bone formation (ossification). The first to develop is the primary ossification centre in the shaft, followed by those at the ends.

3 Child
In childhood, most of the shaft consists of hardened compact and spongy bone. Growth plates (epiphyses) at the ends allow lengthening. Bone is still soft and can bend on impact to form a greenstick fracture.

4 Teenager
At puberty, a surge in sex hormones causes a rapid growth spurt. Increases in height occur when new bone is laid down at the cartilage growth plates (epiphyses) to lengthen the bone shaft.

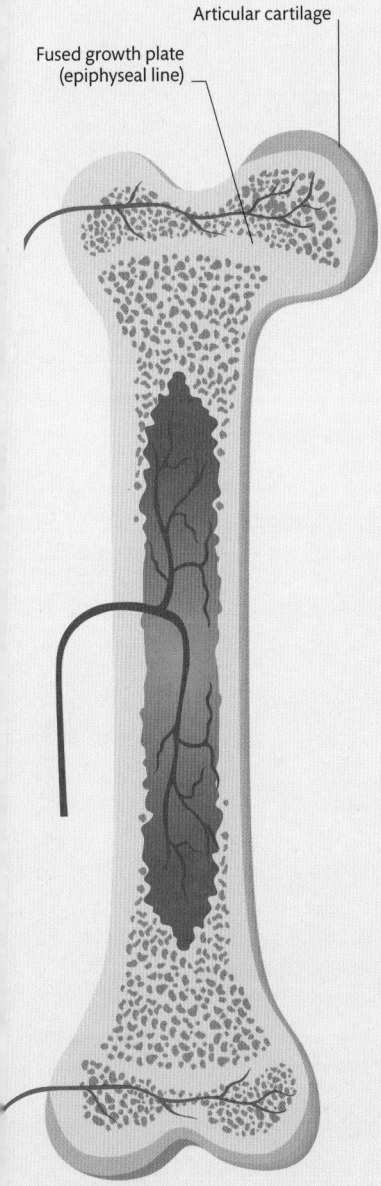

5 Adult
After puberty, the cartilage growth plates are converted into bone (calcified) and fuse. This leaves a hardened area called the epiphyseal line. Bones can still increase in diameter, but can no longer increase in length.

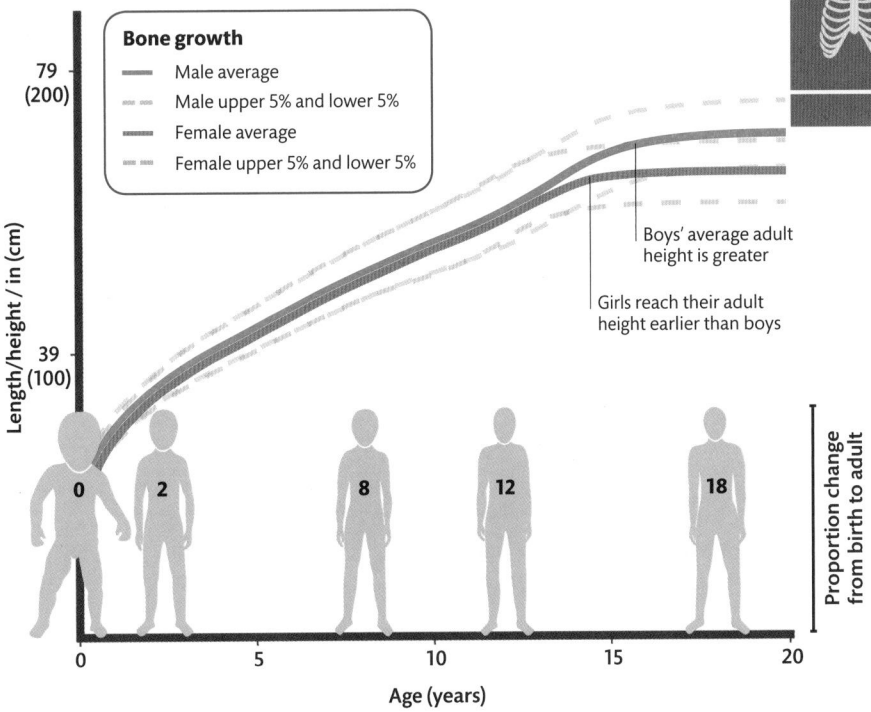

Growth patterns

A baby's head is one-quarter of his or her total body length. Changes in relative growth means that by age two that ratio is down to one-sixth. An adult's head is only one-eighth of body length. Girls enter puberty earlier than boys and reach their adult height around 16–17 years of age. Males only reach their final height between the ages of 19 and 21.

HOW TO CALCULATE YOUR FINAL HEIGHT

Assuming both parents are of normal stature, a child's potential adult height can be calculated as follows. Add father's height to mother's height. For a boy, add 5 in (13 cm) and for a girl deduct 5 in (13 cm). Then divide the total by two. Most children will have a final adult height within 4 in (10 cm) of this estimate.

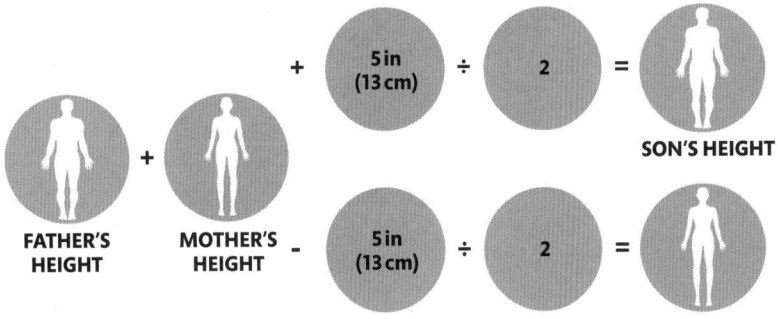

Flexibility

Your joints allow you to move your body and manipulate objects. Movements can be small and controlled, such as when writing your name, or large and powerful, such as when throwing a ball.

Joint structure

A joint forms where two bones come into close contact. Some joints are fixed, with the bones locked together, such as the suture joints in an adult skull. Some joints have a limited range of movement, such as the elbow, while others can move more freely, such as the shoulder.

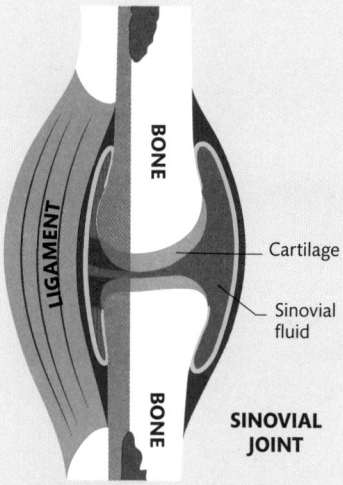

Inside a joint
The bone ends within a mobile joint are coated with slippery cartilage and oiled with synovial fluid to reduce friction. These synovial joints are held together by bands of connective tissue, called ligaments. Some joints, such as the knee, also have internal stabilizing ligaments to stop the bones from sliding apart while bending.

Ellipsoidal
These complex joints involve a bone with a rounded, convex end fitting into a bone with a hollow or concave shape. This allows a variety of movements, including sideways tilting, but not rotation.

Ball and socket
Found in the shoulders and hips, this type of joint allows the widest range of movement, including rotation. The shoulder joint is the most mobile joint in the body.

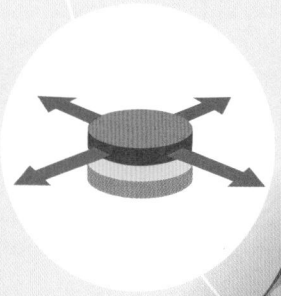

Gliding
These allow one bone to slide over another in any direction within one plane. Gliding joints allow the vertebrae to slide over each other when you flex your back. They are also found in your feet and hands.

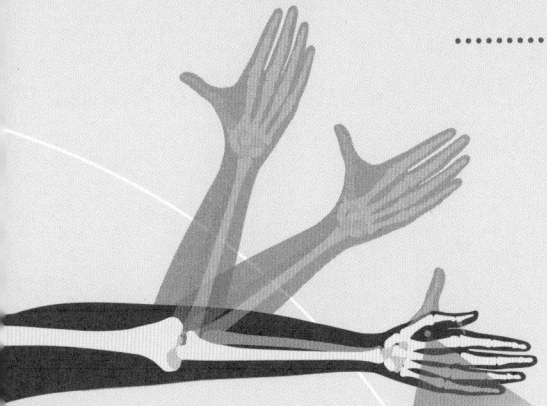

Types of joints

Although your body as a whole moves in complex ways, each individual joint has only a limited range of movement. A few joints have a very limited amount of movement so that they can absorb shock, such as where the two long bones in your lower leg (tibia and fibula) meet or some of the joints in the feet. The temporomandibular joints (see pp.44–45) between your jawbone and each side of the skull are unusual in that they each contain a disk of cartilage that allows the jaw to glide from side to side and protrude forward and backward during chewing and grinding your food.

Saddle
This is only found at the base of the thumb and allows a similar but wider range of movement to ellipsoidal joints, including a circular motion, but without rotation.

Pivot
This allows one bone to rotate around another, for example when you move your forearm to twist your palm to face up or down. A pivot joint in your neck allows your head to turn from side to side.

Hinge
This type of joint mainly allows movement in one plane, rather like a door opening and closing. Good examples are found in the elbow and knee.

THE **SMALLEST JOINTS** ARE FOUND BETWEEN THE **THREE TINY BONES OF THE MIDDLE EAR** THAT HELP **TRANSMIT SOUND WAVES** TO THE **INNER EAR**

DOUBLE-JOINTED PEOPLE

People who are said to be double-jointed have the same number of joints as everyone else, but their joints have a wider than normal range of movement. This trait is usually due to inheriting unusually elastic ligaments or a gene that codes for the production of a weaker type of collagen (a protein found in ligaments and other connective tissues).

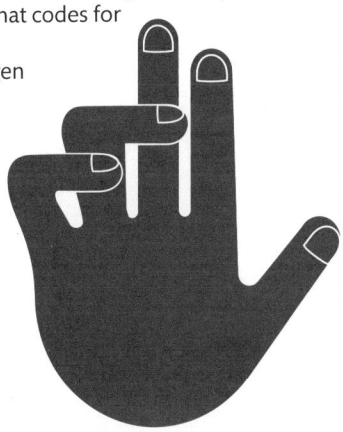

Biting and chewing

Humans struggle to swallow large pieces of food so your teeth break down food as part of the first stage of digestion. Teeth also play a role in speech—it would be difficult to make the sound "tutt" without any teeth, for example.

From baby to adult

Your teeth are all present at birth as tiny buds deep within each jawbone. The first "milk" teeth need to be small to fit within an infant's mouth. These teeth are shed during childhood as the mouth enlarges, leaving more room for adult-sized teeth.

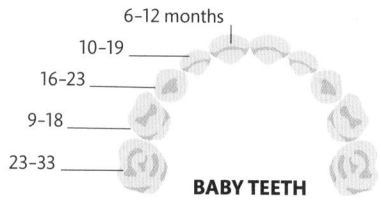

Eruption of milk teeth
The 20 milk teeth usually start to appear between the ages of 6 months and 3 years, although some infants have to wait a year.

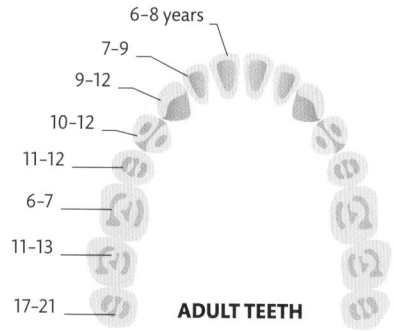

Eruption of adult teeth
The 32 adult teeth appear between the ages of 6 and 20 years and should last for the rest of your life—even if you live to be 100.

MUCH LIKE A **FINGERPRINT,** EACH PERSON HAS A **UNIQUE BITE IMPRESSION**

SECOND INCISOR
CANINE
FIRST PREMOLAR
SECOND PREMOLAR
FIRST MOLAR
SECOND MOLAR
Third molar, or "wisdom tooth"
Crown
Pulp cavity
ENAMEL
DENTINE
GUM
TOOTH ROOT
Cementum, a natural glue holding the tooth in its socket
Root canal
Blood vessel
Periodontal ligaments help anchor tooth in jawbone

Tooth structure
Each tooth has a crown, above the gum, which is coated in hard enamel. This protects the softer dentine forming the tooth root. The central pulp cavity contains blood vessels and nerves.

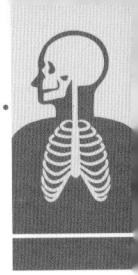

FIRST INCISOR

SECOND INCISOR

CANINE

WHAT ARE WISDOM TEETH?

The last set of molars usually appear between the ages of 17 and 25. It is thought that they are called wisdom teeth because they appear after childhood.

Different types

Your teeth differ in shape and size depending on their use. Sharp-edged incisors cut and bite, canines tear, and molars and premolars have flattened, ridged surfaces that chew and grind food into tiny pieces.

ARE YOU A GRINDER?

One in twelve people grind their teeth while asleep, and as many as one in five clench their jaws while awake. Known as bruxism, this weakens your teeth. You could be a grinder if your teeth look worn down, flattened or chipped, if your teeth are increasingly sensitive, or if you wake with jaw pain, a tightness in your jaw muscles, earache, or a dull headache—especially if you also chewed the inside of your cheeks. Worn-down teeth may be reshaped with crowns.

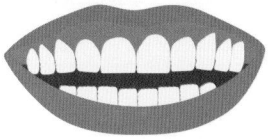

FLATTENED TEETH

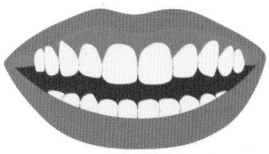

AFTER TREATMENT

Infection

Tooth enamel is the hardest substance in the body, but readily dissolves in acid, exposing the underlying parts of the tooth to bacteria and infection. Acid can come from some foods, juices, and sodas, or from bacterial plaque, which breaks down sugar to form lactic acid.

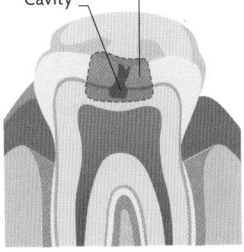

DECAYING TOOTH

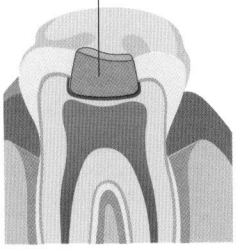

TOOTH WITH FILLING

Decay and filling
When the hard enamel dissolves, it allows infection to rot the softer dentine beneath. A cavity forms as the weakened enamel overhead collapses.

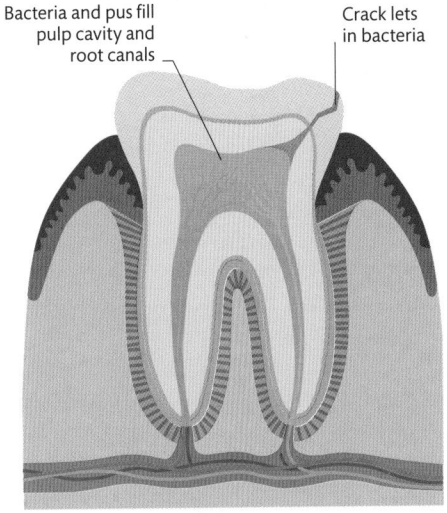

TOOTH WITH AN ABSCESS

Abscess
If bacteria reach the pulp cavity, they may set up an infection in a place that is difficult for the immune system to tackle and lead to an abscess that can spread to the jawbone.

The grinder

Your jaws are powered by strong muscles that produce considerable pressure as you cut and grind food with your teeth. The lower jaw can withstand these forces because it is the hardest bone in your body.

How we chew

Chewing is a complex motion in which the temporalis and masseter muscles control movement of the jaw back and forth, up and down, and side to side. This grinds food between the back molars like a pestle and mortar. The flexibility of the joints in our jaws allow us to slide effortlessly between chewing movements, depending on what we are eating.

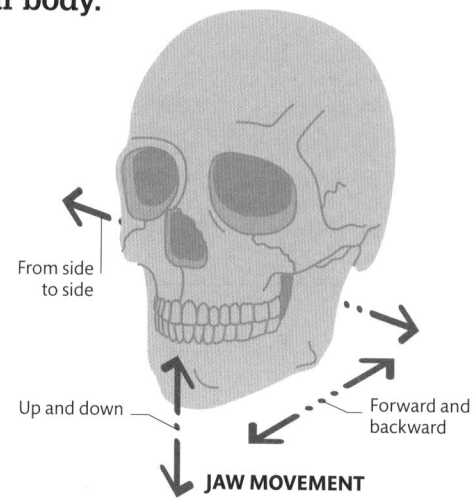

JAW MOVEMENT

WHEN WE ATE LEAVES

Once, our primitive ancestors had smaller skulls and a chewier diet, rather like today's gorilla, pictured. Their powerful jaw muscles were anchored by a tall, sagittal crest along the top of the skull. This acted in a similar way to the breastbone of a bird, which anchors its giant flight muscles.

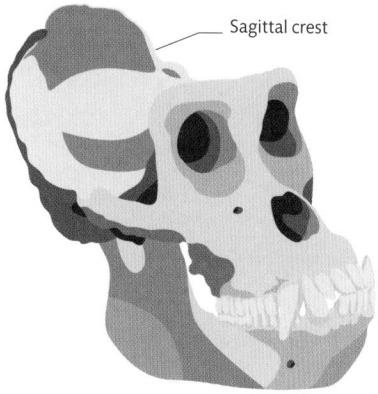

GORILLA SKULL

How the jaw works

The two temporomandibular joints between the lower jawbone and the skull each contain a disk of cartilage that provides a wider range of movement than is possible in other hinged joints, such as the elbow and knee. This disk is what allows the jaw to glide from side to side and move forward and backward when talking, chewing, or yawning.

WHAT CAUSES A CLICKING JAW?

If the protective disk of cartilage is displaced forward, you may have a clicking jaw. The lower jawbone clicks against the zygomatic arch as you chew.

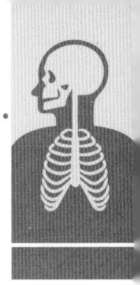

975 lb

(442 KG) THE POUNDS OF FORCE THAT THE MASSETER MUSCLE CAN EXERT DURING A BITE

Temporalis tendon attaches to the cranium with hundreds of extensions of the tendon's collagen fibers, which perforate the bone and anchor the muscle

Temporalis muscle forms a thin sheet over the side of the skull

Chewing muscles attach to the front and back of the cheek bone

CRANIUM

TEMPORALIS TENDON

TEMPORALIS MUSCLE

ZYGOMATIC ARCH

Cartilage disk

MASSETER MUSCLE

MAXILLA, OR UPPER JAWBONE

Pterygoid muscle pulls hinge joint open when using jaw

Masseter muscle can close the jaw with great force

MANDIBLE, OR LOWER JAWBONE

Jaw muscles
The chewing muscles are attached to the skull. The strong temporalis and masseter muscles control the jaw as it grinds, snaps, and closes.

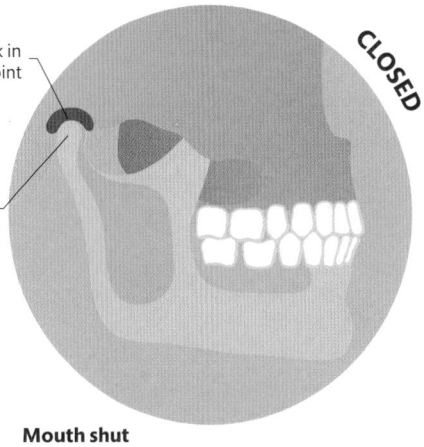

Mouth shut
The cartilage disk within the temporomandibular joint sits in a socket in the skull and wraps around a knob on the lower jawbone called the condyloid process. The disk cushions the joint and prevents the jawbone from grinding against the skull bones when you chew.

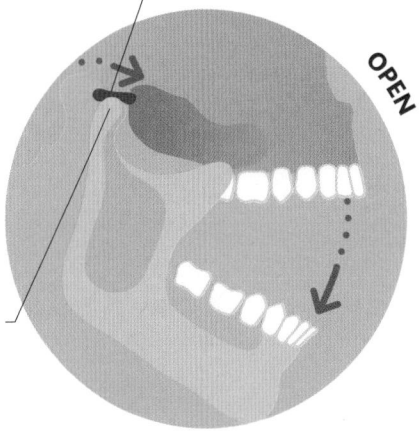

Mouth gaping
Both the lower jaw and the cushioning disk of cartilage can rock forward out of their socket, allowing your lower jawbone to hang open. Three fingers should fit between your upper and lower teeth.

Skin damage

Damaged skin, whether it is a superficial scrape or a cut that penetrates deeper into the skin, lets infection enter the body. It is therefore important for healing to occur quickly, to prevent infections from spreading.

WHY DO SCABS ITCH?

During healing, when the cells move around the base of the wound, they begin to contract, which helps stitch the skin back together. As the tissues shrink, they stimulate specialized itch-sensitive nerve endings. Try not to scratch the scab off, though!

Wound healing

When the skin is breached, the first important step for the circulatory system is to stem bleeding from a cut, or weeping fluid loss from a burn or blister. Some wounds need medical attention to seal them more firmly with stitches, butterfly bandages, or tissue glue. Covering the wound with a dressing will aid healing and reduce the chance of infection.

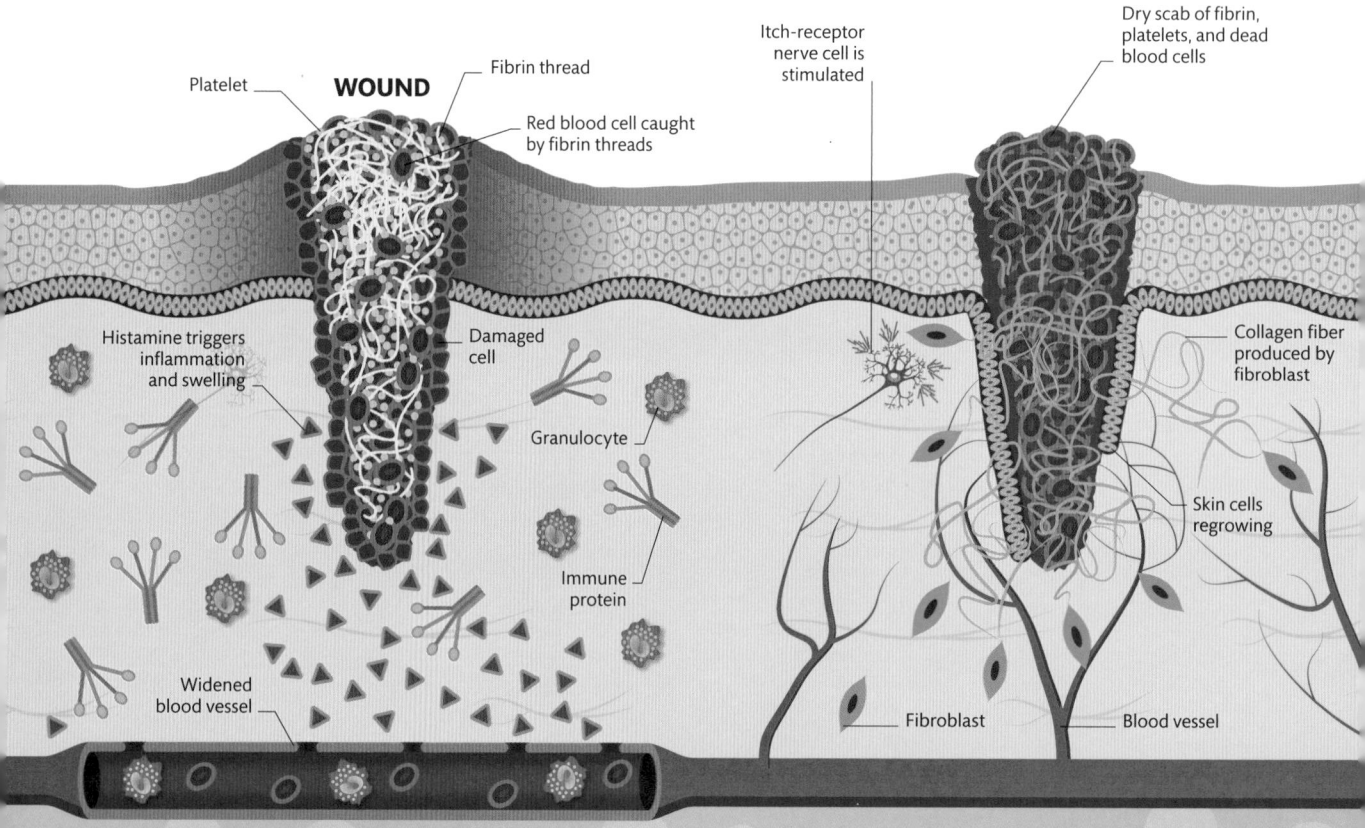

1 Clotting and inflammation
Platelets, which are fragments of blood cells, clump together to form a clot. Clotting factors form fibrin threads, which hold the clot in place. Inflammation floods the area with granulocytes and other cells and proteins of the immune system, which attack invading microbes.

2 Skin cells proliferate
Proteins called growth factors attract fiber-producing cells (fibroblasts), which move into the wound. They make granulation tissue, which is rich in tiny new blood vessels that grow into the area. Skin cells multiply to heal the wound from the base and sides.

WET AND DRY HEALING

When exposed to the air, a scab hardens so new skin cells have to burrow underneath and dissolve it away. Modern dressings help keep a wound moist so skin cells can leapfrog across the moist wound surface. This helps wounds heal more quickly, with less pain, less risk of infection, and less scarring.

DRY HEALING

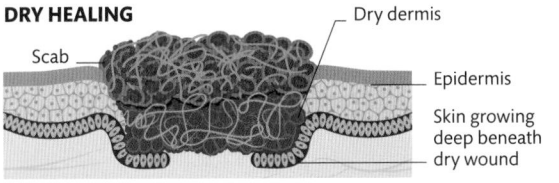

WET HEALING

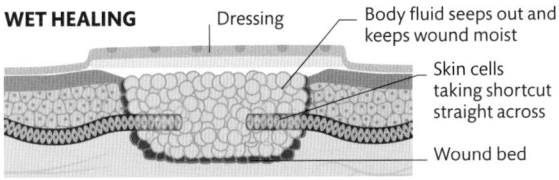

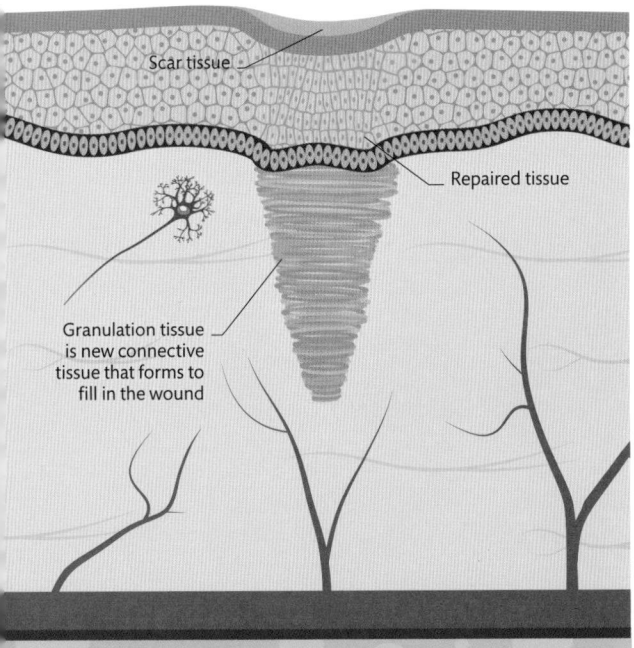

3 Remodeling
The surface skin cells have completed their job of growing over the damaged area and converting the scab into scar tissue. The scar shrinks to leave a red area that slowly becomes paler. Granulation tissue remains for a while.

Burns

If skin is heated above 120°F (49°C), its cells are damaged to cause a burn. Burns can also result from contact with chemicals and electricity.

1st degree burn
Only the top layer of skin is injured, causing reddening and pain. Dead cells may peel after a few days.

2nd degree burn
Cells in the deeper layers are destroyed and large blisters form. Enough live cells may remain to prevent scarring.

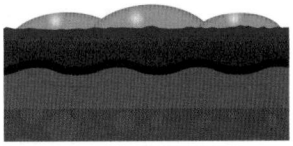

3rd degree burn
The full skin thickness is burned and skin grafts may be needed. There is a risk of scarring.

Blisters

A combination of heat, moisture, and friction may cause layers of skin to separate from each other and form a fluid-filled bubble, which protects the damaged skin. Covering them with a hydrocolloid gel blister bandage will soak up the fluid and form a cushioning, antiseptic environment so that the blister can heal faster.

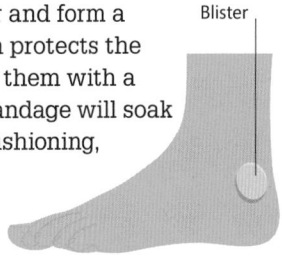

Acne

Sebaceous glands release oil (sebum) onto the skin and hair. When the glands produce an excessive amount of sebum, the hair follicle can become clogged with sebum and dead skin cells to form a blackhead. Skin bacteria can infect the plug to cause a pimple or cyst, which can leave a scar when it heals.

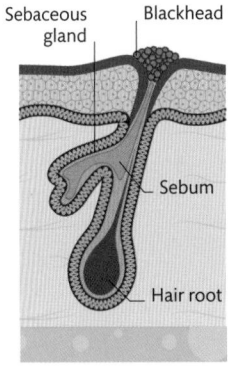

Breaking and mending

A fracture is a break in a bone, which commonly results from an accident such as a fall, a traffic accident, or a sports injury. Some fractures are relatively minor dents or hairline cracks that heal quickly, while severe impacts can shatter a bone into more than three pieces.

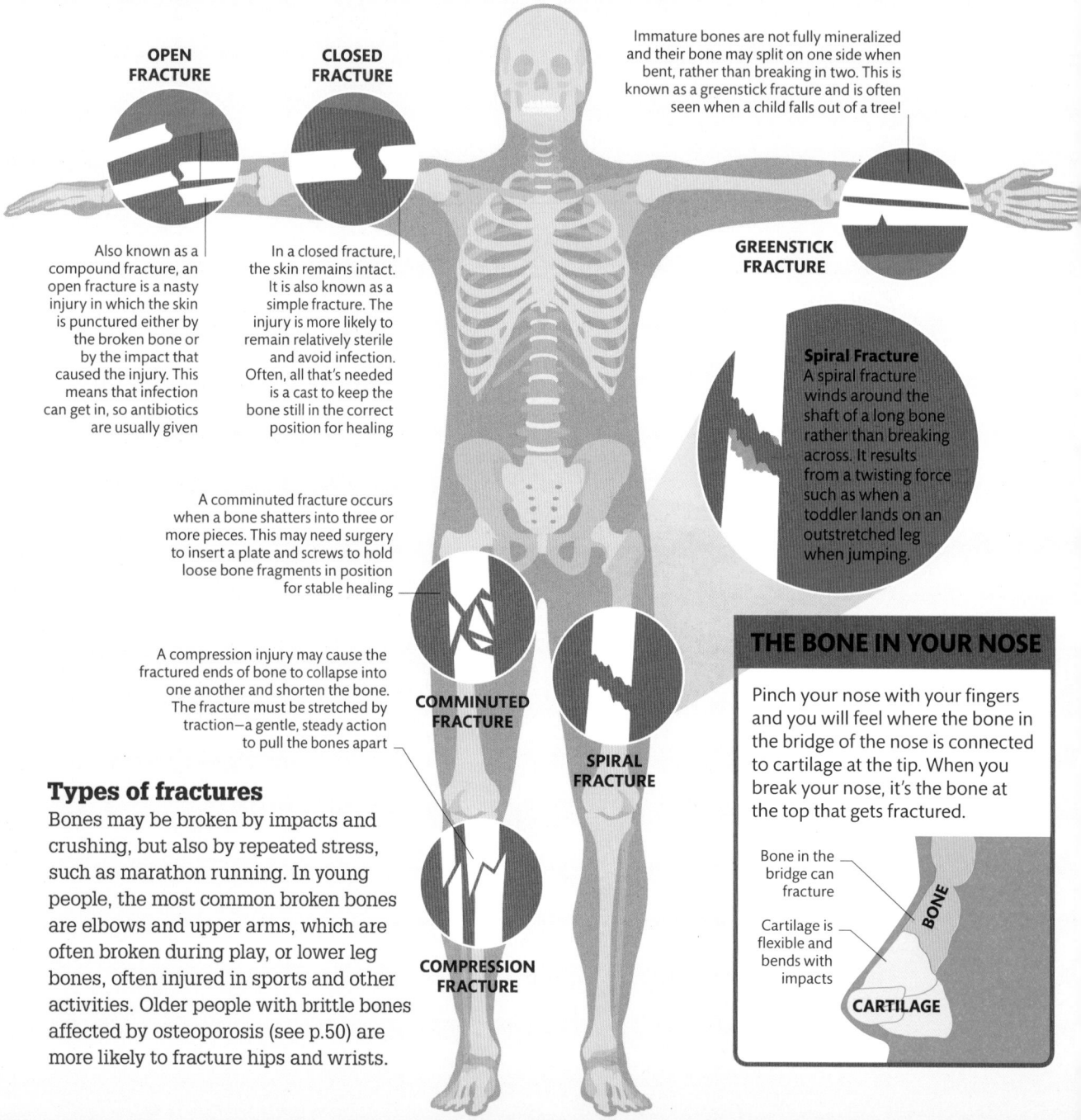

Types of fractures

Bones may be broken by impacts and crushing, but also by repeated stress, such as marathon running. In young people, the most common broken bones are elbows and upper arms, which are often broken during play, or lower leg bones, often injured in sports and other activities. Older people with brittle bones affected by osteoporosis (see p.50) are more likely to fracture hips and wrists.

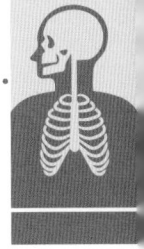

Dislocation

If the ligaments supporting a mobile joint are stretched during a wrenching accident, the bones can slip out of place, causing a joint dislocation. It is most common in the shoulder, finger, and thumb joints. To treat a dislocation, orthopedists fit the bones back into place and keep the joint still with a cast or a sling, so that the ligaments can heal. Some joints, such as the shoulder, can dislocate again and again if the ligaments remain slack.

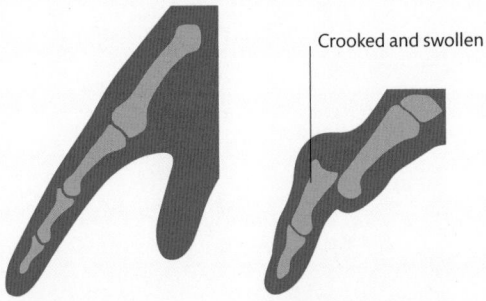

Dislocated joint
The finger joints may dislocate if you catch a ball awkwardly. It causes pain, swelling, and an obviously abnormal shape. Once the dislocated bones are repositioned (after an X-ray to rule out a fracture) the fingers are splinted together to allow healing.

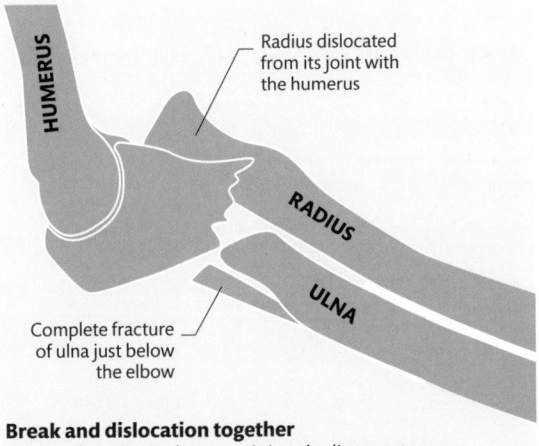

Break and dislocation together
When a fracture is close to a joint, the ligaments may give way so both a fracture and a dislocation occur. This is commonly seen at the elbow when the ulna fractures, and the head of the radius is displaced.

Healing

Bones can heal like any other living tissue, but the process takes longer as minerals must be laid down until the bone is strong again. A broken bone is immobilized by a rigid cast around the body part. If it needs firmer support, surgical screws or a metal plate may be inserted. The fracture then heals in several stages.

1 Immediate response
The fracture site quickly fills with blood to form a massive clot. The tissue around the injury forms a bruise-like swelling. The area is painful, inflamed, and some bone cells die due to poor circulation.

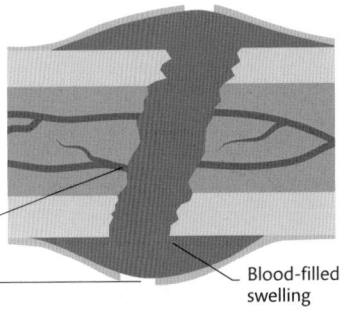

2 Three days later
Blood capillaries grow into the blood clot and the damaged tissue is slowly broken down, absorbed, and removed by scavenger cells. Specialized cells move into the area and start laying down collagen fibers that act like scaffolding for bone cells.

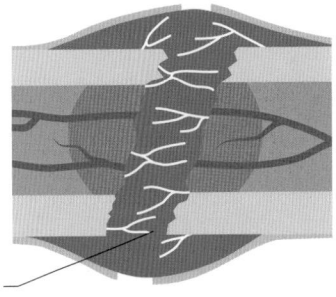

3 Three weeks later
Collagen fibers join up across the fracture to link the bone ends. The repair process forms a swelling, called a callus, which is initially formed of cartilage. This provides weak support which can easily refracture if moved too early.

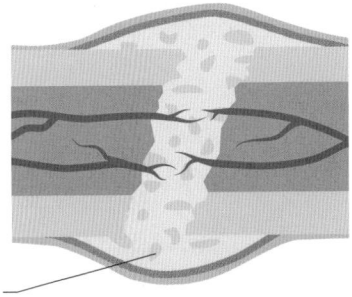

4 Three months later
Cartilage within the repair tissue is replaced with strong spongy bone and compact bone forms around the outer edge of the fracture. As the fracture heals, bone cells remodel the bone, removing the excess callus and eventually straightening out the swelling.

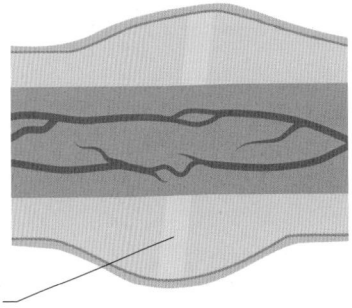

Wearing thin

Cells in our bones are constantly remodeling our skeletons by dissolving old bone and laying down new bone. However, sometimes this process becomes unbalanced, leading to a variety of problems, not all of which are easily resolved.

When bones wear out

The brittle bone disease, osteoporosis, develops when not enough new bone is made to replace the old. This imbalance can happen if you don't eat enough calcium-rich foods or if you don't top up your vitamin D—either in your diet or by getting enough sun (see p.33)—which the body needs to absorb calcium efficiently. It can also result from hormone changes in later life, such as when women's estrogen levels fall after menopause. Osteoporosis produces few symptoms, but the first indication is often when a fracture of the hip or wrist occurs after a minor fall.

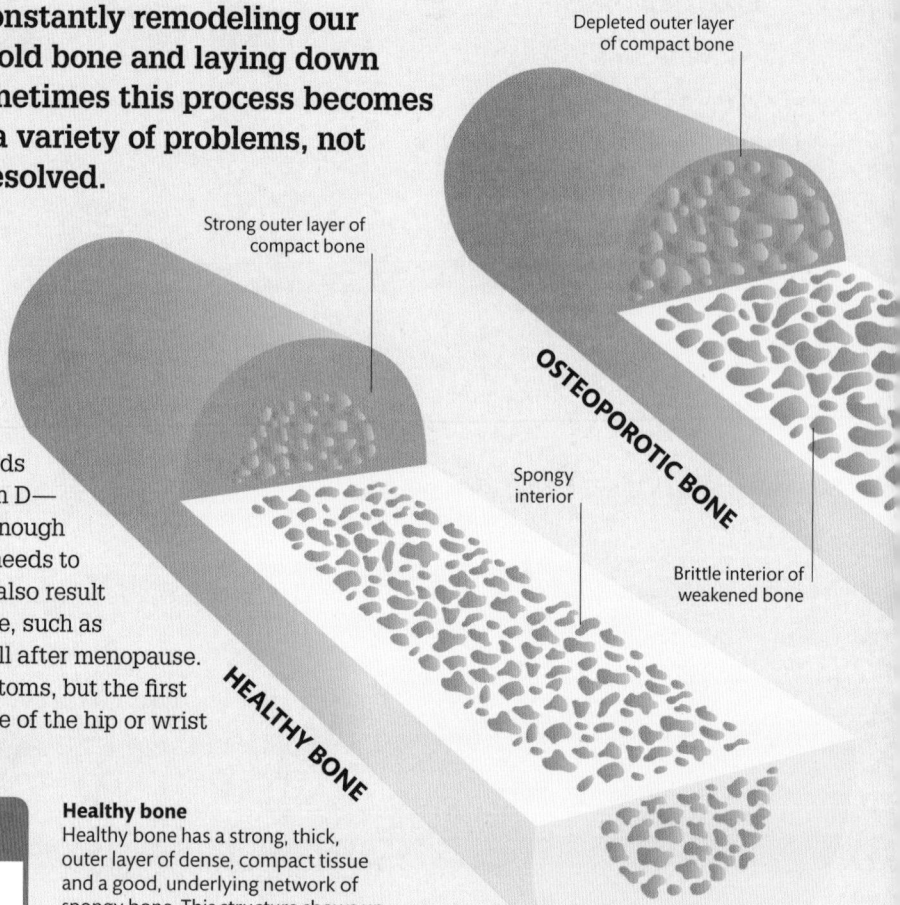

Healthy bone
Healthy bone has a strong, thick, outer layer of dense, compact tissue and a good, underlying network of spongy bone. This structure shows up clearly on X-rays and is strong enough to withstand minor blows, such as a fall onto your outstretched hands.

BONE EXERCISE

Regular exercise stimulates the production of new bone tissue. High-impact exercises, such as aerobics, jogging, or racquet sports are best, but any weight-bearing exercise, including gentle yoga or tai chi, helps stimulate strengthening at areas where bone is stressed.

In this yoga exercise, the tibia (shinbone) is under stress

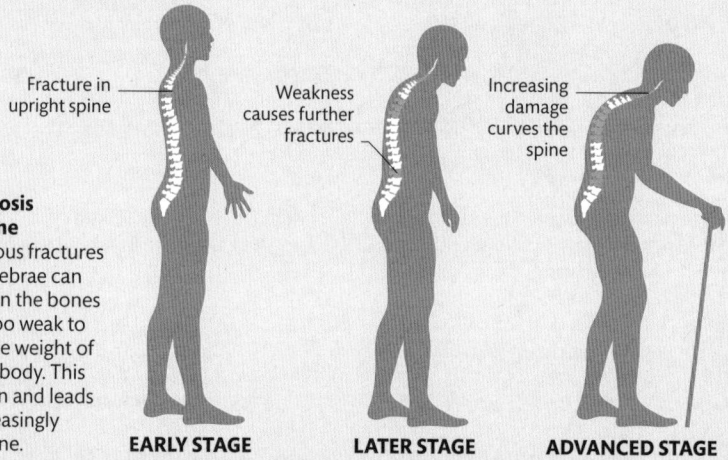

Osteoporosis in the spine
Spontaneous fractures of the vertebrae can occur when the bones become too weak to support the weight of the upper body. This causes pain and leads to an increasingly curved spine.

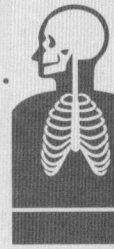

HOW COMMON IS OSTEOPOROSIS ?

Worldwide, one in three women and one in five men over the age of 50 experience an osteoporotic bone fracture. Smoking cigarettes, alcohol, and lack of exercise increase the risk of injury.

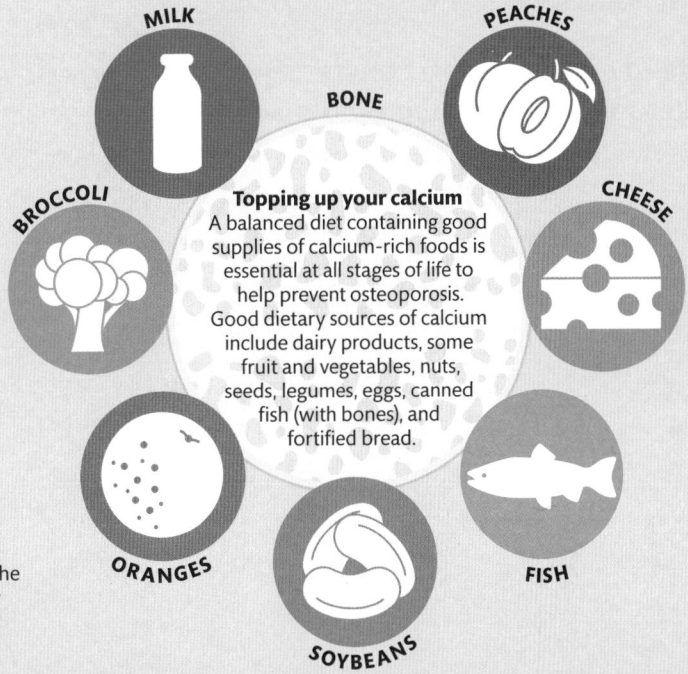

Topping up your calcium
A balanced diet containing good supplies of calcium-rich foods is essential at all stages of life to help prevent osteoporosis. Good dietary sources of calcium include dairy products, some fruit and vegetables, nuts, seeds, legumes, eggs, canned fish (with bones), and fortified bread.

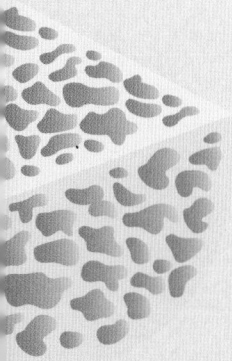

Osteoporotic bone
Brittle bones have only a thin outer layer of dense, compact bone and fewer struts within the underlying network of spongy bone. Thin bones barely show on X-rays and may fracture in a simple fall.

When joints becomes weak

Joints are subject to a lot of wear and tear, which leads to a type of inflammation called osteoarthritis. This is especially common in weight-bearing joints, such as the knee and the hip, causing increasing pain, stiffness, and restricted movements. The joint cartilage weakens and flakes away, leaving the bone ends to rub together and form bony outgrowths.

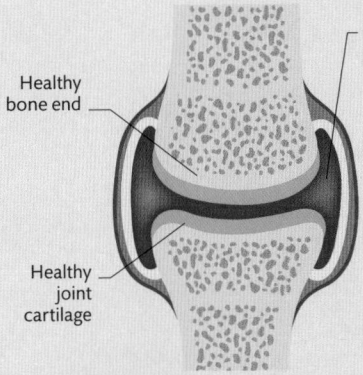

Healthy joint
In a healthy joint, the two bones are cushioned with cartilage and are separated by a film of lubricant called synovial fluid.

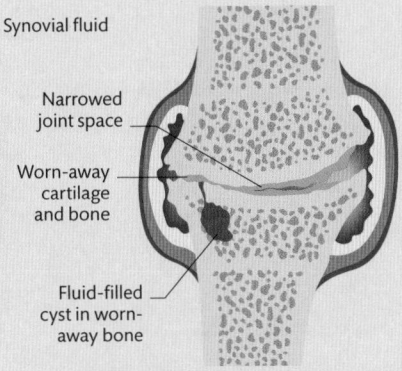

Arthritic joint
In an arthritic joint, the joint cartilages are eroding. The bones grind together and the synovial fluid is unable to lubricate the joint.

JOINT REPLACEMENT

Osteoarthritis is treated simply with analgesics, but when symptoms interfere with a person's quality of life, a better solution is to replace the worn-out joint with an artificial one made of metal, plastic, or ceramic. However, even artificial joints eventually wear out, and may need to be replaced every 10 years or so. A commonly replaced joint is the hip joint.

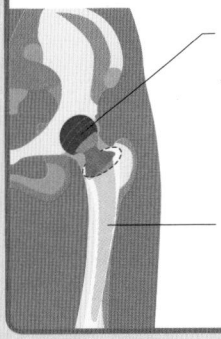

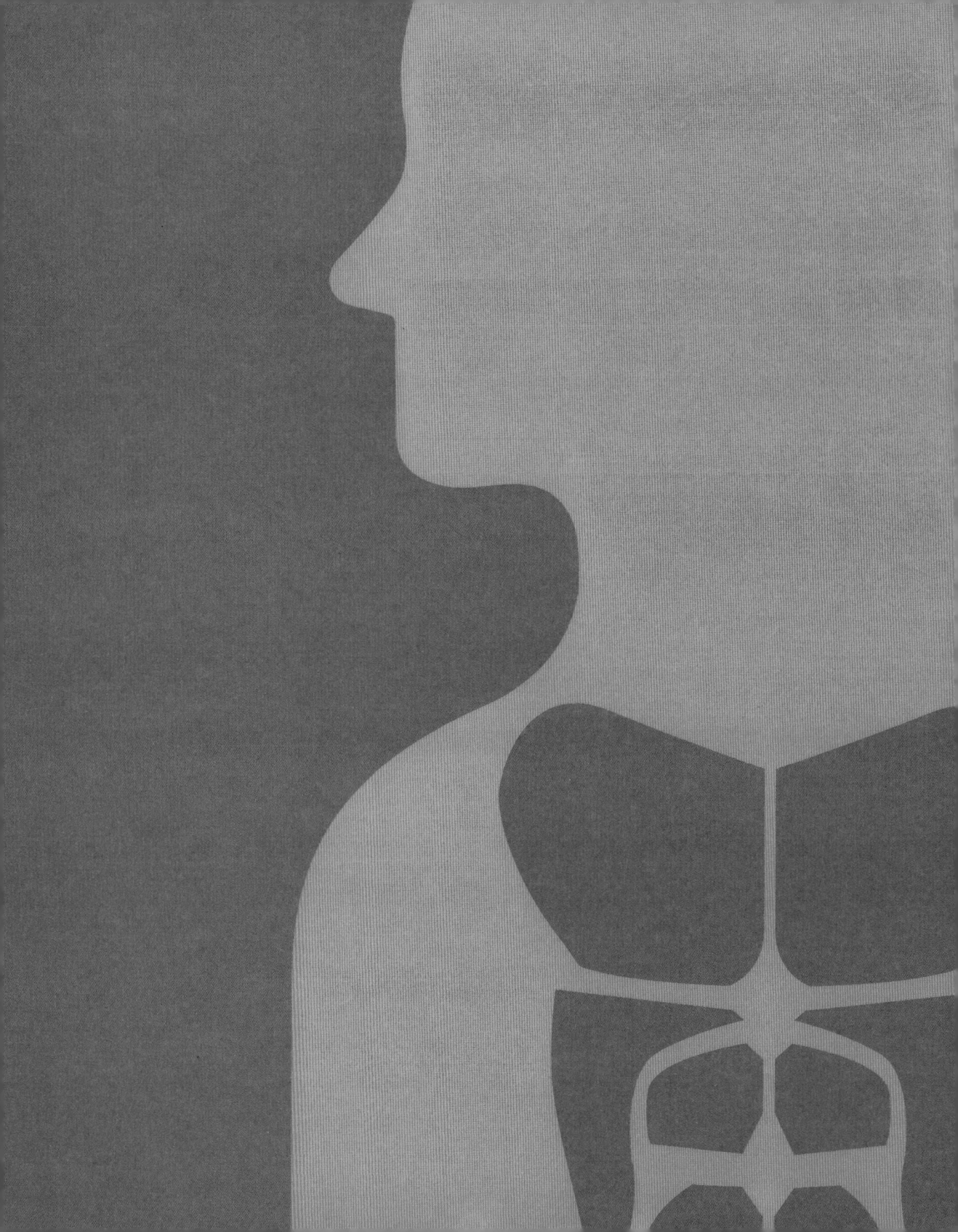

ON THE MOVE

Pulling power

Muscles carry out all the body's movements and are attached to bones by tendons. The tendons are made of strong connective tissue that can stretch to help deal with the forces produced during movement.

Teamwork

Muscles can only pull, they cannot push. They therefore work in pairs or teams that work in opposition to each other. When one set of muscles contracts to bend a joint, the other relaxes. They swap roles to straighten the joint again. For example, contraction of the biceps bends the elbow, while contraction of the triceps straightens it as the biceps relaxes. Muscle can only "push" indirectly, via levers.

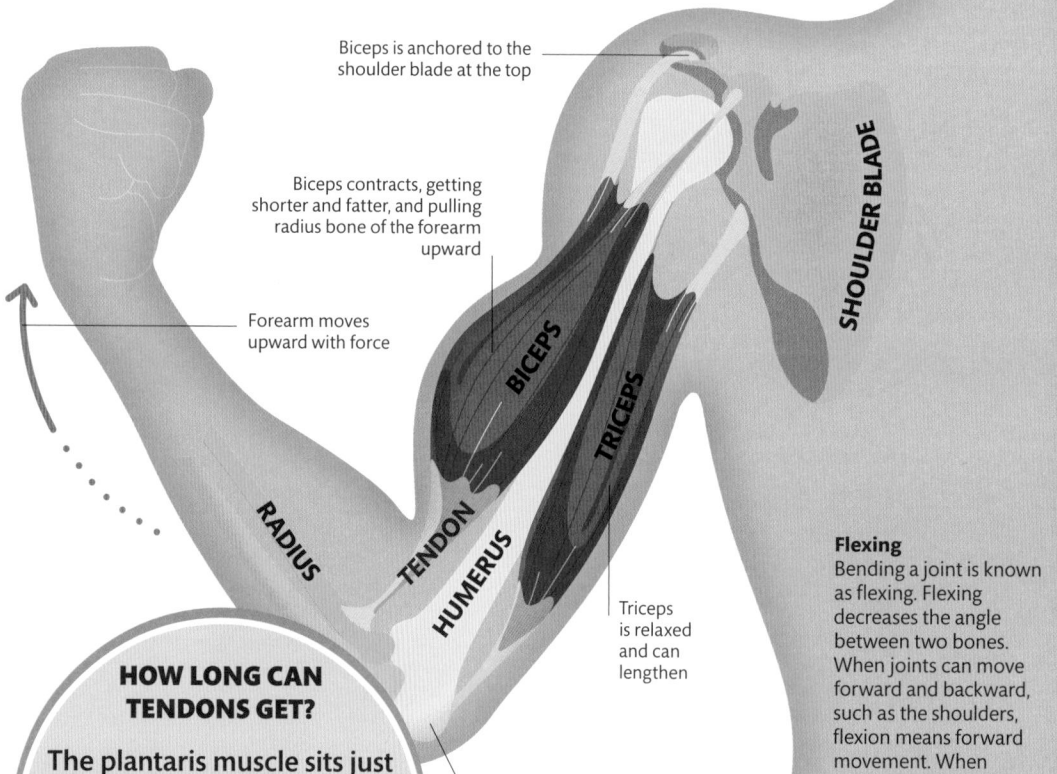

Flexing
Bending a joint is known as flexing. Flexing decreases the angle between two bones. When joints can move forward and backward, such as the shoulders, flexion means forward movement. When sitting down, both your hips and your knees flex.

Extending
Extension is the opposite of flexion and increases the angle between two bones. When joints can move forward and backward, such as the hips, extension means backward movement. When standing, both your hips and your knees extend.

HOW LONG CAN TENDONS GET?

The plantaris muscle sits just above the back of the knee and pulls on the heel bone via a tendon 20 in (50 cm) long. The Achilles tendon is the strongest and thickest in the body.

Body levers

A lever allows movement to occur around a point called a fulcrum. A first-class lever has the fulcrum in the middle. A second-class lever places the load between the effort and fulcrum. In a third-class lever, the effort occurs between the load and fulcrum—like using a pair of tweezers.

Lever action
Fulcrum · Direction of force · Movement of load

First-class lever
Neck muscles work like first-class levers. When they contract, they force your chin up on the opposite side of the fulcrum (a joint between your skull and spine).

NECK MUSCLE

Third-class lever
The biceps acts as a third-class lever. Pulling close to the fulcrum—the elbow—it moves the bones only a little, but creates a lot of movement for the hand at the end of the lever. A small effort translates into a big movement.

BICEPS

Fulcrum is the elbow joint

CALF MUSCLE

Body rises a little way, but with great force

Second-class lever
The calf muscle can act as a second-class lever by pulling when the foot is on the ground. The foot then bends at the base of the toe so the entire weight of the body is raised on tiptoe.

THE ACHILLES TENDON IS **STRONG ENOUGH** TO SUPPORT MORE THAN **10 TIMES YOUR BODY WEIGHT** WHEN RUNNING

BICEPS

TRICEPS

Triceps is anchored to the shoulder blade and humerus at the top end

Biceps is relaxed, and can lengthen, allowing triceps to extend the elbow

Forearm moves down

ULNA

Elbow extends (straightens)

Triceps contracts, pulling on the ulna bone of the forearm

Finger extender muscle is anchored to the upper arm bone at one end

Muscle's tendon splits to pull on four fingertips

Remote control

Muscles pull on bones via tendons. However, the tendons can be very long, and the muscles far from the joints they are operating. Amazingly, there are no muscles at all in the fingers. All of their movement is made by remote control—by muscles in the hand and arm.

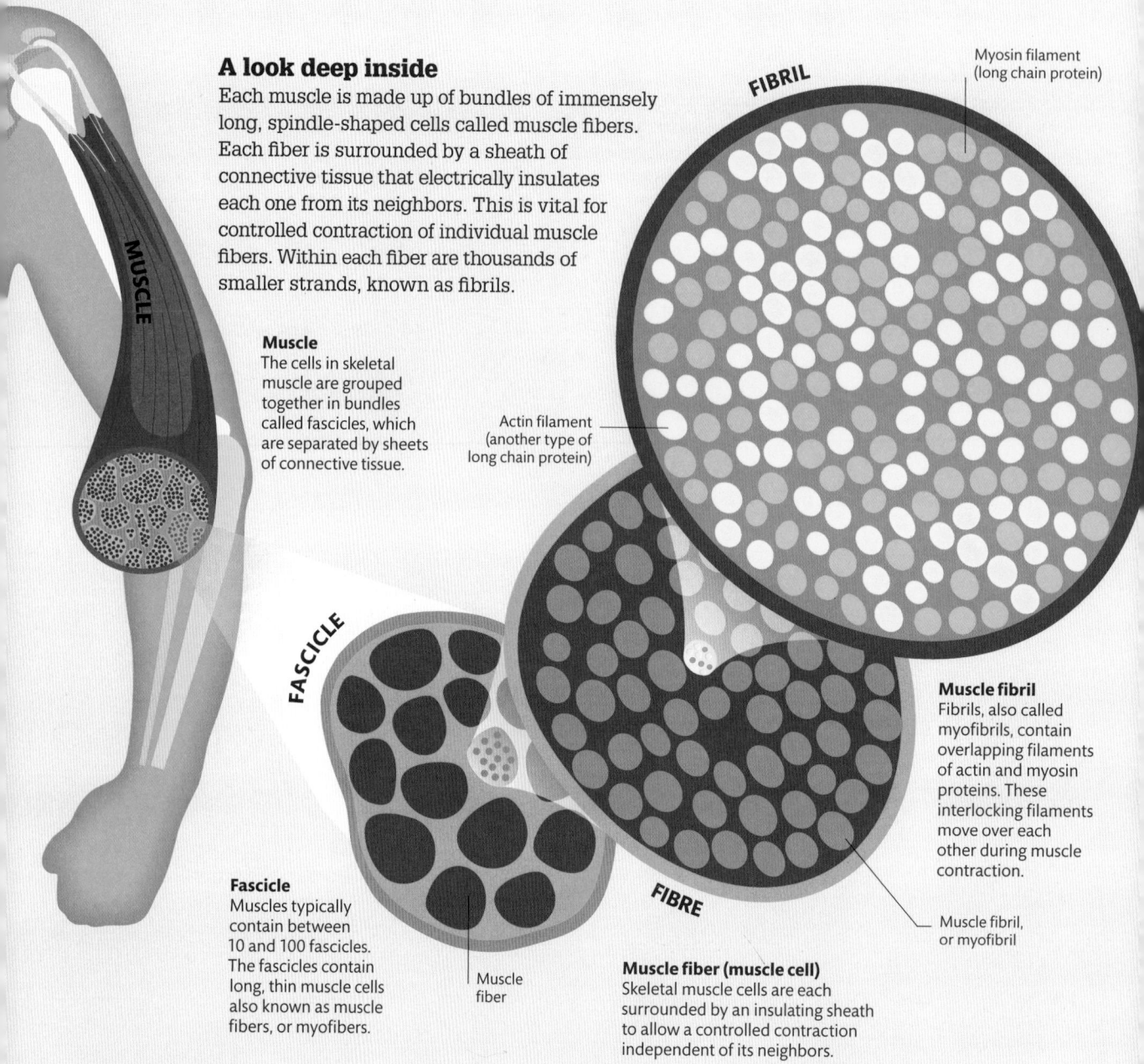

A look deep inside

Each muscle is made up of bundles of immensely long, spindle-shaped cells called muscle fibers. Each fiber is surrounded by a sheath of connective tissue that electrically insulates each one from its neighbors. This is vital for controlled contraction of individual muscle fibers. Within each fiber are thousands of smaller strands, known as fibrils.

Muscle
The cells in skeletal muscle are grouped together in bundles called fascicles, which are separated by sheets of connective tissue.

Muscle fibril
Fibrils, also called myofibrils, contain overlapping filaments of actin and myosin proteins. These interlocking filaments move over each other during muscle contraction.

Fascicle
Muscles typically contain between 10 and 100 fascicles. The fascicles contain long, thin muscle cells also known as muscle fibers, or myofibers.

Muscle fiber (muscle cell)
Skeletal muscle cells are each surrounded by an insulating sheath to allow a controlled contraction independent of its neighbors.

How do muscles pull?

Muscle cells carry out all body movements. Some muscles are under voluntary control and only contract when you want them to. Others contract automatically to keep your body working smoothly. Muscle cells are able to contract due to actin and myosin molecules.

Miracle molecules

Actin and myosin filaments are arranged in units called sarcomeres. When a muscle receives a signal to contract, the myosin filaments repeatedly pull the actin filaments along so that they slide closer and closer together. This makes the muscle shorten. They slide apart when the muscle relaxes again.

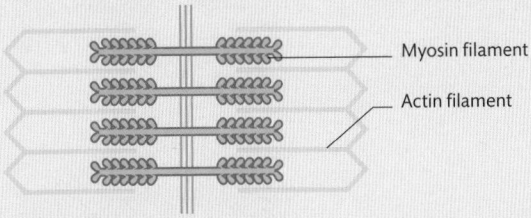

SARCOMERE OF RELAXED MUSCLE

1 Myosin energized
Myosin head is energized by ATP molecule (produced from sugars and oxygen).

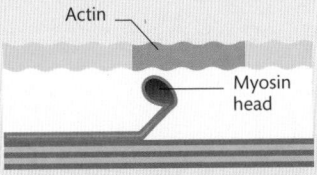

2 Myosin head sticks to actin
The energized myosin head sticks to the actin filament, forming a cross bridge.

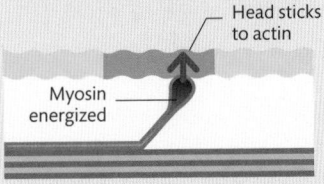

3 Head pivots
The myosin head releases energy and pivots, sliding the actin filament over. The cross bridge weakens.

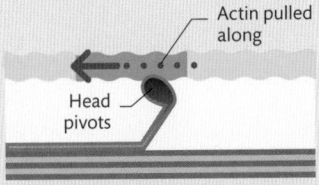

4 Reenergizing
The cross bridge releases and the myosin head is reenergized. These steps happen many times during a single contraction.

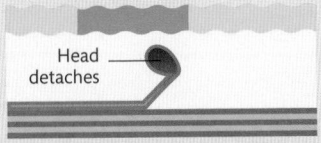

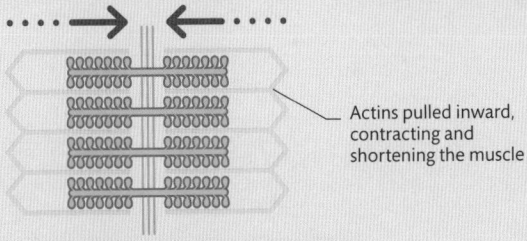

SARCOMERE OF CONTRACTED MUSCLE

FAST AND SLOW TWITCHING

Muscles have two types of fibers. Fast-twitch fibers reach peak contraction—the peak of their power output—in 50 milliseconds, but fatigue after a few minutes. Slow-twitch fibers take 110 milliseconds to reach peak contraction but do not tire. The explosive power needed by sprinters is made possible by having more fast-twitch fibers. Long-distance runners usually have more slow-twitch fibers, which don't fatigue as quickly as fast-twitch fibers.

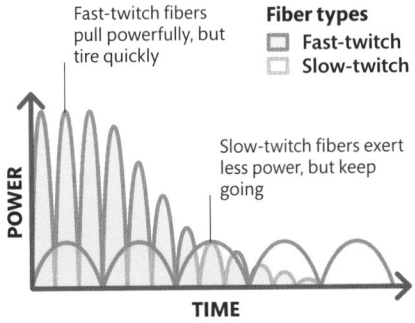

CRAMP

Sometimes voluntary muscle may contract involuntarily, causing painful cramping. This occurs when chemical imbalances—for example, when poor circulation leads to low oxygen levels and a buildup of lactic acid—interfere with the release of the cross bridges. Gently stretching and rubbing the contracted muscle stimulates circulation and helps muscle relaxation.

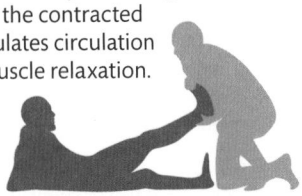

FAST-TWITCH FIBERS CAN CONTRACT AT A RATE OF 30–50 TIMES PER SECOND.

Working, stretching, pulling, braking

Muscles shorten and pull on bones to bend joints and create movement. However, they also contract without any movement to create power and tension, which can hold a weight steady. If the weight is too great to hold, muscles can even contract and lengthen as they brake the movement of the weight.

Pulling and shortening

If you contract your biceps muscle when lifting a gym weight during a "biceps curl," the muscle shortens, producing a movement in the direction of the contraction. The force generated by the muscle is greater than the weight or force it pulls against. The muscles contain both contractile fibers, which shorten, and elastic fibers, which stretch if tension increases. During a shortening contraction, the contractile fibers cause the muscle length to change, but the tension in the elastic fibers remain unchanged.

WHY WARM UP BEFORE EXERCISE?

Doing exercise to loosen muscles and increase blood flow helps to limit muscle injuries, such as tears and strains, which can occur with sudden vigorous movement.

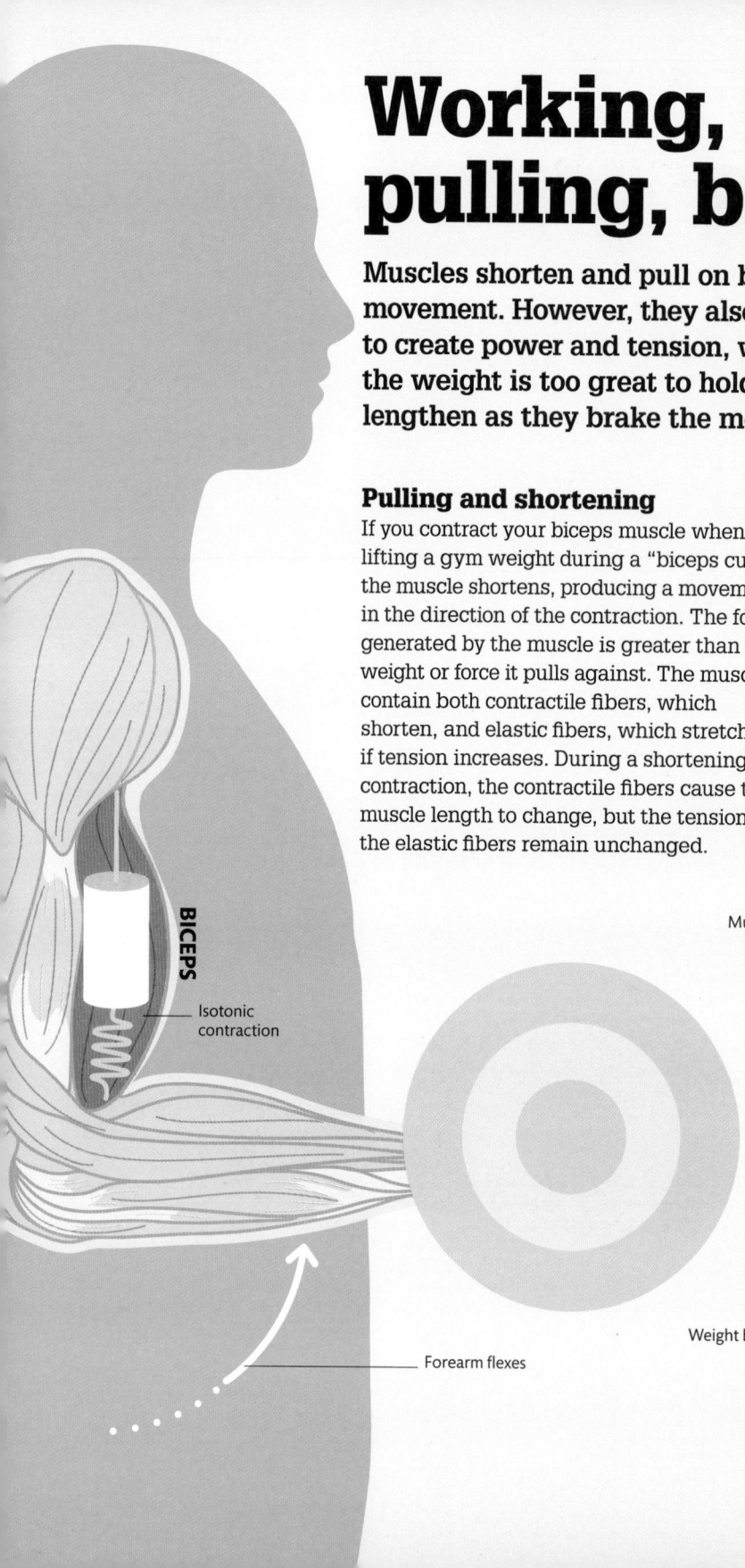

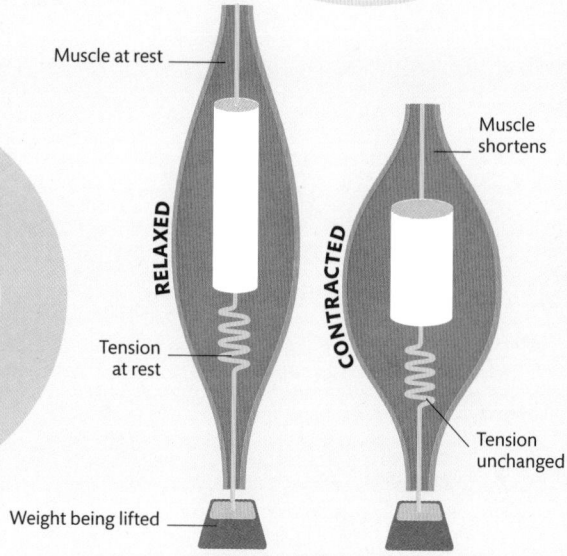

Same tension, different length
A muscle contraction is isotonic when the muscle length changes, but the tension is unchanged. If the muscle shortens, the contraction is also called concentric.

Pulling without shortening

If you hold a weight steady, without dropping it, the muscle does not change in length or generate movement. Instead of shortening, it produces a strong pulling force, or tension. In fact, many of your muscles are always slightly contracted to offset the effects of gravity on the body.

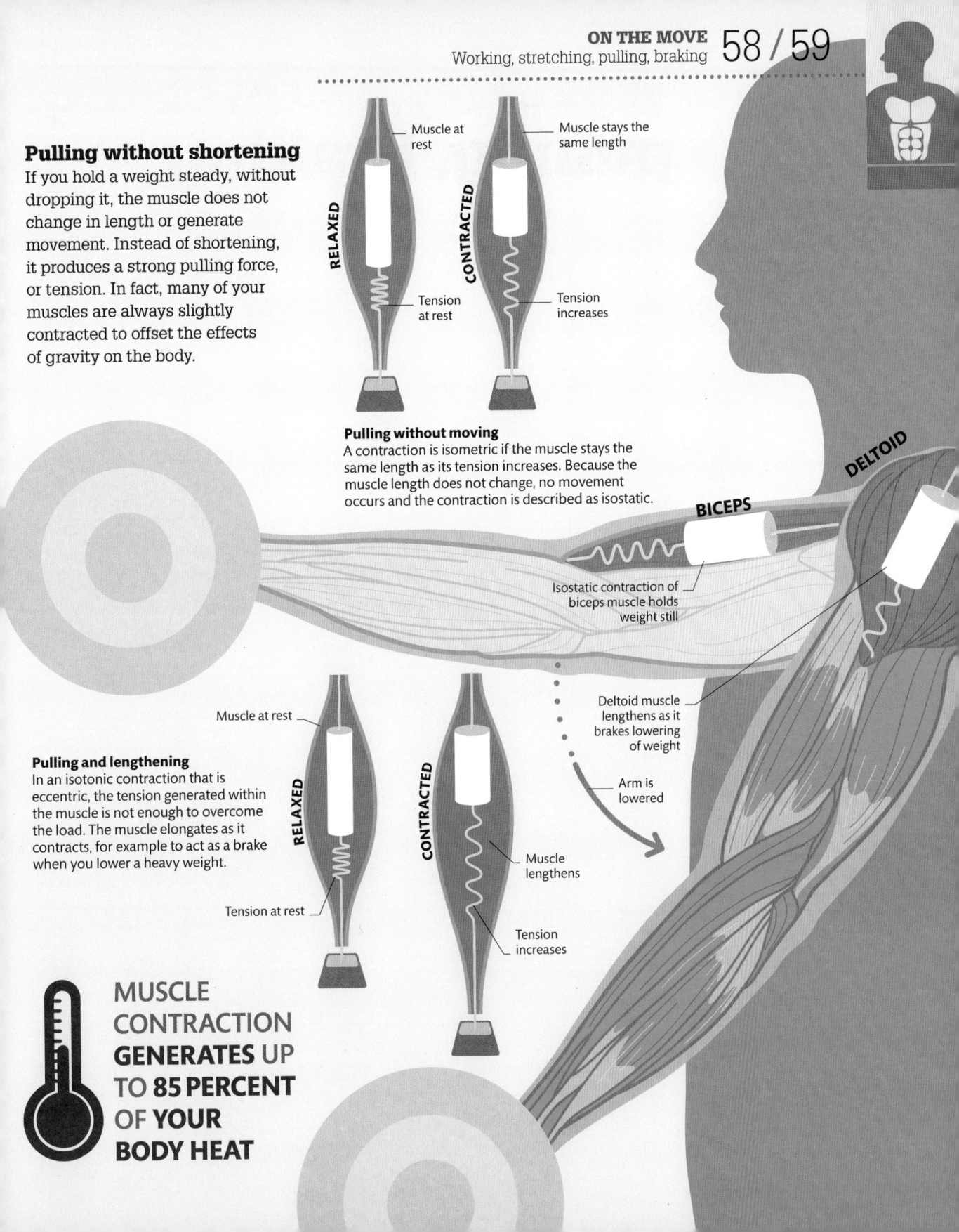

Pulling without moving
A contraction is isometric if the muscle stays the same length as its tension increases. Because the muscle length does not change, no movement occurs and the contraction is described as isostatic.

Pulling and lengthening
In an isotonic contraction that is eccentric, the tension generated within the muscle is not enough to overcome the load. The muscle elongates as it contracts, for example to act as a brake when you lower a heavy weight.

MUSCLE CONTRACTION **GENERATES** UP TO **85 PERCENT** OF **YOUR BODY HEAT**

Sensory input, action output

The brain and spinal cord form the central nervous system. They receive sensory input from all over the body via a vast network of "sensory" nerve cells. In response to the sensory information, the brain and spinal cord send instructions down "motor" nerve cells to control your actions.

IT CAN TAKE YOUR BRAIN UP TO **400 MILLISECONDS** TO **PROCESS INCOMING INFORMATION** BEFORE YOU **BECOME CONSCIOUS** OF IT

HOW FAST?

Reflex reactions are much faster than reaction times routed via the brain. This is true of reactions to visual, hearing, or touch sensations.

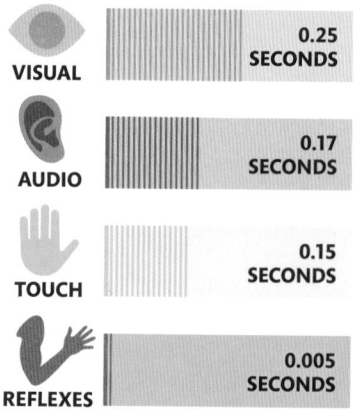

Consulting the brain

If a movement requires conscious thought, such as listening for a starter gun, the sensory signal travels up the spinal cord to the brain for processing before the body takes action. Some conscious actions become relatively automatic and are performed on "autopilot," without thinking. In fact, most nerve signals sent to and from the brain, just to keep the body functioning properly, occur subconsciously.

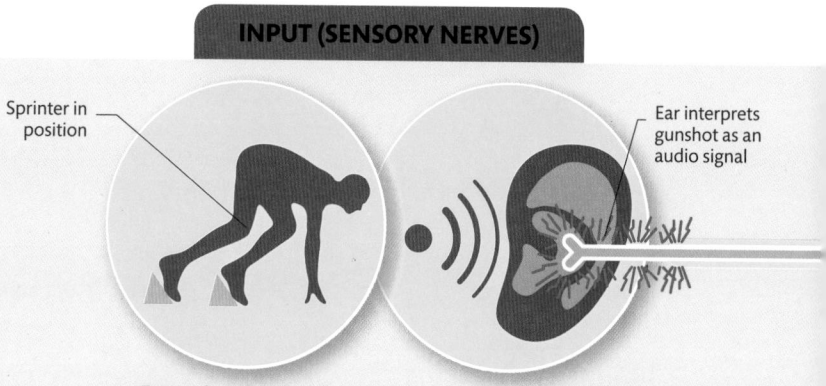

Expecting the signal
A sprinter is poised at the start line, waiting for the gunshot to start running.

Audio cue
The starter gun sounds. Audio waves reach the ear, which sends sensory messages to the brain.

Taking the brain out of the loop

Survival sometimes requires instant responses that bypass the brain and happen as automatic reflexes. Reflex pathways are routed via the spinal cord to avoid the delays that would occur if the messages traveled via the brain. When a reflex action is performed, the brain may be informed immediately afterward.

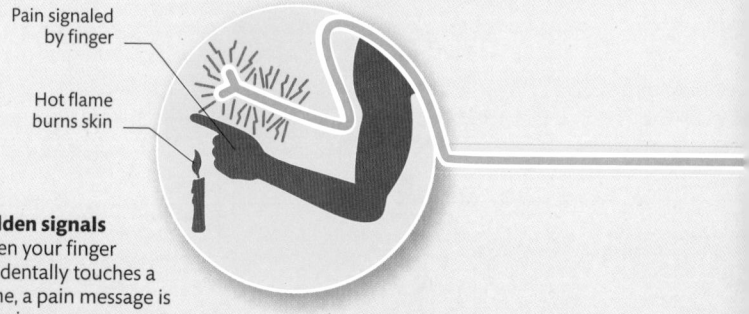

Sudden signals
When your finger accidentally touches a flame, a pain message is sent via a sensory nerve to the spinal cord.

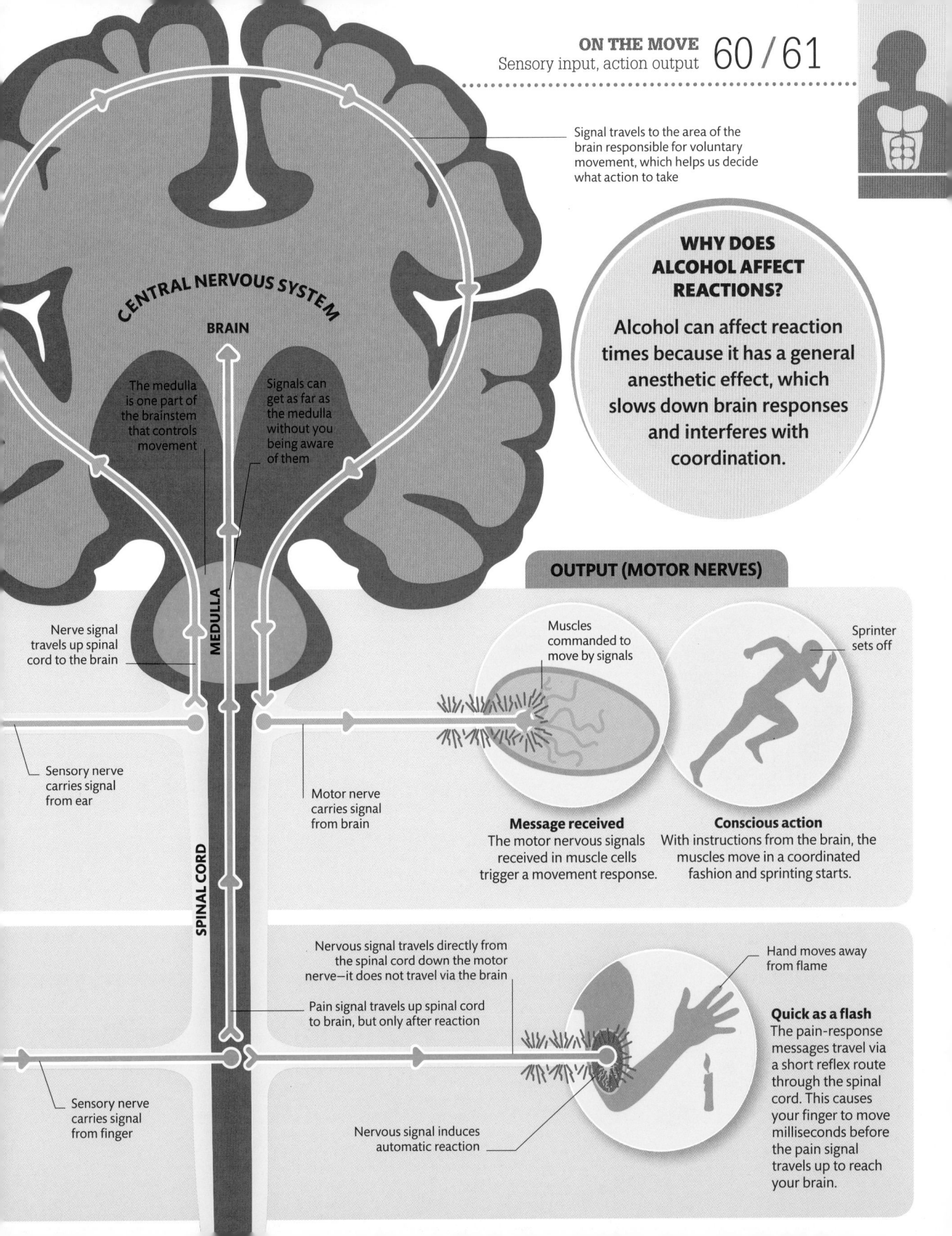

WHY DOES ALCOHOL AFFECT REACTIONS?

Alcohol can affect reaction times because it has a general anesthetic effect, which slows down brain responses and interferes with coordination.

OUTPUT (MOTOR NERVES)

The control center

The brain coordinates all body functions. It contains billions of nerve cells whose interconnections make it the most complex of all your organs. The brain can process thoughts, actions, and emotions simultaneously. Despite popular belief, you use all of your brain although the exact function of some areas remains elusive.

Inside the brain

The brain is divided into two main parts—the higher brain and the primitive brain. The higher brain is the largest and consists of the cerebrum, which is divided into two halves called the left and right hemispheres. The higher brain is the part where conscious thoughts are processed. The more primitive part of the brain, which connects with the spinal cord, is where your body's automatic functions, such as breathing and blood pressure, are controlled.

Gray matter
The darker outer layer of the brain is composed mainly of nerve cell bodies, some of which cluster together to form nerve ganglia.

White matter
The fine nerve filaments, or axons, which carry electrical impulses away from each nerve cell, form the paler tissue beneath the gray matter.

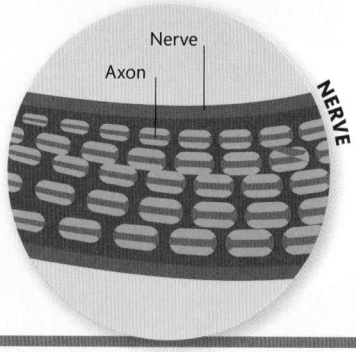

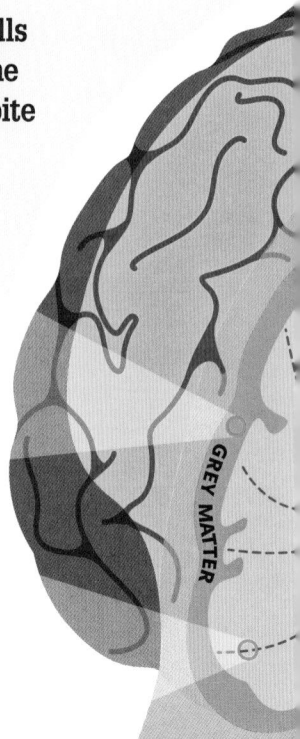

Primitive brain
The cerebellum, thalamus, and brainstem deal with instinctive responses and automatic functions, such as body temperature and sleep-wake cycles. This part of the brain also generates primitive emotions, such as anger and fear. The cerebellum coordinates muscle movements and balance.

The brain at work

When you learn a skill, new connections form between the brain cells that are used. This means that unfamiliar actions start to become automatic. The amount of practice a golfer does is reflected in the active areas of the brain when the club is swung.

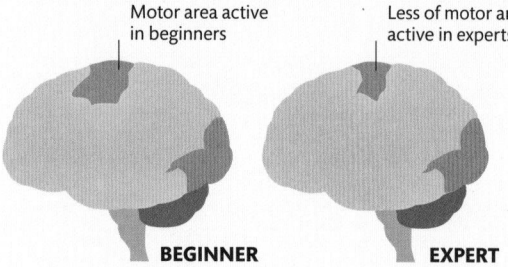

Outer cerebral activity
As you practice your shots, less of your motor area will be stimulated as the once unfamiliar action becomes more refined. Areas devoted to coordination and visual processing in both beginners and experts remain the same.

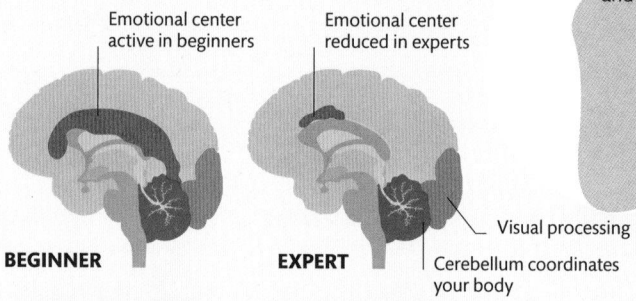

Inner cerebral activity
A cross section of the brain reveals that the brain's emotional center is active in beginners, who may deal with anxiety or embarrassment. Expert golfers learn to control their emotions and concentrate solely on taking the shot.

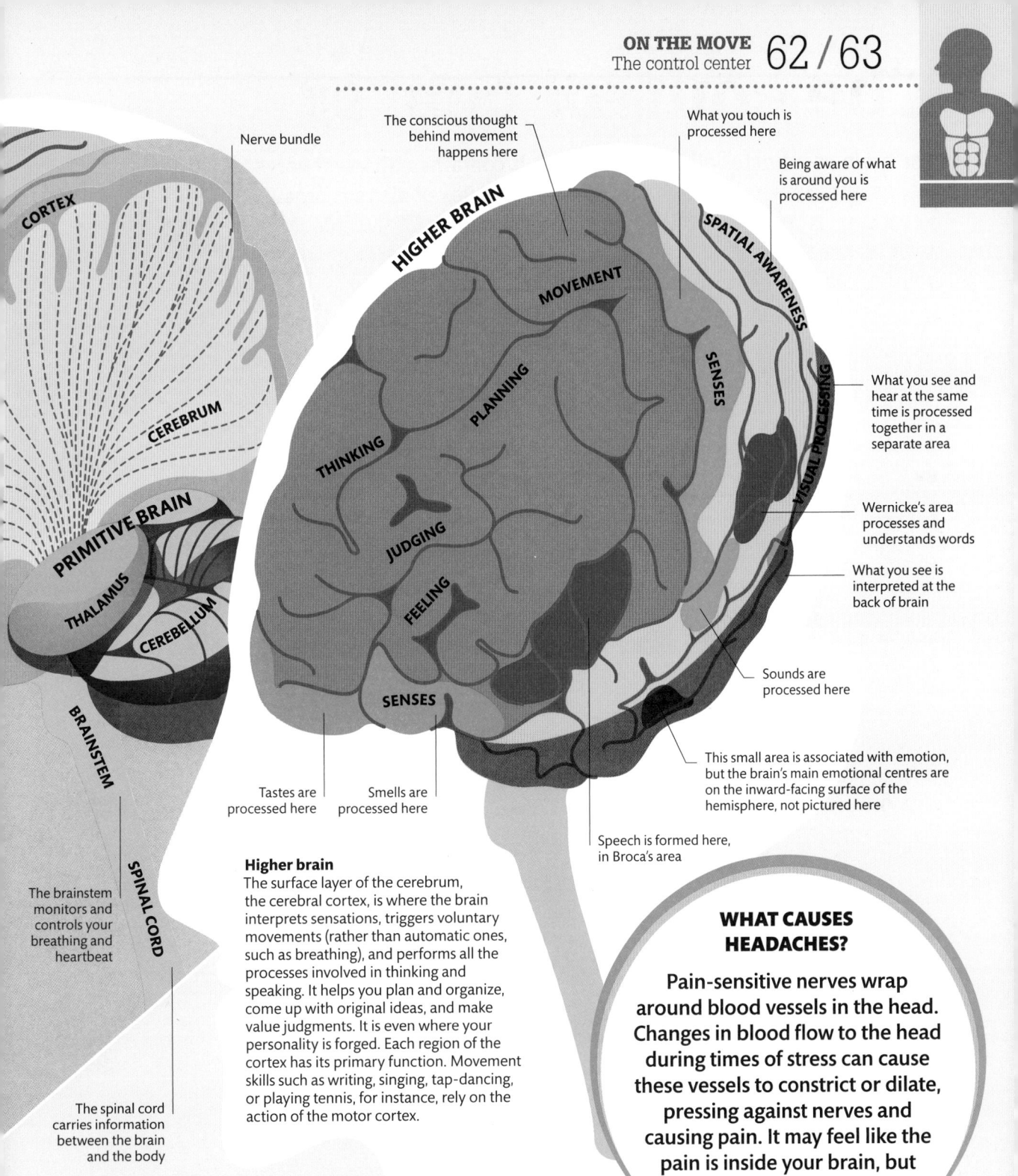

Higher brain
The surface layer of the cerebrum, the cerebral cortex, is where the brain interprets sensations, triggers voluntary movements (rather than automatic ones, such as breathing), and performs all the processes involved in thinking and speaking. It helps you plan and organize, come up with original ideas, and make value judgments. It is even where your personality is forged. Each region of the cortex has its primary function. Movement skills such as writing, singing, tap-dancing, or playing tennis, for instance, rely on the action of the motor cortex.

WHAT CAUSES HEADACHES?

Pain-sensitive nerves wrap around blood vessels in the head. Changes in blood flow to the head during times of stress can cause these vessels to constrict or dilate, pressing against nerves and causing pain. It may feel like the pain is inside your brain, but no pain-sensitive nerves are there!

Communication hub

When you think or act, it is not a single region of the brain that becomes active, but rather a network of cells spanning several brain regions. It is these patterns of activity that command your mind and body.

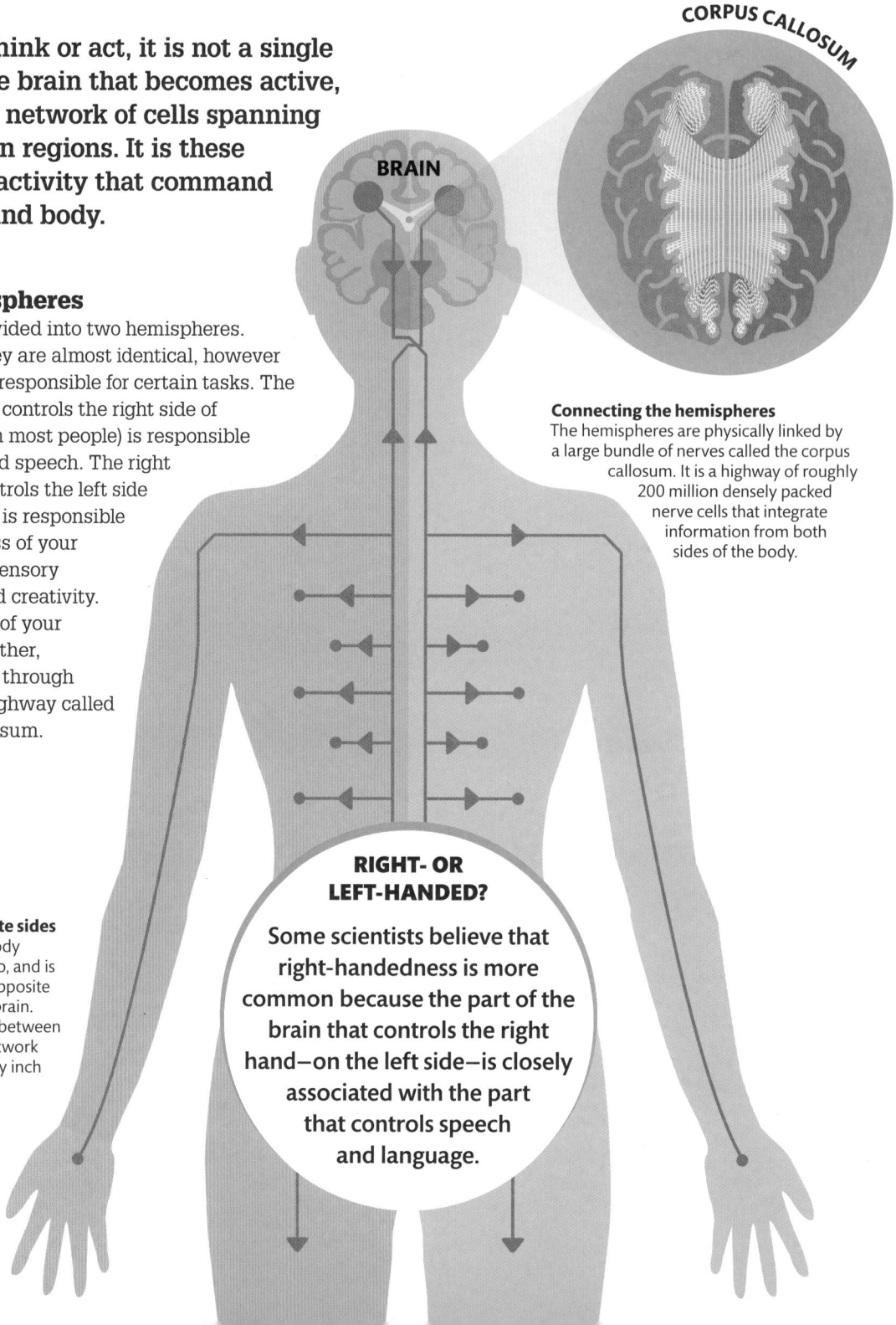

Brain hemispheres

Your brain is divided into two hemispheres. Structurally, they are almost identical, however each of them is responsible for certain tasks. The left hemisphere controls the right side of the body and (in most people) is responsible for language and speech. The right hemisphere controls the left side of the body and is responsible for an awareness of your surroundings, sensory information, and creativity. The two halves of your brain work together, communicating through a nerve superhighway called the corpus callosum.

Connecting the hemispheres
The hemispheres are physically linked by a large bundle of nerves called the corpus callosum. It is a highway of roughly 200 million densely packed nerve cells that integrate information from both sides of the body.

Controling opposite sides
Each side of your body sends information to, and is controlled by, the opposite hemisphere of the brain. Information travels between them by a nerve network that spreads to every inch of your body.

RIGHT- OR LEFT-HANDED?

Some scientists believe that right-handedness is more common because the part of the brain that controls the right hand—on the left side—is closely associated with the part that controls speech and language.

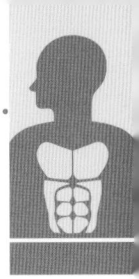

THE **BRAIN** CONTAINS **86 BILLION NERVE CELLS** JOINED BY **100 TRILLION CONNECTIONS**—MORE THAN THE NUMBER OF **STARS IN THE MILKY WAY**

Nerve pathway that connects brain regions

One of many nodes in brain active while playing chess

Multiple areas at work
When you play chess, you use many regions of your brain. Not only do you use your visual processing region, you may also activate your memory and planning areas to recall previous games and establish a strategy.

Networks in the brain

To performing the simplest action, such as walking, or a complex maneuver, such as a dance, you rarely use just one area of your brain. In fact, networks of connected areas all over the brain are activated as you go about your day. By looking for regions consistently activated together, researchers can track the flow of information around the brain. These networks can change during your lifetime as you learn new skills and information, and as a result new nerve pathways are made. Unused nerve pathways may be pruned as you grow older.

This nerve cell is connected to four others, forming a network across the brain

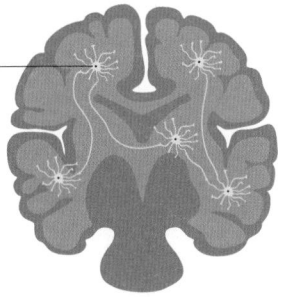

Physical connections
Scientists can trace the physical connections between nerve cells in the brain. The density of nerve pathways indicates which brain regions communicate the most.

Nerve activity is shown as areas that light up on some brain scans

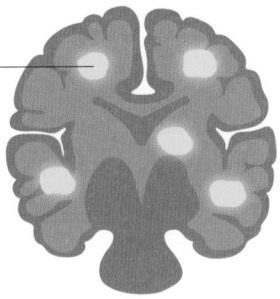

Active brain areas
The electrical activity that nerve cells generate can be picked up on certain types of brain scans. Looking at these scans can shed light on which brain regions are most active during particular tasks.

DEFAULT MODE

When you are relaxed and not focusing on the world around you, your brain shows a specific pattern of activity; this is called the default mode network. It is thought that this network helps generate thoughts as your mind wanders, and may be linked with creativity, self-reflection, and moral reasoning.

DAYDREAMER

CREATIVE THOUGHTS

Sparking into life

Nerves transmit electrical messages around the body in milliseconds. Each nerve is like a cable of insulated wires, and each wire is called a nerve fiber, or axon. An axon is the main part of a single, immensely long cell—called a neuron—whose job it is to pass on the signal.

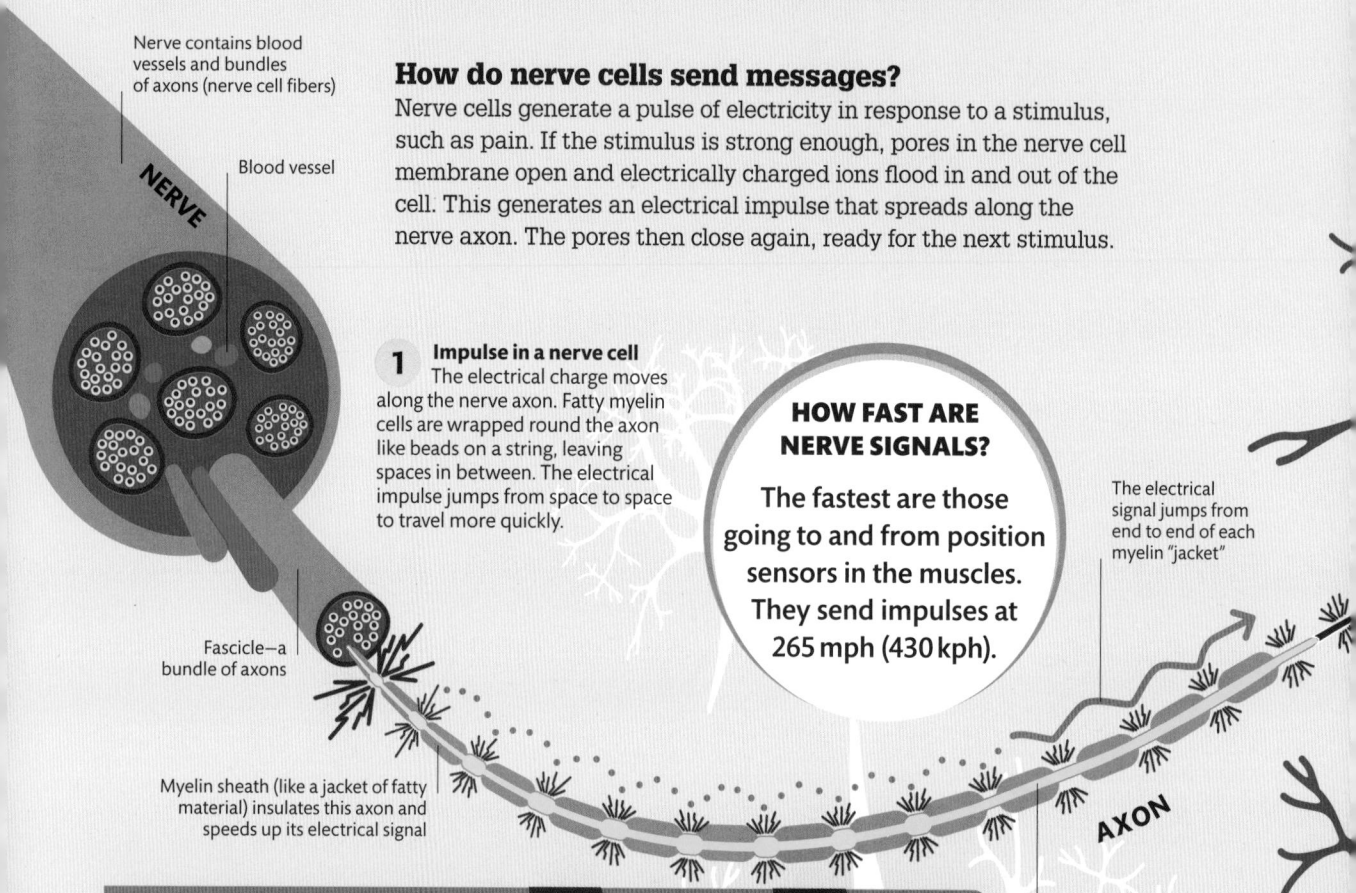

How do nerve cells send messages?

Nerve cells generate a pulse of electricity in response to a stimulus, such as pain. If the stimulus is strong enough, pores in the nerve cell membrane open and electrically charged ions flood in and out of the cell. This generates an electrical impulse that spreads along the nerve axon. The pores then close again, ready for the next stimulus.

1 Impulse in a nerve cell
The electrical charge moves along the nerve axon. Fatty myelin cells are wrapped round the axon like beads on a string, leaving spaces in between. The electrical impulse jumps from space to space to travel more quickly.

HOW FAST ARE NERVE SIGNALS?

The fastest are those going to and from position sensors in the muscles. They send impulses at 265 mph (430 kph).

PINS AND NEEDLES

Pressure over a nerve, such as from a tight sock, can cut off its blood supply. This causes numbness by preventing the nerve from sending messages. When pressure is relieved, blood flow returns. As the nerve and its receptors become active again, a tingling sensation, which can be unpleasant, occurs.

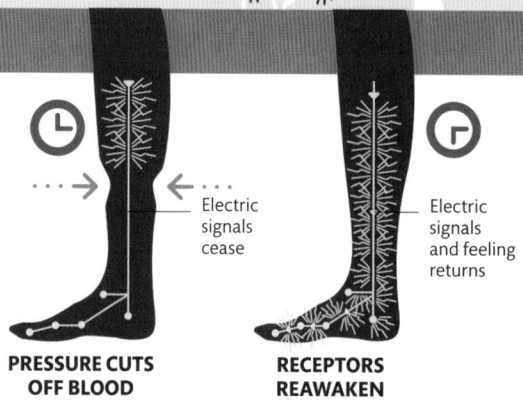

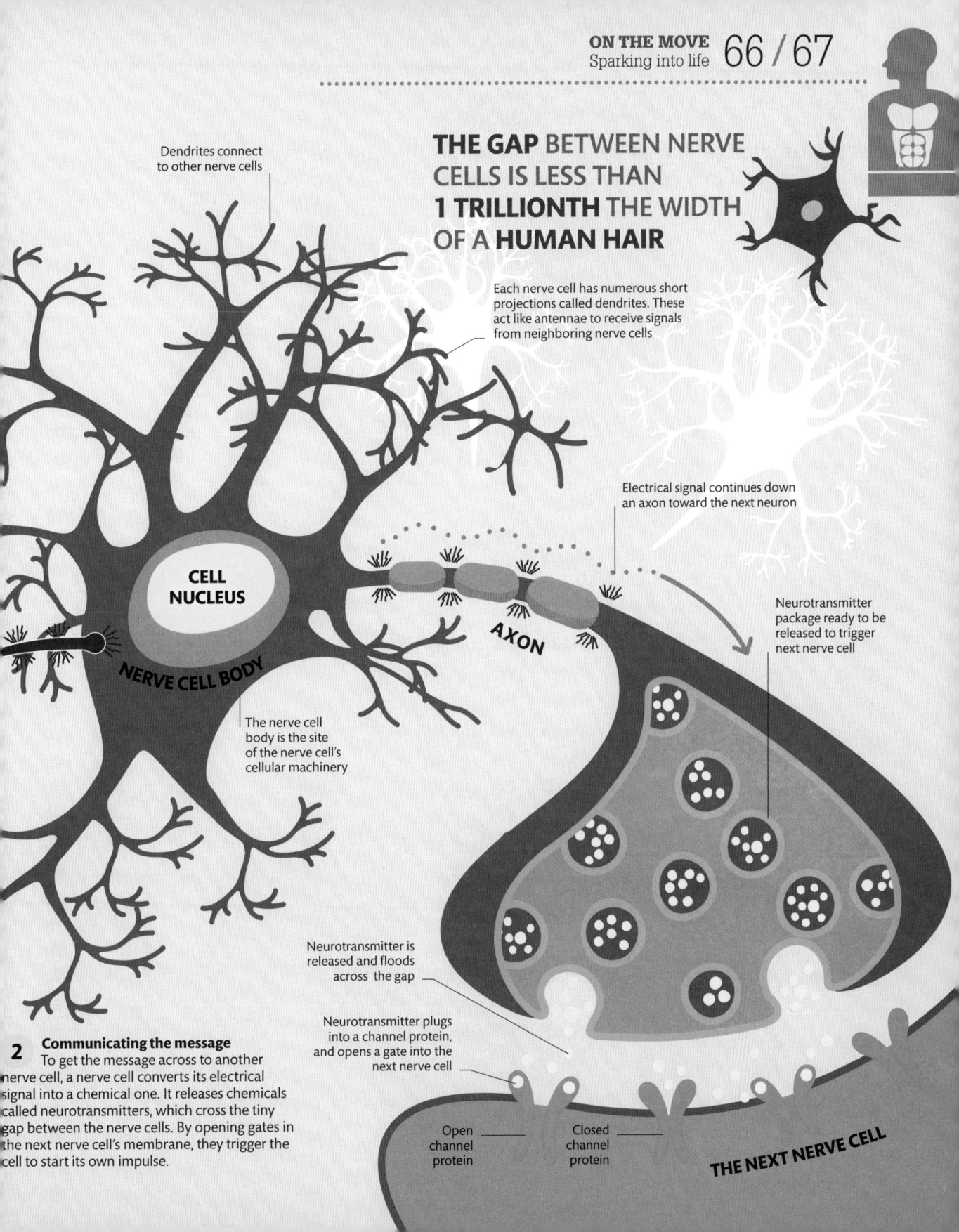

THE GAP BETWEEN NERVE CELLS IS LESS THAN 1 TRILLIONTH THE WIDTH OF A HUMAN HAIR

2 Communicating the message
To get the message across to another nerve cell, a nerve cell converts its electrical signal into a chemical one. It releases chemicals called neurotransmitters, which cross the tiny gap between the nerve cells. By opening gates in the next nerve cell's membrane, they trigger the cell to start its own impulse.

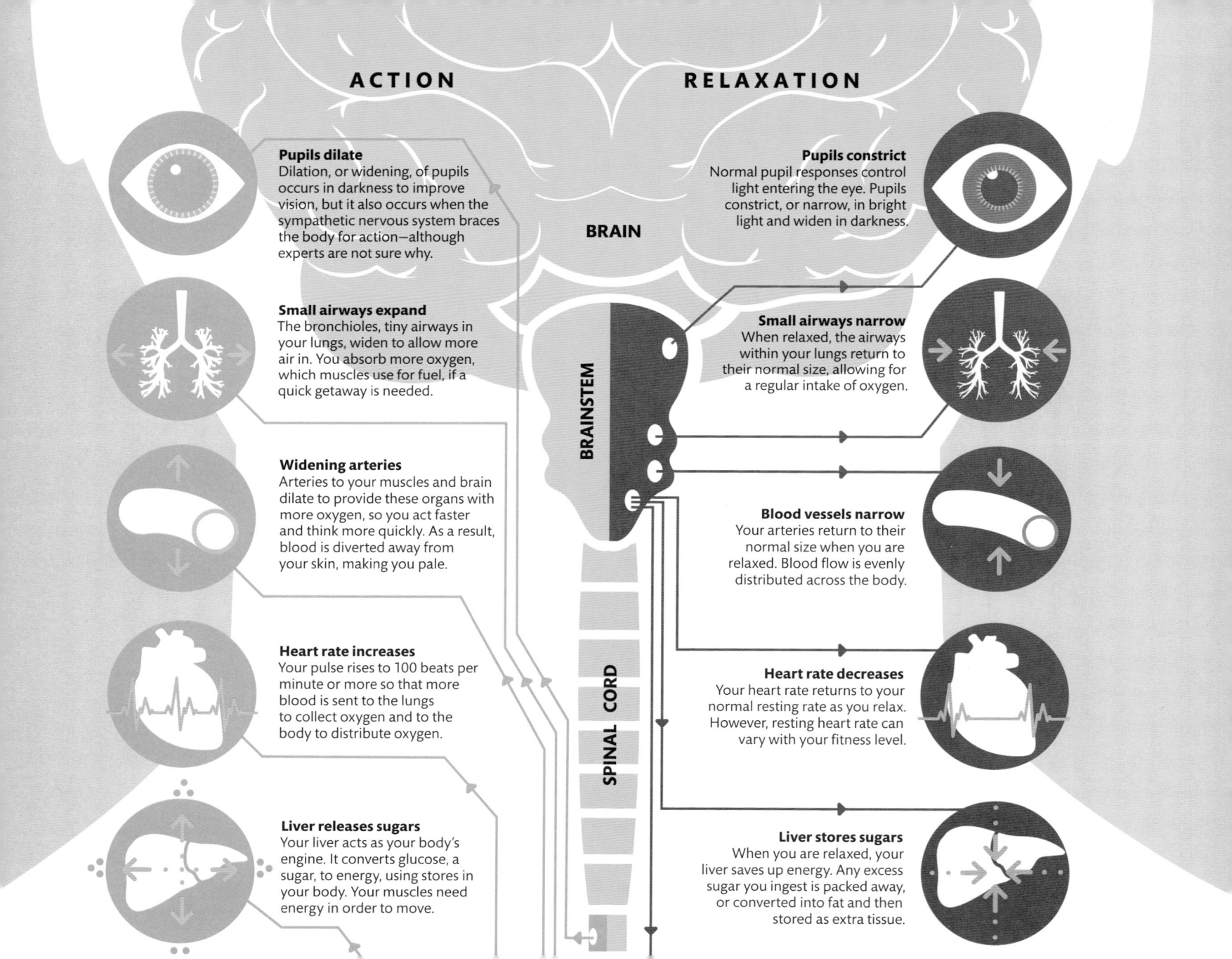
ACTION
RELAXATION
BRAIN
BRAINSTEM
SPINAL CORD
Pupils dilate
Dilation, or widening, of pupils occurs in darkness to improve vision, but it also occurs when the sympathetic nervous system braces the body for action—although experts are not sure why.
Pupils constrict
Normal pupil responses control light entering the eye. Pupils constrict, or narrow, in bright light and widen in darkness.
Small airways expand
The bronchioles, tiny airways in your lungs, widen to allow more air in. You absorb more oxygen, which muscles use for fuel, if a quick getaway is needed.
Small airways narrow
When relaxed, the airways within your lungs return to their normal size, allowing for a regular intake of oxygen.
Widening arteries
Arteries to your muscles and brain dilate to provide these organs with more oxygen, so you act faster and think more quickly. As a result, blood is diverted away from your skin, making you pale.
Blood vessels narrow
Your arteries return to their normal size when you are relaxed. Blood flow is evenly distributed across the body.
Heart rate increases
Your pulse rises to 100 beats per minute or more so that more blood is sent to the lungs to collect oxygen and to the body to distribute oxygen.
Heart rate decreases
Your heart rate returns to your normal resting rate as you relax. However, resting heart rate can vary with your fitness level.
Liver releases sugars
Your liver acts as your body's engine. It converts glucose, a sugar, to energy, using stores in your body. Your muscles need energy in order to move.
Liver stores sugars
When you are relaxed, your liver saves up energy. Any excess sugar you ingest is packed away, or converted into fat and then stored as extra tissue.

Digestion slows
Your stomach is instructed to bring digestion to a halt. In times of true terror, you may vomit to stop digestion. A full stomach can slow you down if running.

Intestine slows
Blood is diverted from the intestine, since it is an unimportant organ in times of stress, and movements in your gut slow down or stop altogether.

Bladder opening contracts
The sympathetic nervous system causes the muscle around the opening to close and keeps your bladder shut. This may be overridden at times of stress.

Braced for action

The job of igniting and stimulating your body ready for action lies with the sympathetic nervous system, which uses different nerves. Once it has served its purpose, the parasympathetic system kicks in, counteracting the sympathetic effects to wind your body back down into a relaxed state.

BUTTERFLIES IN THE STOMACH

The sensation of butterflies before a stage performance or big interview is due to the reduction of blood flow to the stomach when readying the body for danger. The stomach has a dense network of nerves, and some of these nerves signal nervous, fluttery feelings, or even nausea, as blood flow drops.

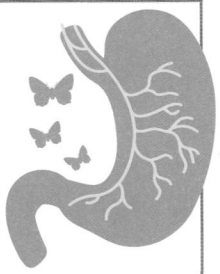

Digestion stimulated
In the absence of stress, your stomach churns away to start the digestion process. This could be why you can hear rumbling stomachs in quiet rooms.

Act or relax?

Automatic, unconscious functions of the body are managed by the "primitive" parts of the central nervous system—the spinal cord and brainstem. However, they use two different networks of nerves to control our body parts depending on whether we need to get moving or put our feet up.

Calming the nerves

Our twin automatic nervous systems are called the sympathetic and the parasympathetic. Together they form what is called the autonomic nervous system. The parasympathetic nerves tend to slow things down and start digestion. You don't tend to notice their effects.

Bladder opening relaxes
The parasympathetic system causes the bladder wall to contract and the opening to relax, resulting in the bladder emptying.

Intestine speeds up
Nutrients are absorbed from the small intestines, and bowel movements push undigested waste onward. This process works best when you are still and relaxed.

Bumps, sprains, and tears

Soft tissues of the body, such as nerves, muscles, tendons, and ligaments, are susceptible to injury, leading to bruising, swelling, inflammation, and pain. Some injuries result from sports, while others can occur from overuse or accidents. Injuries are more common with age and poor fitness.

WHY DOES HITTING YOUR "FUNNY BONE" FEEL FUNNY?

Knocking your elbow compresses the ulnar nerve, which runs down the outside of your elbow, against bone, causing an electric shock sensation.

Nerve problems

Nerves stretch for long distances and often travel through narrow spaces between bones. These tunnels guide and protect the nerve, but can also trap it to cause pain, numbness, or tingly feelings. Pinching can occur when repetitive movements cause tissues to swell, from maintaining an awkward position for a long time (such as keeping an elbow bent during sleep), or when surrounding tissues move out of alignment, which occurs with a slipped disk.

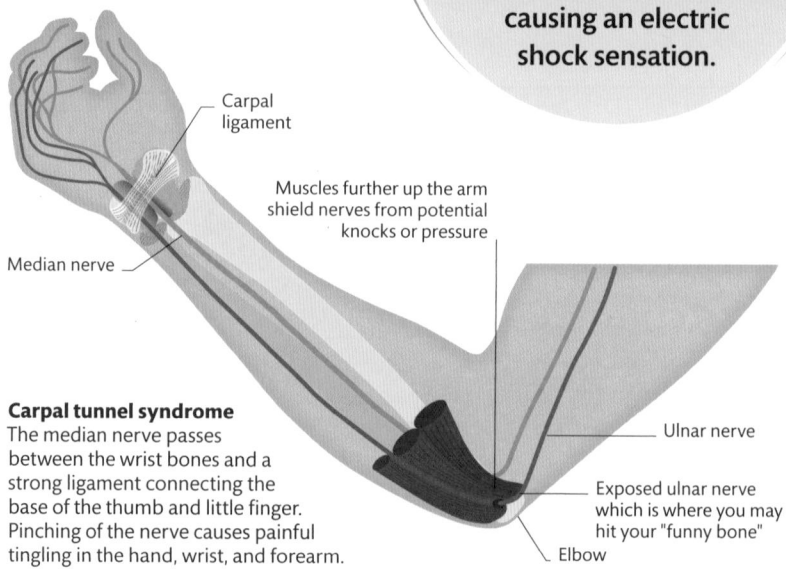

Carpal tunnel syndrome
The median nerve passes between the wrist bones and a strong ligament connecting the base of the thumb and little finger. Pinching of the nerve causes painful tingling in the hand, wrist, and forearm.

Whiplash

This injury to the neck occurs when the head is suddenly whipped backward and then forward or vice versa. This commonly happens to those traveling in a car that is hit from behind by another vehicle.

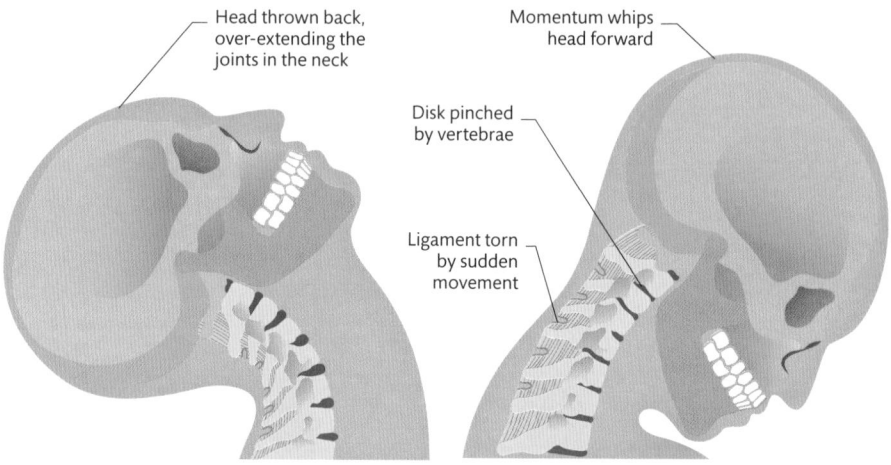

Squashed disks and torn ligaments
The sudden whiplash movement jars the neck. This motion can injure bones in the spine, compress discs between the vertebrae, tear ligaments and muscles, and stretch nerves in the neck.

Back pain

Back pain most commonly occurs in the lower spine, which is vulnerable as it supports most of the body's weight. Many cases result from heavy lifting without protecting the back by keeping it straight. Excessive strain can lead to tearing and spasms of the muscles, the stretching of ligaments, and even a dislocation of one of the tiny gliding joints (see p.40) between the vertebrae. Pressure may cause the soft, jellylike center of an intervertebral disk to rupture through its fibrous coat and press on a nerve. Treatment involves analgesics, manipulation, and remaining as mobile as possible.

Muscle tears in your back are difficult to heal because blood flow is limited

Muscle strain
When you are unfit, muscles have poor tone. They are easily strained from lifting, carrying, bending awkwardly, or even prolonged sitting in one position.

Slipped disk
A damaged spinal disk presses on a nerve root causing pins and needles, spasm, and back pain. Sciatic nerve irritation causes shooting pain down one leg.

Slipped spinal disk

Bone spurs
As aging vertebrae start to wear out, mild inflammation and the bone's attempt to heal can produce spurlike growths that press against nerve roots causing pain.

Bone growth

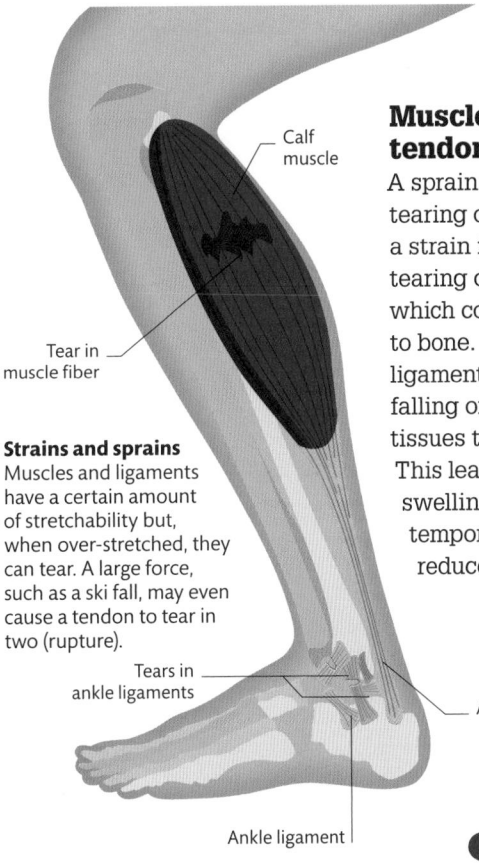

Strains and sprains
Muscles and ligaments have a certain amount of stretchability but, when over-stretched, they can tear. A large force, such as a ski fall, may even cause a tendon to tear in two (rupture).

Muscle and tendon strain

A sprain is the stretching or tearing of a ligament, while a strain is the stretching or tearing of a muscle or a tendon, which connects the muscle to bone. Muscle strains and ligament sprains occur when falling or twisting causes tissues to stretch or tear. This leads to painful spasms, swelling, and can result in temporary stiffness and reduced mobility.

THE ANKLE IS THE MOST COMMON AREA OF THE BODY TO GET A SPRAIN

"PRICE" TECHNIQUE

The PRICE technique is an effective way to treat a strain or sprain: Protection—use a support, crutch, or sling to relieve pressure. Rest—keep the injured area free from movement. Ice—apply an ice-pack to minimize swelling and bleeding. Compression—an elastic bandage reduces swelling. Elevation—keep the area raised to reduce swelling.

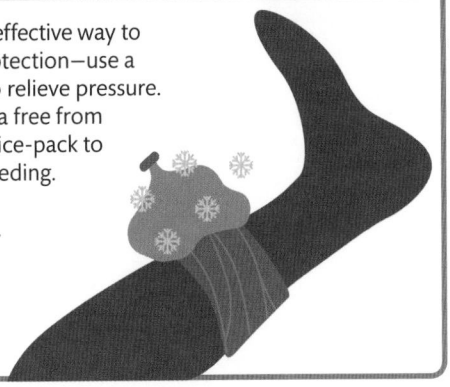

SENSITIVE TYPES

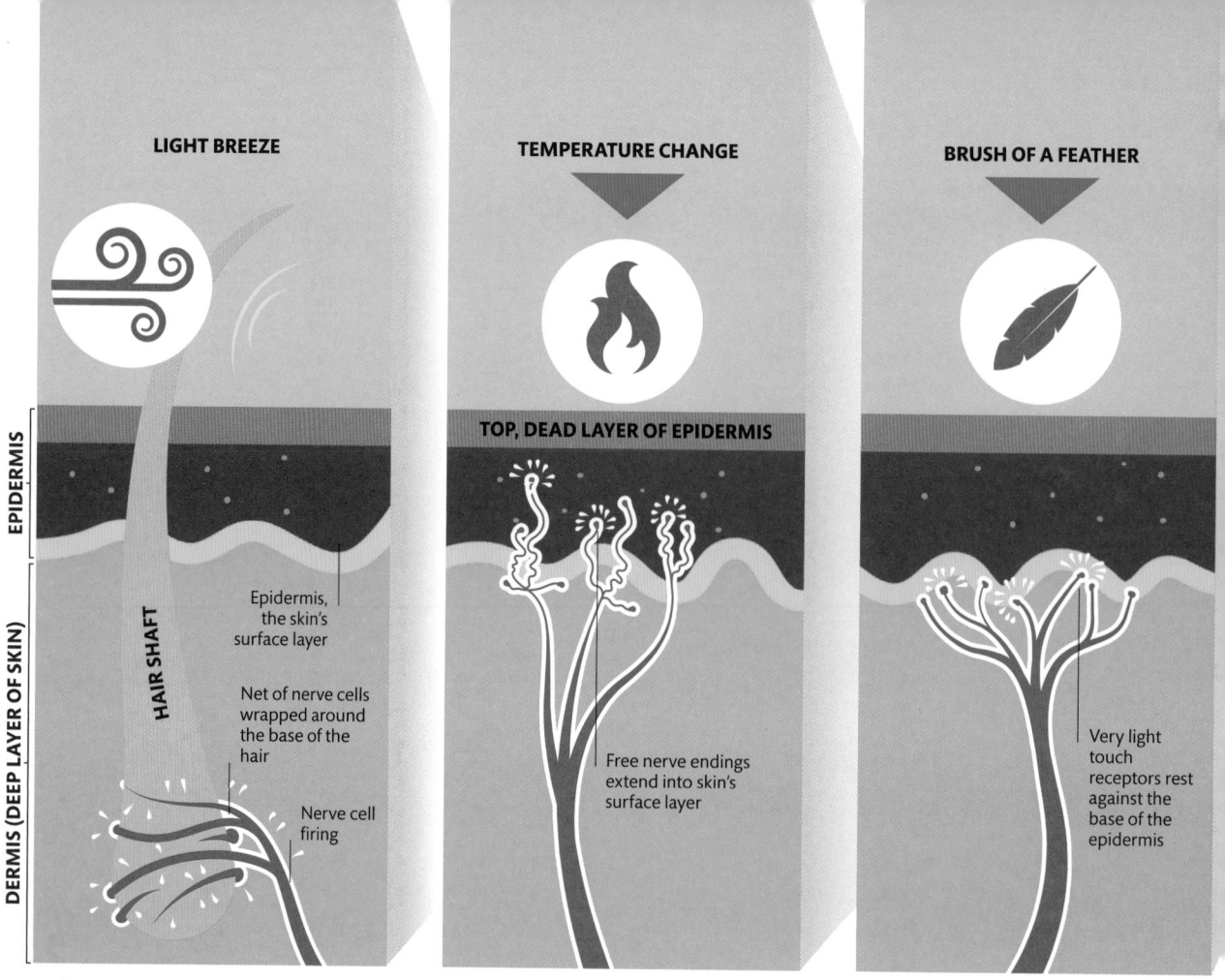

Hair movement
We can sense things that haven't touched our skin. Air currents, or the brushing of hair against objects, distorts and triggers nerves wrapped around a hair's base.

Temperature and pain
Nerves without any special structure around them are sensitive to cold, heat, or pain. They are the shallowest receptors, extending right into the skin's surface layer.

Very light touch
Slightly lower than the free nerve endings are Merkel's cells, which are sensitive to the faintest touch. They are particularly dense in the fingertips.

Feeling the pressure

What we think of as our sense of touch is actually composed of signals from several different receptors in our skin. Some receptors are concentrated in certain areas, such as the sensitive fingertips.

How the skin feels

Our skin is full of microscopic sensors, or receptors, that are buried at different depths and are poised to respond to touches of different kinds—from faint, brief contacts to sustained pressure. In effect, each represents a subtly distinct sense. Receptors work by responding (triggering a nerve impulse) when they are disturbed or distorted.

HOW DO WE FEEL DEEP INSIDE THE BODY?

Nearly all of our touch sense is in the skin and joints. But we also feel discomfort in our guts. This comes from stretch receptors and chemical sensors in and around our intestines.

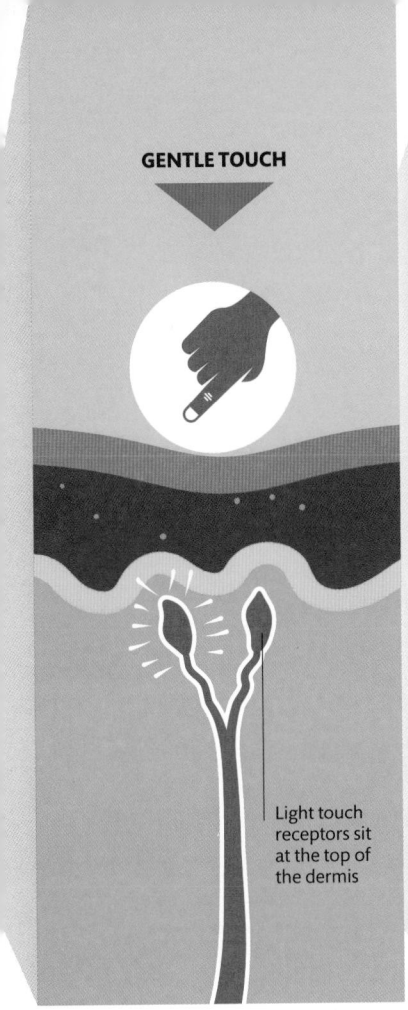

Light touch
Light-touch receptors are good for reading Braille because they are arranged densely and their firing dies away quickly. This gives precise, rapidly updating information.

Pressure and stretch
If the skin is stretched or distorted by pressure, deep receptors fire. They stop firing after a few seconds, so they report rapid changes, not continuous pressure.

Vibration and pressure
The deepest type of touch receptor occurs in joints as well as skin. These sensors don't give up firing, so they respond to sustained pressure as well as vibration.

FROM PALM TO FINGERTIP

Our palms and fingers are very sensitive, but our fingertips have more nerve endings than anywhere else on our skin. Light-touch sensors are packed by the thousands into the pads of our fingers. The pattern in which they fire tell us about the texture of surfaces that we touch.

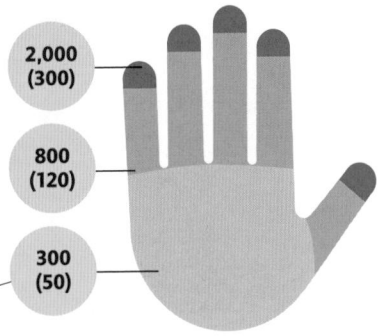

Number of nerve endings per sq in (per sq cm)

EACH OF YOUR **FINGERTIPS** CAN DETECT DIFFERENCES IN TEXTURE **10,000 TIMES SMALLER** THAN THE **WIDTH OF A HAIR**

How do you feel?

From our skin, tongue, throat, joints, and other body parts, microscopic sensors send touch information along sensory nerves to the brain. The destination of these nerve impulses is a part of the brain's outer layer called the sensory cortex, where the touch information is organized and analyzed.

Homunculus
A sensory homunculus is a body pictured in proportion to the area of sensory cortex devoted to it. The colors of this one match those on the large illustration of the brain.

How the brain feels

We can tell where something touches us because the brain contains a map of the body. The map is on a strip of the brain's outer layer called the sensory cortex, but it is a distorted. Because some body parts are so much more sensitive, with closely packed nerve endings, those parts occupy a hugely exaggerated area of the map. The cortex needs such a great area to record precisely the detailed touch data. It combines the information to calculate whether an object is hard or soft, rough or smooth, warm or cold, stiff or flexible, wet or dry, and much more.

Touch-sensitive brain
Viewed from the side, the part of the brain's surface that receives touch information is a narrow strip. It continues down the inside into the deep canyon between the brain's two halves.

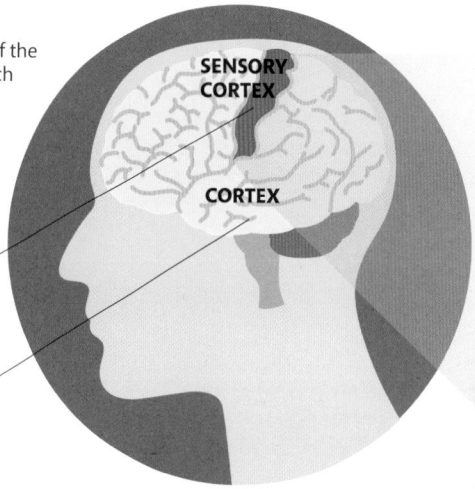

This pink band is the sensory cortex—the part of the cortex that receives touch information

The cortex, in yellow, is the outer layer of the cerebrum—the giant, folded structure that forms most of the human brain

Sensitive bits
The cortex reserves a disproportionate amount of space for the body parts that deliver the most detailed touch information—the lips, palms, tongue, thumb, and fingertips.

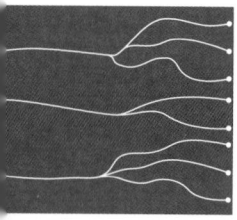

5 MILLION
THE TOTAL AMOUNT OF **SENSORY NERVE ENDINGS** IN **THE SKIN**

LEFT HEMISPHERE receives touch information from the right side of the body

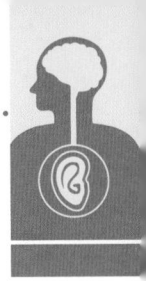

HOW DO WE SENSE TEMPERATURE?

Specific skin nerve endings are sensitive to hot or cold. In the range 41–113°F (5–45°C), both types fire all the time, but at different rates, giving the brain an idea of how hot or cold it is. Outside this range, different nerve endings take over. These register not heat, but pain.

LEG
TRUNK
HEAD
ARM
HAND
FOOT
TOES
GENITALS
FINGERS AND THUMB
EYE
FACE
LIPS
TONGUE

RIGHT HEMISPHERE receives touch information from the left side of the body

Why can't we tickle ourselves?

When we try to tickle ourselves, our brain takes a copy of the intended movement pattern of our fingers and sends it to the body part about to be tickled, warning it and dampening its tickle response. This works because unlike tickles from other people, our brain can predict the precise movement of our own hands and filter it out. This is an example of the brain's vital ability to filter unwanted sensory data.

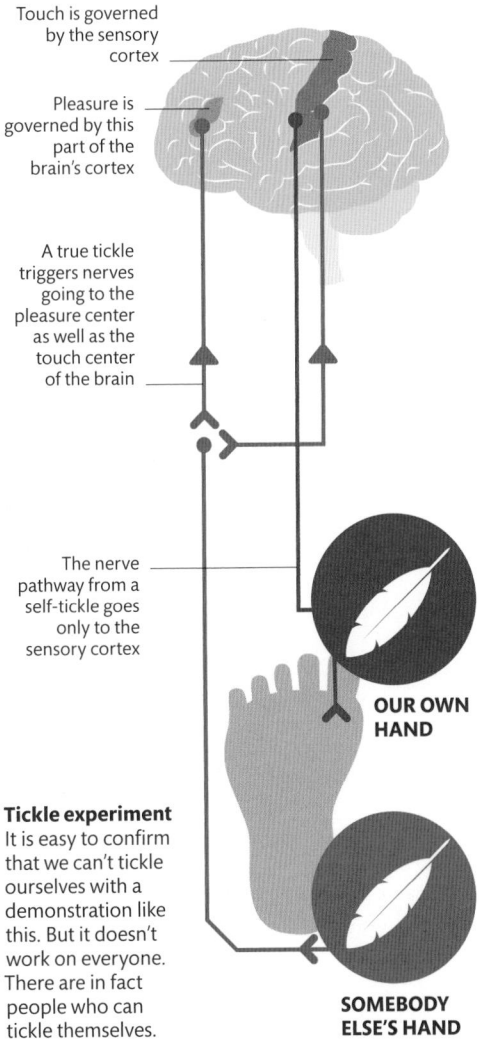

Tickle experiment
It is easy to confirm that we can't tickle ourselves with a demonstration like this. But it doesn't work on everyone. There are in fact people who can tickle themselves.

Pain's pathway

Pain, while unpleasant, is actually incredibly helpful. It tells you when your body is damaged, and the level of pain you feel helps you act accordingly.

Feeling the pain

Pain signals travel from nerve cell receptors at the site of injury along nerves to the spinal cord, and then to the brain, which tells you that you are in pain. Man-made or natural analgesic (painkilling) chemicals work by stopping this flow of information.

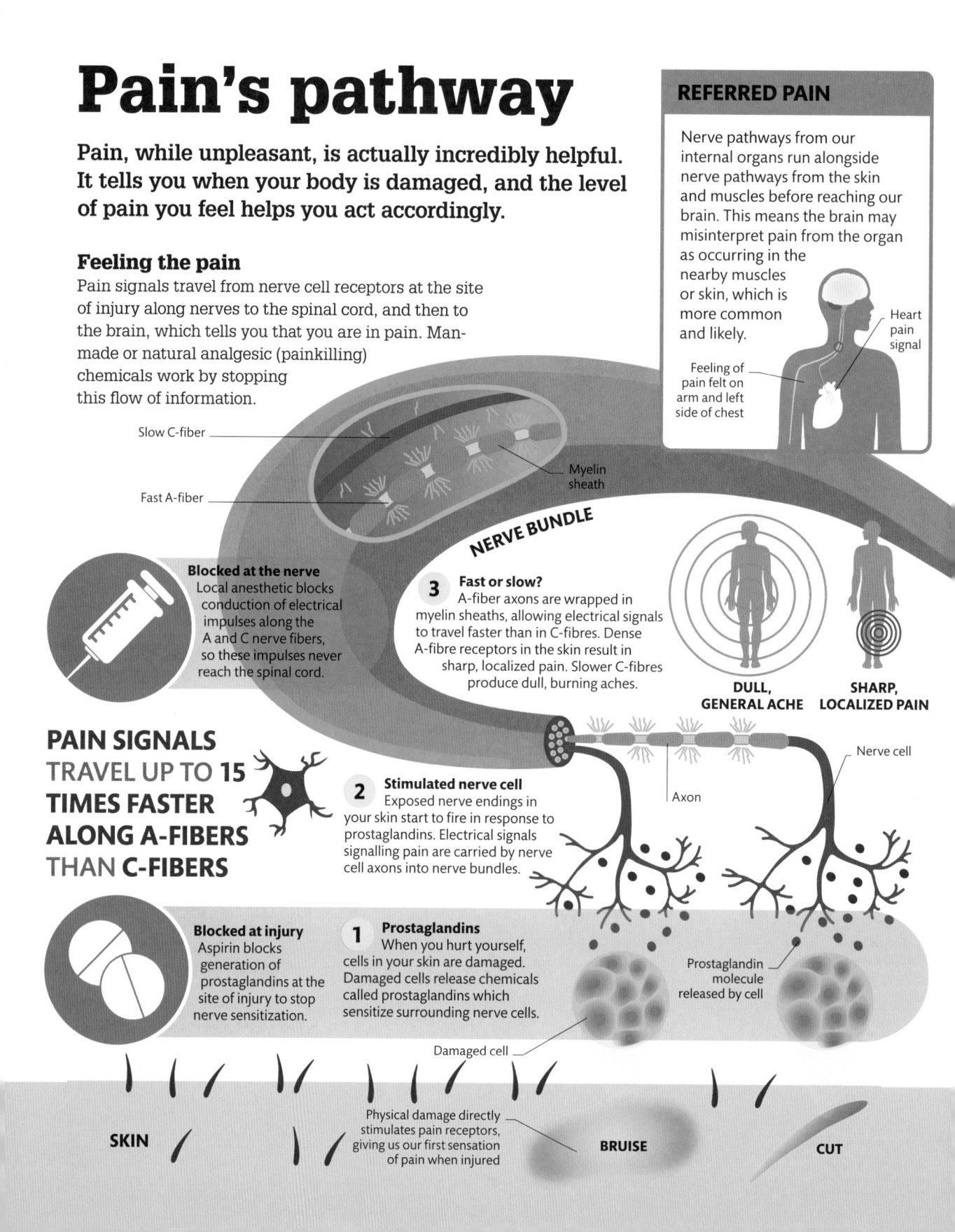

Nerve cell synapse-junction passing on chemical signal for pain

Receiving nerve cell

Chemical message for pain

Painkillers
Opioids such as morphine mimic your body's natural opioids, binding to the nerve cells in order to reduce or even block pain's chemical message. It can erase the feeling of pain altogether, which is useful during medical emergencies.

4 Traveling message
Just like any nervous signal, the electrical impulse is converted into a chemical message to reach the next nerve cell on the path to the brain. The brainstem can release natural opioid painkillers that inhibit some of the chemical message from crossing the gap, dampening the feeling of pain.

Higher cortex registers chemical messages as pain

BRAIN

Thalamus distributes pain signals to various areas of cortex

Nerve traveling up to brain

5 Reaching the brain
The signal continues to the conscious part of the brain, the cortex. Feeling pain requires activity in cortex areas involved in emotion, attention, and accessing significance. People can experience pain due to this activity, even if there is no cause.

Nerve in the spinal cord

DORSAL HORN

SPINAL CORD

Dorsal horn of spinal cord
The dorsal horn is one of the four main columns of nerves found in the spinal cord. It is responsible for processing touch and related senses, including pain.

Nerve connecting to spinal cord

Why do we itch?

Itches arise when our skin is irritated by something on its surface, or by chemicals released by the body when parts of the skin have become inflammed due to disease. It is likely to have evolved to protect us against biting insects. Itch receptors are separate from touch or pain receptors. When they are stimulated, a signal travels through the spinal cord to the brain where the scratch response is initiated. Scratching an itch stimulates both touch and pain receptors, blocking signals from the itch receptor and distracting you from the urge to scratch.

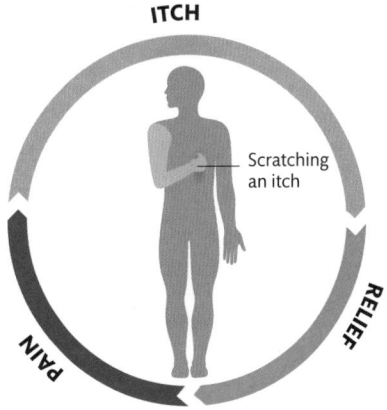

Itching cycle
Scratching can irritate the skin further, which makes the itch signal ever more persistant. Scratching also causes the brain to release serotonin to dampen the pain caused, providing temporary relief. However, once it wears off, the urge to itch returns stronger than before.

How the eye works

Our visual capabilities are amazing. We can see detail and color, see near and far objects clearly, and judge speed and distance. The first stage in the visual process is image capture—a sharp image forms on the eye's light receptors. The image then needs to be converted into nerve signals (see pp.82–83) so that it can be processed by the brain (see pp.84–85).

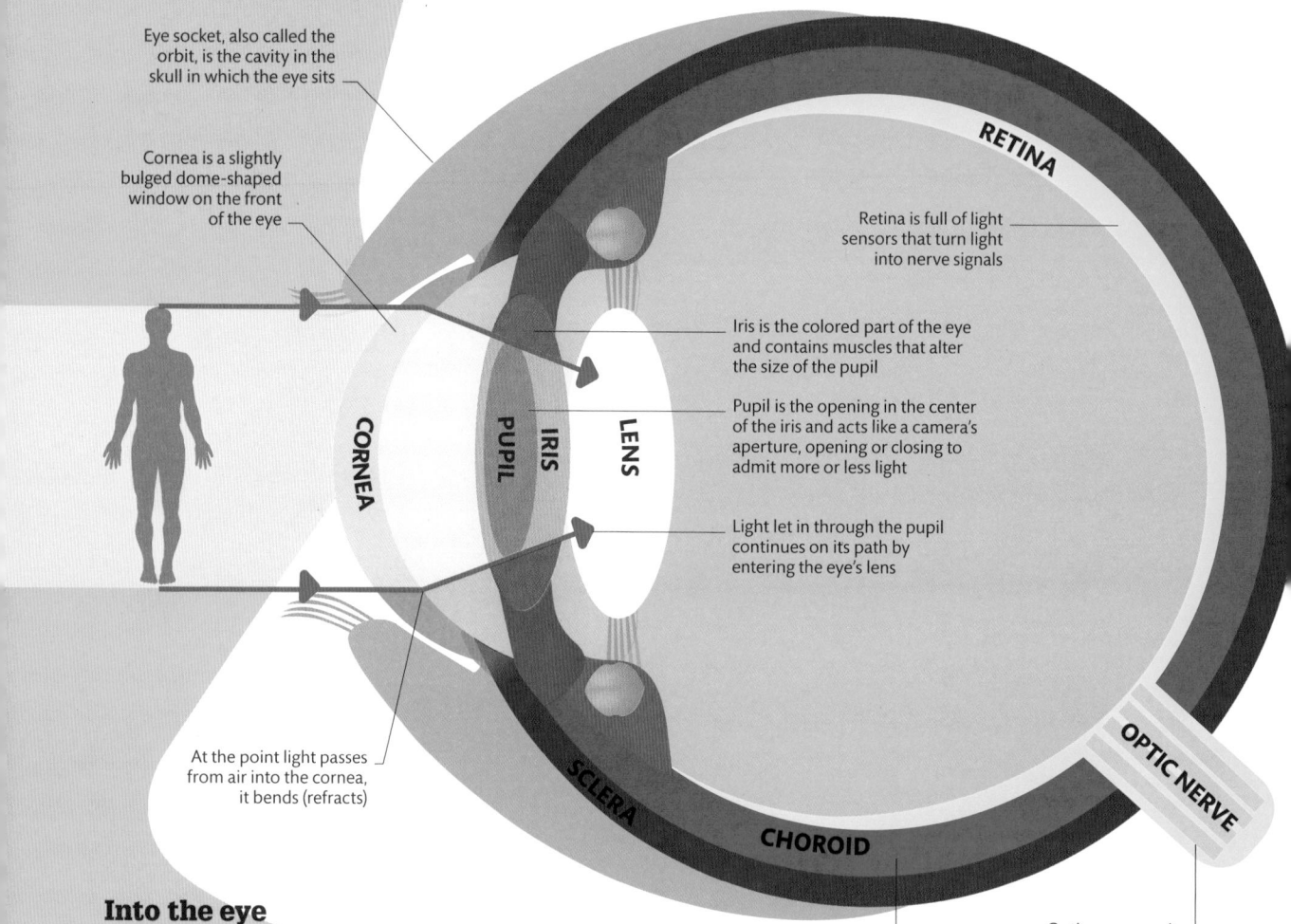

Into the eye

The eyes scan our environment constantly, taking in rays of light produced by, or reflected, from objects. The rays enter the eye first through a clear, bulging window called the cornea. Light is bent by the cornea, passing through the pupil—which controls light intensity—and is then fine-focused by the adjustable lens onto the retina, whose millions of photoreceptor cells form an image to be sent to the brain.

1 Bending light
Due to the cornea's domed shape, light refracting through it bends inward through the pupil toward a focal point within the eye. The pupil, which is a hole in the iris, lets a controlled amount of light through.

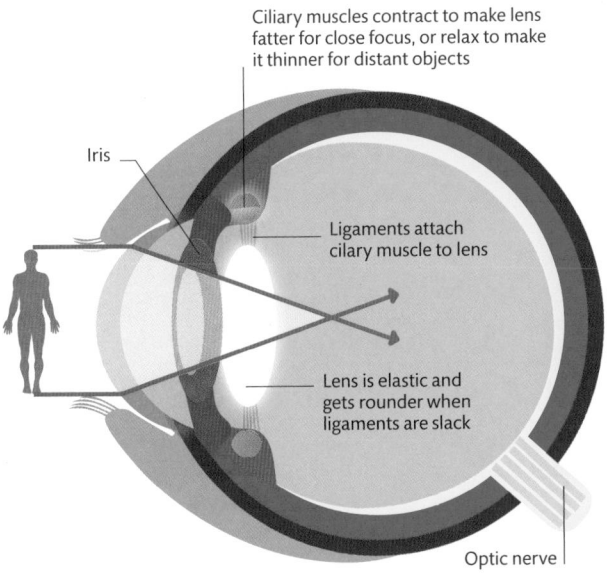

2 Autofocusing
As we look at nearby and distant objects, we adjust the focus of our eyes without thinking. For close work, the muscles that pull on the lens contract, the ligaments go slack, and the lens bulges to increase its focusing power.

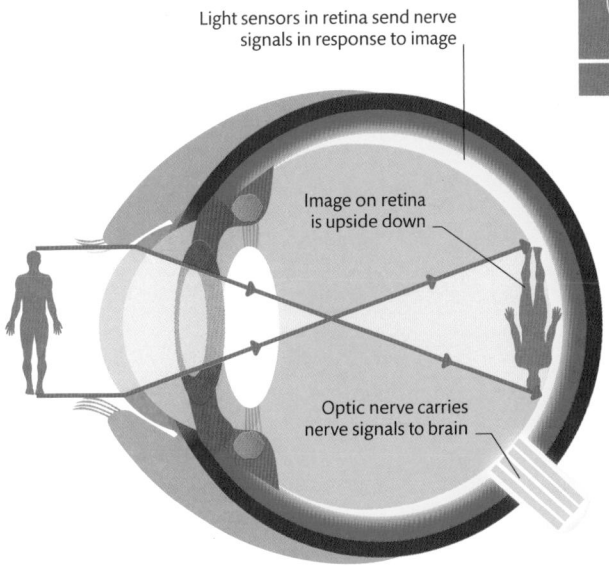

3 Image on the retina
When light hits the retina, more than 100 million light receptors are stimulated, like the pixels on a digital camera's sensor. The pattern of light intensity and color in the image is preserved as an electrical signal in the optic nerve, which sends it to the brain.

Bright light

The iris is the colored part of the eye with a central opening called the pupil. It contains muscles that contract or relax to alter the size of the pupil and so let more or less light into the eye.

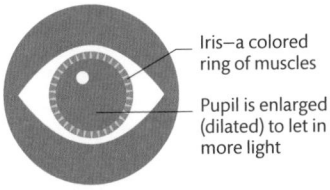

DIM LIGHT

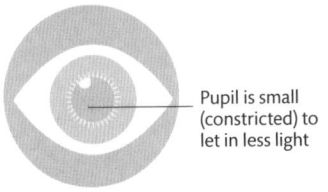

BRIGHT LIGHT

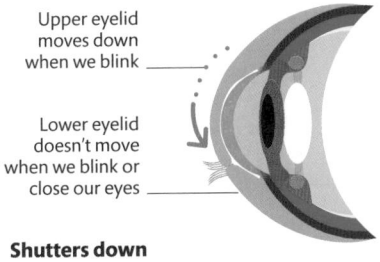

Shutters down
Our eyes are extremely delicate. The eyelids close by reflex action if we are in danger of getting something in our eyes.

First line of defense

The eyelashes and eyelids help protect our eyes. The eyelashes prevent dust and other small particles from getting into the eyes. The eyelids help protect against larger objects and irritant substances in the air. The eyelids also spread tears across the surface of the eye.

Lubrication
Produced by tear glands under the upper eyelid, tears moisten and lubricate the eye and wash away small particles from the eye's surface. Tears are produced continually, although we only notice when we cry or our eyes water.

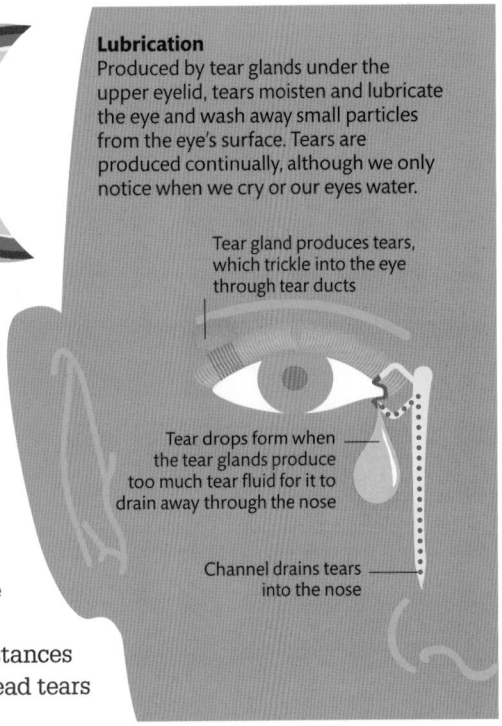

Forming an image

The part of our eye that creates images, the retina, is only the size of a thumbnail, but can produce an incredibly sharp and detailed image. We rely on cells inside the retina to convert light rays into images.

WHAT ARE LIGHT SPOTS?

The gel-like fluid that fills the inner part of your eye can break loose, blocking incoming light rays and casting shadows on your retina. These shadows appear as flashing dots or shapes in your vision.

How we see

Images are formed at the back of the eye in a layer called the retina. Cells inside the retina are sensitive to light. When light rays strike them, they trigger nerve signals, which then travel to the brain to be processed as an image. The retina contains two types of light sensor cells; cone cells, or cones, detect color (wavelength) of light rays, whereas rod cells, or rods, do not.

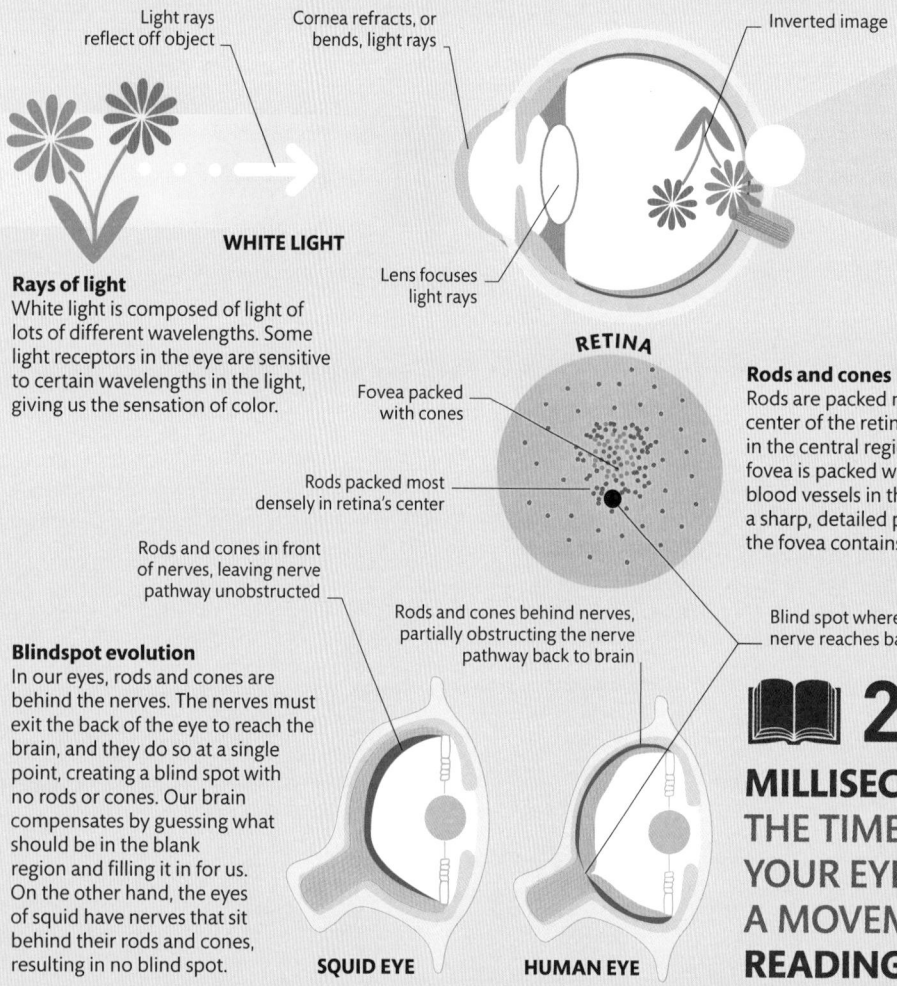

Rays of light
White light is composed of light of lots of different wavelengths. Some light receptors in the eye are sensitive to certain wavelengths in the light, giving us the sensation of color.

Rods and cones
Rods are packed most densely around the center of the retina, although none are found in the central region, known as the fovea. The fovea is packed with cones, and there are no blood vessels in this small area, so it produces a sharp, detailed picture. The very center of the fovea contains only red and green cones.

Blindspot evolution
In our eyes, rods and cones are behind the nerves. The nerves must exit the back of the eye to reach the brain, and they do so at a single point, creating a blind spot with no rods or cones. Our brain compensates by guessing what should be in the blank region and filling it in for us. On the other hand, the eyes of squid have nerves that sit behind their rods and cones, resulting in no blind spot.

20-100

MILLISECONDS— THE TIME TAKEN FOR YOUR EYES TO MAKE A MOVEMENT WHEN **READING QUICKLY**

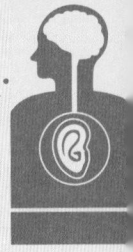

Reaching the retina
Once focused by the lens, light rays travel through the inner eye toward the retina, where our light receptors—rods and cones—are located. Light rays then hit the rods and cones, and a nearby nerve cell fires a nervous signal which travels along nerve fibers back in the opposite direction toward the brain.

Cones send nervous signals in response to green, red, or blue light

Connecting nerve cell

Rod sends nervous signal in response to any color of light; it works in dim light

NERVE SIGNAL

Nerve signals travel along nerve fibers

Blind spot

NERVE SIGNAL

NERVE CELLS

LIGHT RECEPTOR CELLS

Light ray traveling through inner eye to retina at back of eye

Wall of cells forming back of retina

In dim light, flower may seem black and white

SHADES OF GRAY

Grayscale vision
Rods are very sensitive to light and enable us to see in dim conditions, but they cannot distinguish between different colors. Cones are not stimulated at low light levels, so what you see may appear "grayscale."

Cones are responsible for seeing a flower's full color

FULL COLOR

Color vision
Cones provide color vision but work only in bright light. There are three types of cone, each one sensitive to red, blue, or green light. Combining these three colors allows us to see millions of different colors.

Light and nerve signals
The white arrows show the direction of light rays. Green and blue arrows refer to nervous signals traveling through the eye.

Light rays

Black and white

Colour

AFTERIMAGE

If you stare at an image steadily, the rods and cones it stimulates start to "fatigue" and fire less often. When you look away, these rods and cones remain fatigued, while those sensitive to different wavelengths of light are still fresh, so begin to fire rapidly. This leads to an afterimage forming on your retina in a contrasting colour. You can prove this by staring at this bird for 30 seconds, then looking at the cage.

Vision in the brain

Our eyes provide basic visual data about the world, but it is our brain that extracts useful information from it. This is done by selectively modifying it, producing our visual perception of the world—deducing movement and depth and taking into account lighting conditions.

Binocular vision

We are able to see in 3-D because of the placement of our eyes. They both point in the same direction, but are spaced apart slightly, so that they see slightly different images when looking at an object. How different these images are depends on the distance of the object relative to where you are fixating, so we use the disparity between the images to judge how far away an object is.

Visual pathways
Information from the eyes is carried to the back of the brain, where it is processed and turned into conscious vision. Along the way, signals converge at the optic chiasm, where half of the signals cross over to the opposite hemisphere of the brain.

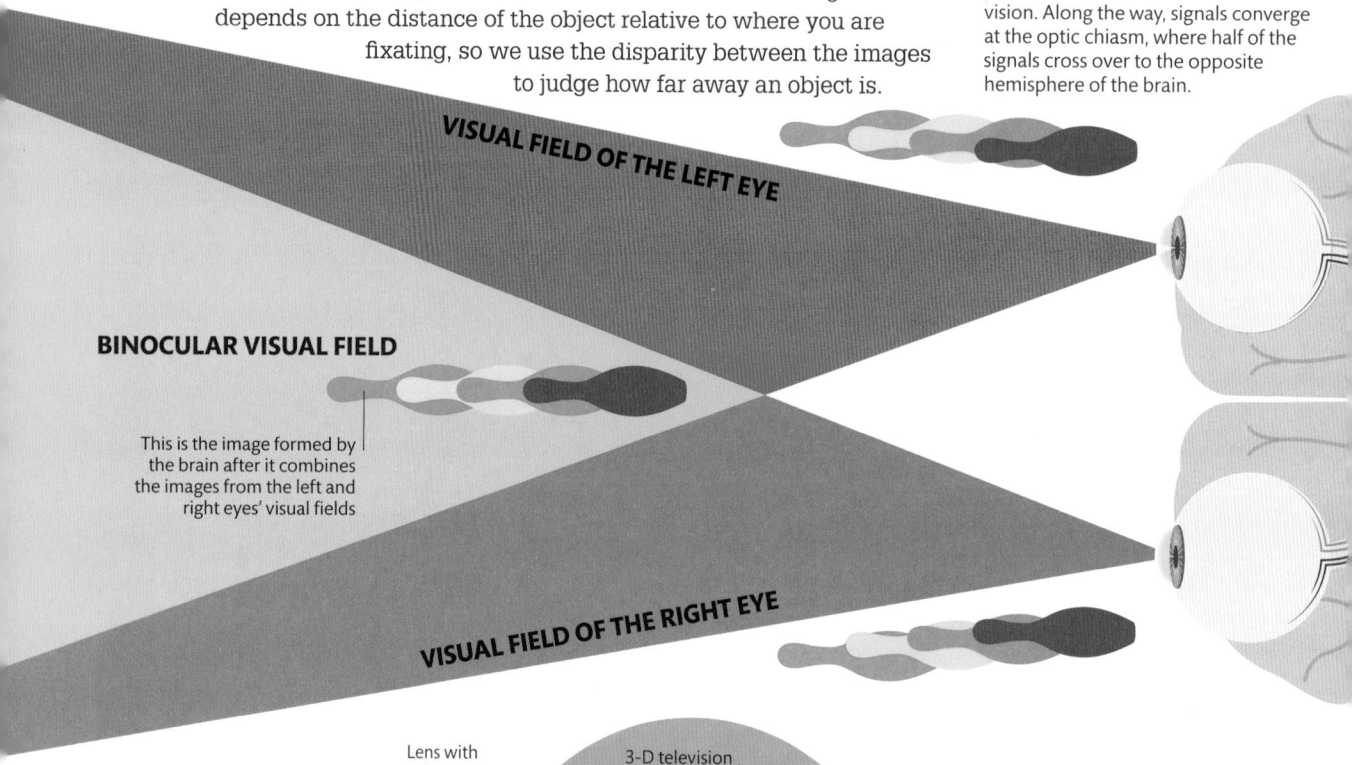

Seeing in 3-D
The way our brains have evolved to perceive depth can be used to produce 3-D movies. Filmmakers film one image out of polarized light waves that are oscillating up and down, and a different image, filmed from a different angle, from light oscillating from side to side. By providing each eye with these slightly different images, they trick the brain into thinking it is seeing in 3-D.

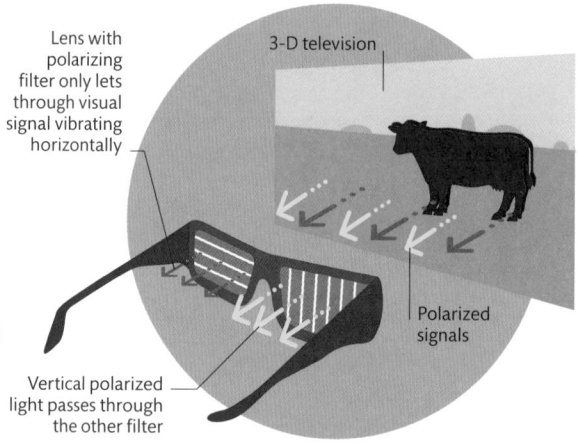

24

THE NUMBER OF **FRAMES PER SECOND** AT WHICH **FILM IS RECORDED**

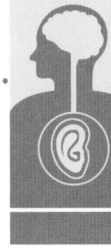

Perspective

Experience tells us that two straight lines, such as railroad tracks, appear to converge in the distance. We use this to estimate depth from an image—by combining this with other cues, such as changes in texture and comparisons to objects of known size, we can estimate distances. The image to the right creates an illusion because we interpret converging lines as distance and compare the cars' sizes to lane width.

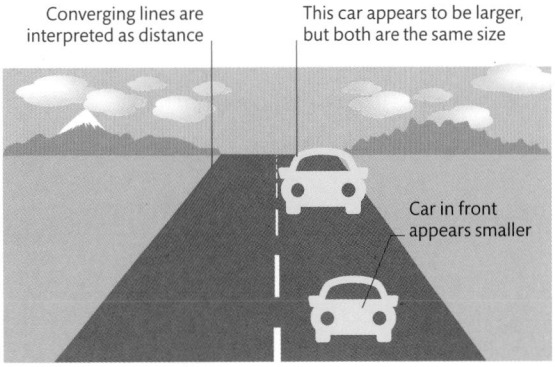

PERSPECTIVE ILLUSION

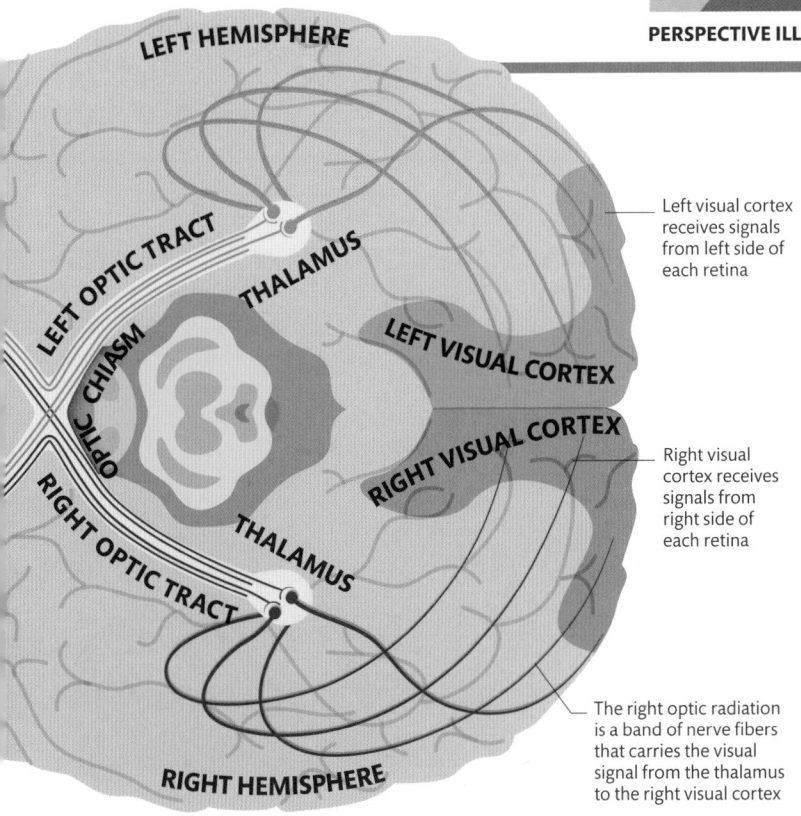

COLOR CONSTANCY

We are used to seeing objects in a variety of lighting conditions and our brain takes this into account to cancel out the effects of shadows and lighting. This means we always see a banana as yellow, no matter how it is illuminated. But sometimes our brain sees only what it expects.

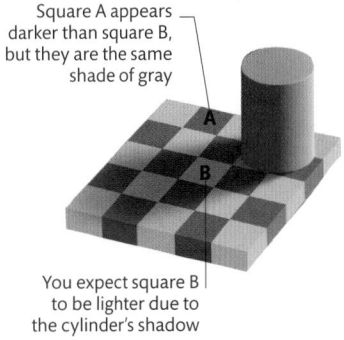

Moving pictures

Surprisingly, our eyes don't provide a smooth stream of moving visual information. They deliver a series of snapshots to the brain, just like film or video. The brain creates the perception of movement from the images, which is why we find it easy to blend the frames of film and TV into the impression of smooth motion. The process can go wrong, however, because a sequence of still frames can be misleading.

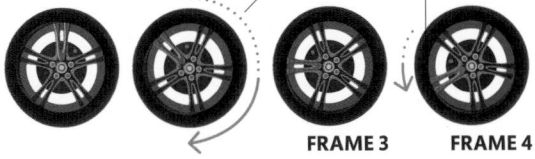

Apparent motion
When the wheels of cars on TV seem to go backward, it is because they make a little less than one rotation between frames. Our brain wrongly reconstructs a slow backward motion.

Eye problems

Your eyes are complex, delicate organs and therefore vulnerable to disorders caused by damage or natural degeneration as you get older. Eye problems affect most people at some point in their lives, but luckily, many eye conditions are easily treatable.

Why do you need glasses?

You see sharp, clear images when light from an object is bent by your lens and cornea and focused on the retina (see pp.80–81). If this system is slightly off, images appear blurred. Glasses can correct for too much or too little bending of the light, bringing the image back into focus. The prevalence of nearsightedness appears to be increasing—possibly because modern day life, especially in urban environments, requires us to focus more on nearby objects than those far away.

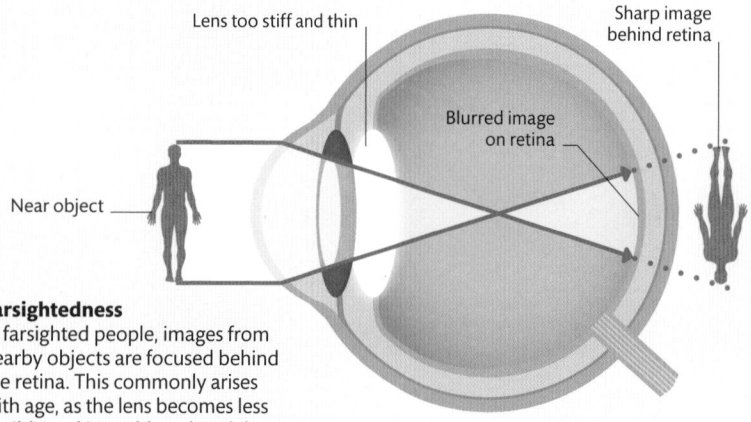

Farsightedness
In farsighted people, images from nearby objects are focused behind the retina. This commonly arises with age, as the lens becomes less flexible and is unable to bend the light enough to focus the image on the retina.

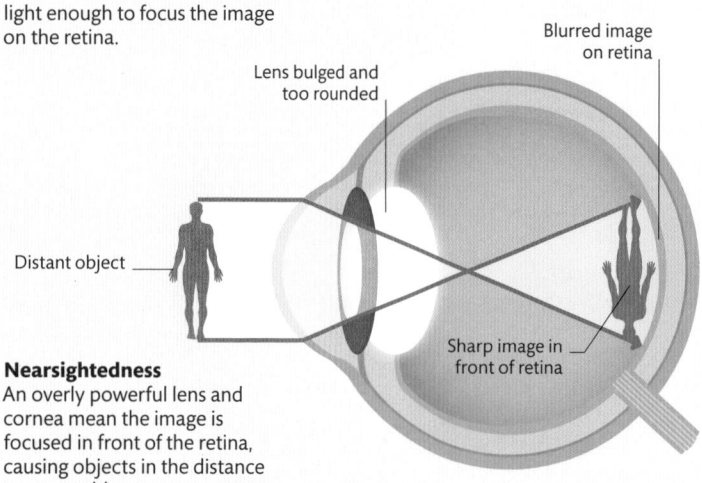

Nearsightedness
An overly powerful lens and cornea mean the image is focused in front of the retina, causing objects in the distance to appear blurry.

90%

THE PROPORTION OF 16–18 YEAR OLD **CHILDREN** WITH **NEARSIGHTEDNESS** IN SOME **CITIES**

Astigmatism

The most common type of astigmatism is caused by a cornea or lens shaped more like a football than a soccer ball. This means that while the image may be focused on the retina horizontally, the vertical aspect could be focused in front of or behind the retina (or vice versa). It can be corrected using glasses or contact lenses, or through laser eye surgery.

What you see
People with astigmatism may see vertical or horizontal lines blurred, but the other in focus. Sometimes, both axes are distorted—one can be farsighted and the other nearsighted.

HEALTHY VISION

NO FOCUS

VERTICAL FOCUS

HORIZONTAL FOCUS

Cataracts

Cataracts are cloudy lenses that disturb vision and are the cause of half the cases of blindness worldwide. They are common in older people, but can also be caused by environmental factors, such as exposure to ultraviolet (UV) light, or injury. They can be treated through surgery, during which the lens is removed and replaced with an artificial one.

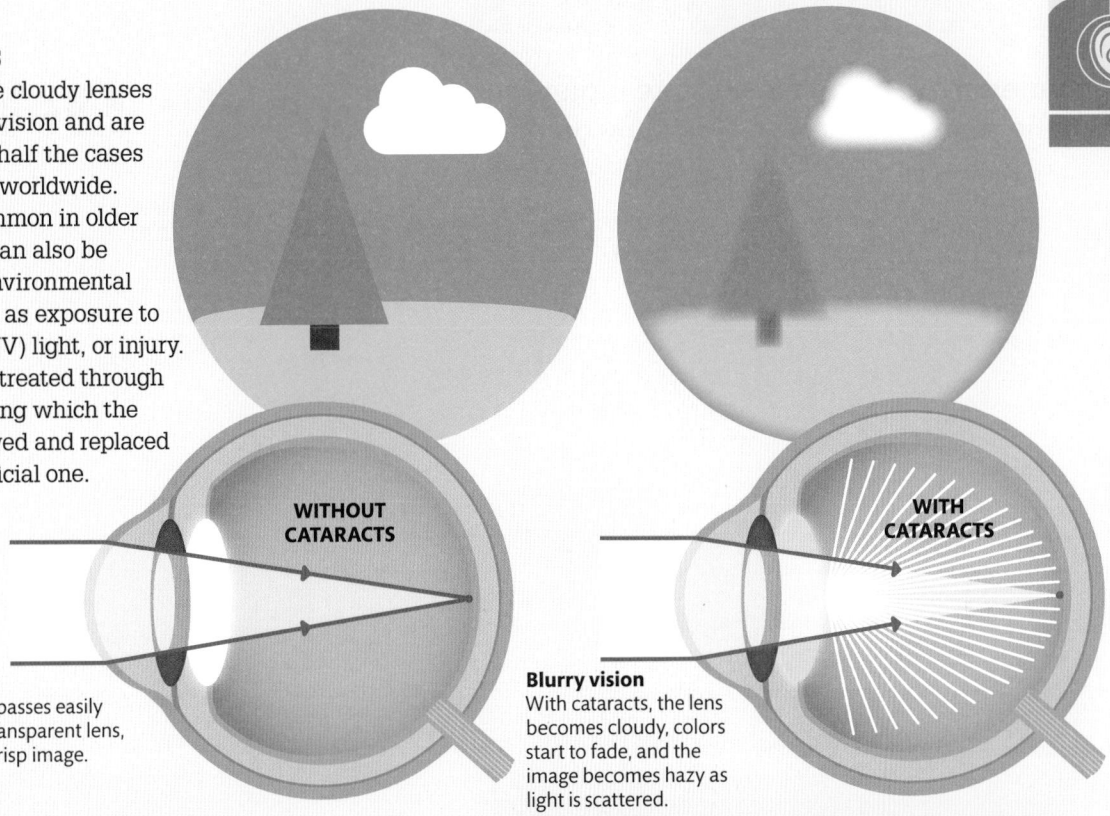

Healthy vision
Normally, light passes easily through your transparent lens, and you see a crisp image.

Blurry vision
With cataracts, the lens becomes cloudy, colors start to fade, and the image becomes hazy as light is scattered.

Glaucoma

Normally, excess fluid in your eyes drains harmlessly into the blood. Glaucoma occurs when blocked drainage channels cause fluid to buildup in the eye. Causes of glaucoma aren't well understood, although genetics plays a part.

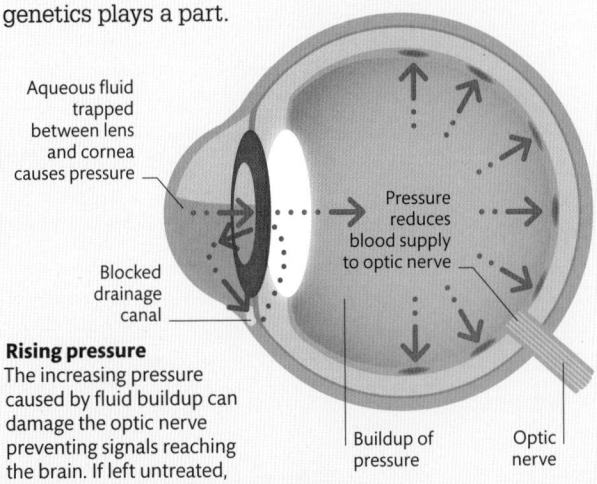

Rising pressure
The increasing pressure caused by fluid buildup can damage the optic nerve preventing signals reaching the brain. If left untreated, it can cause total blindness.

TESTING YOUR VISION

Vision tests allow ophthalmologists to examine your ability to see at far and near distances and to check that your eyes are working together and the muscles are healthy. They also inspect the eye inside and out, which can pick up illnesses such as diabetes and vision problems such as glaucoma or cataracts. Another type of vision problem that may be detected is color blindness. Color vision deficiencies are caused by missing or defective cone types, so sufferers rely on fewer than the three cone types most people have. This means they confuse certain colors—most commonly reds and greens.

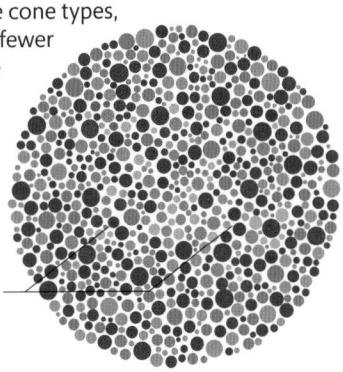

Some people see the number 74, some 21, and some neither

How the ears work

Our ears have the tricky job of converting sound waves in the air into nerve signals for our brains to interpret. The series of steps used ensures that as much of the information as possible is preserved. Ears can also amplify faint signals, and determine where sounds are coming from.

Getting sound into the body

When sound waves travel from air to liquid, as they must to enter the body, they are partially reflected, so they have less energy and sound quieter. Our ear prevents the sound from bouncing off by easing the wave energy in, step by step. When the eardrum vibrates, it pushes on the first of three tiny bones called ossicles, which move in turn, pushing on the oval window and setting up waves in the cochlea's liquid. As the sound passes through the ossicles, they amplify it by 20–30 times.

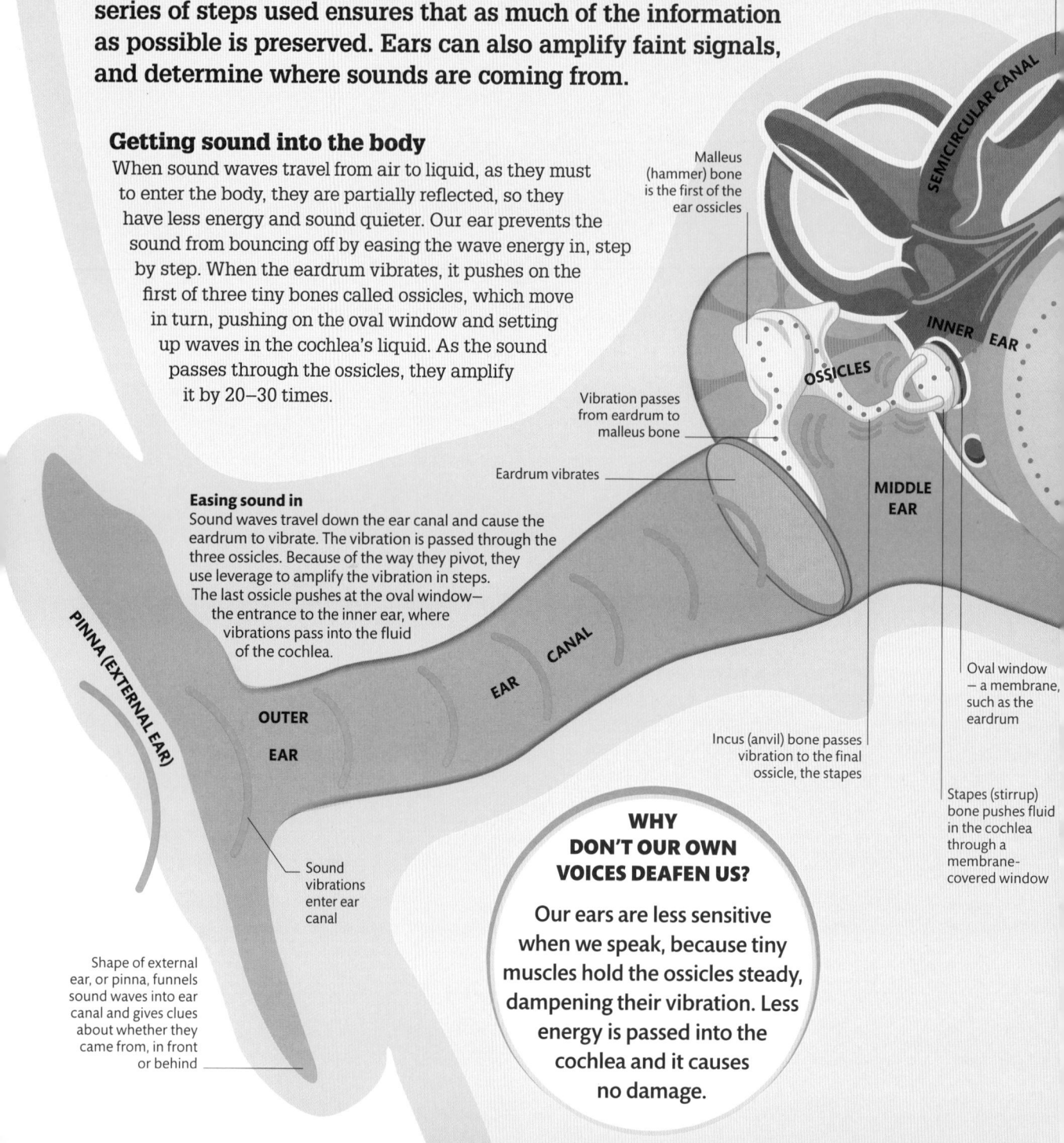

Easing sound in
Sound waves travel down the ear canal and cause the eardrum to vibrate. The vibration is passed through the three ossicles. Because of the way they pivot, they use leverage to amplify the vibration in steps. The last ossicle pushes at the oval window—the entrance to the inner ear, where vibrations pass into the fluid of the cochlea.

WHY DON'T OUR OWN VOICES DEAFEN US?

Our ears are less sensitive when we speak, because tiny muscles hold the ossicles steady, dampening their vibration. Less energy is passed into the cochlea and it causes no damage.

THE WORD **COCHLEA** COMES FROM THE **GREEK FOR SNAIL,** BECAUSE OF ITS **COILED SHAPE**

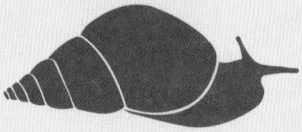

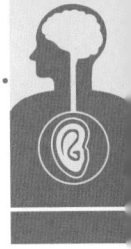

Sounds of different pitches

Inside the cochlea is the basilar membrane, which is connected to sensitive hair cells. Each section of the membrane vibrates most at a particular frequency, because its stiffness changes along its length. Different sounds, therefore, cause deflection of different hair cells. The brain deduces the pitch of the sound using the position of the disturbed cells.

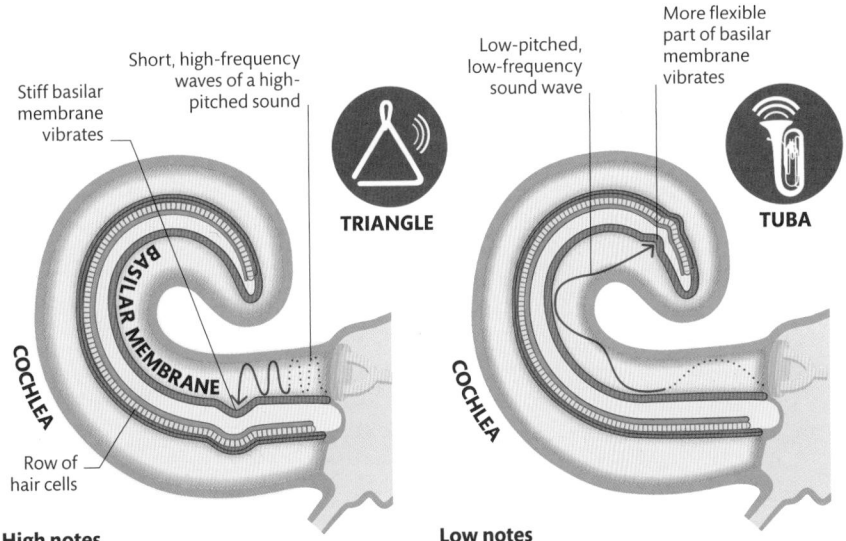

High notes
High notes are caused by high-frequency waves. These activate the basilar membrane near its base, where it is narrower and stiffer and vibrates more rapidly.

Low notes
Longer, lower-frequency waves travel farther through the cochlea before causing the basilar membrane to vibrate nearer its tip, where it is floppier and wider.

Sound into electricity

The information in the sound—including its pitch, tone, rhythm, and intensity—is converted into electric signals to be sent to the brain for analysis. Exactly how the information is encoded is still unknown, but it is achieved by the hair cells and auditory nerves.

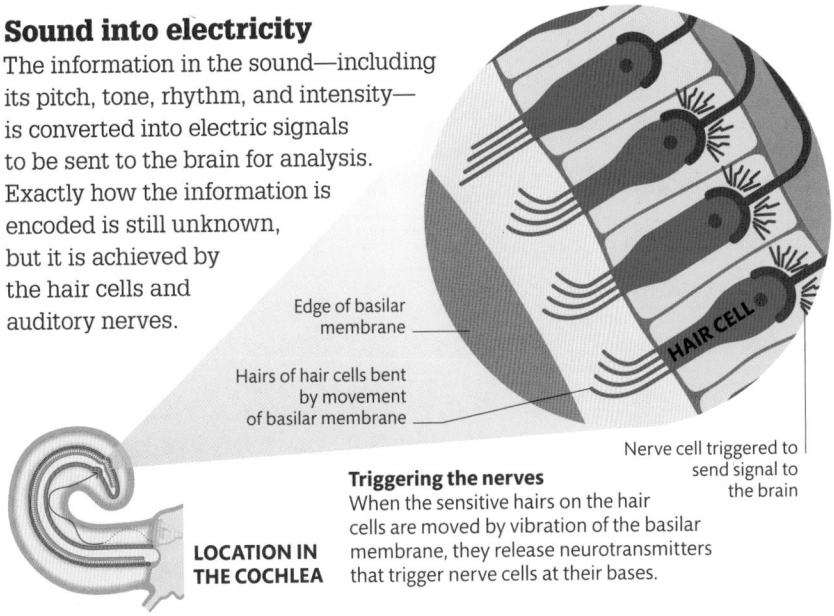

Triggering the nerves
When the sensitive hairs on the hair cells are moved by vibration of the basilar membrane, they release neurotransmitters that trigger nerve cells at their bases.

How the brain hears

Once signals from the ear reach the brain, complex processing is needed to extract information. Our brains determine what the sound is, where it is coming from, and how we feel about it. The brain is able to focus in on one sound over another, and even tune out unnecessary noises completely.

Localizing sound

We use three main cues to find the location of a sound source—its loudness, its frequency pattern, and the difference in arrival time at each ear. We use frequency pattern to tell if the sound is in front or behind us, because our ear's shape means that a sound coming from in front has a different pattern of frequencies than the same sound coming from behind. Our ears don't help much in pinning down the height of sound sources, though. Left and right localization is easier—a sound from the left is louder in the left ear than in the right, particularly at high frequencies. It also reaches our left ear a few milliseconds before our right. The diagrams on the right show how the brain uses this information.

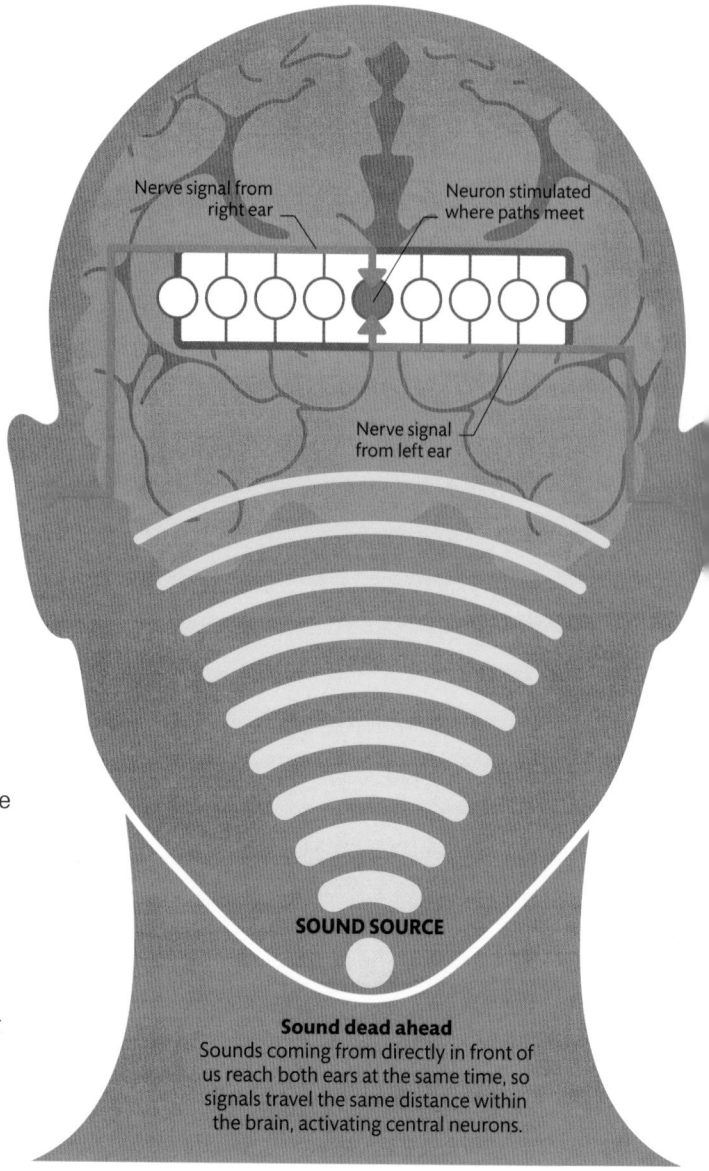

Sound dead ahead
Sounds coming from directly in front of us reach both ears at the same time, so signals travel the same distance within the brain, activating central neurons.

Tuned in

Our brains can "tune in" to a single conversation over the babble of noise at a party by grouping sounds into separate streams, based on frequency, timber, or source. It might seem as though you don't hear any of the other conversations—but you will notice if someone mentions your name. That's because your ears still send signals from the other conversations to the brain, which will override the filtering if something important comes up elsewhere.

WE CAN PICK OUT A CONVERSATION IN NOISY ENVIRONMENTS

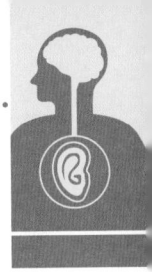

THE BRAIN HAS CELLS THAT **RESPOND** ONLY TO SOME **FREQUENCIES,** JUST LIKE THE DIFFERENT PARTS OF THE **COCHLEA IN THE INNER EAR**

The neuron that fires tells us from how far to the left or right the sound is coming

Signal travels farther from this side before meeting the pathway from other ear

Sound waves reach the closer ear first

Off-center sound source
Different neurons are activated depending on the delay between a sound first reaching the nearer ear and then reaching the farther ear. This tells us what direction the sound comes from.

Sounds from anywhere inside the "cone of confusion" produce identical neural responses, so they can't be told apart

Sounds outside the cone produce unique neural responses, so they are easier to locate

FINDING THE SOURCE

SOUND SOURCE

Cone of confusion
In a cone-shaped region outside each ear, signals are ambiguous and we find it difficult to localize sounds. Tilting or swiveling our heads can move the sound source out of this confusing region and help us locate the sound.

Why does music make us emotional?

Music can cause strong emotional reactions—whether it's the soundtrack heightening fear in a scary movie, or chills created by a haunting melody. We know there are a wide range of brain areas involved in the emotions elicited, but we don't know why or how music creates such dramatic feelings in the listener, or why the same song affects people differently.

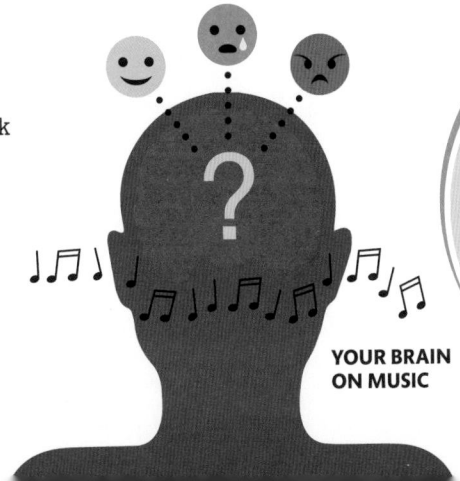

YOUR BRAIN ON MUSIC

WHY DO WE STAND STILL TO LISTEN?

It is easier to listen carefully when we stop moving altogether. This helps us hear better by stopping sounds that are generated by our own movements.

Balancing act

As well as hearing, our ears are responsible for keeping our balance and telling us how and in which direction we are moving. They do this using a set of organs in the inner ear—one on each side of the head.

Turning and movement

Inside each of our ears, three fluid-filled canals sit at roughly 90 degrees to each other. One responds to motions such as forward rolls, the second to cartwheels, and the third to pirouettes. The relative motion of the fluid tells our brains in what direction we are moving. When spinning repeatedly in the same direction, the fluid builds up momentum. Once that matches the rate of spin, it stops deflecting the hair cells and you no longer feel motion. After you stop, however, the liquid continues, giving the feeling that you are still moving, known as dizziness.

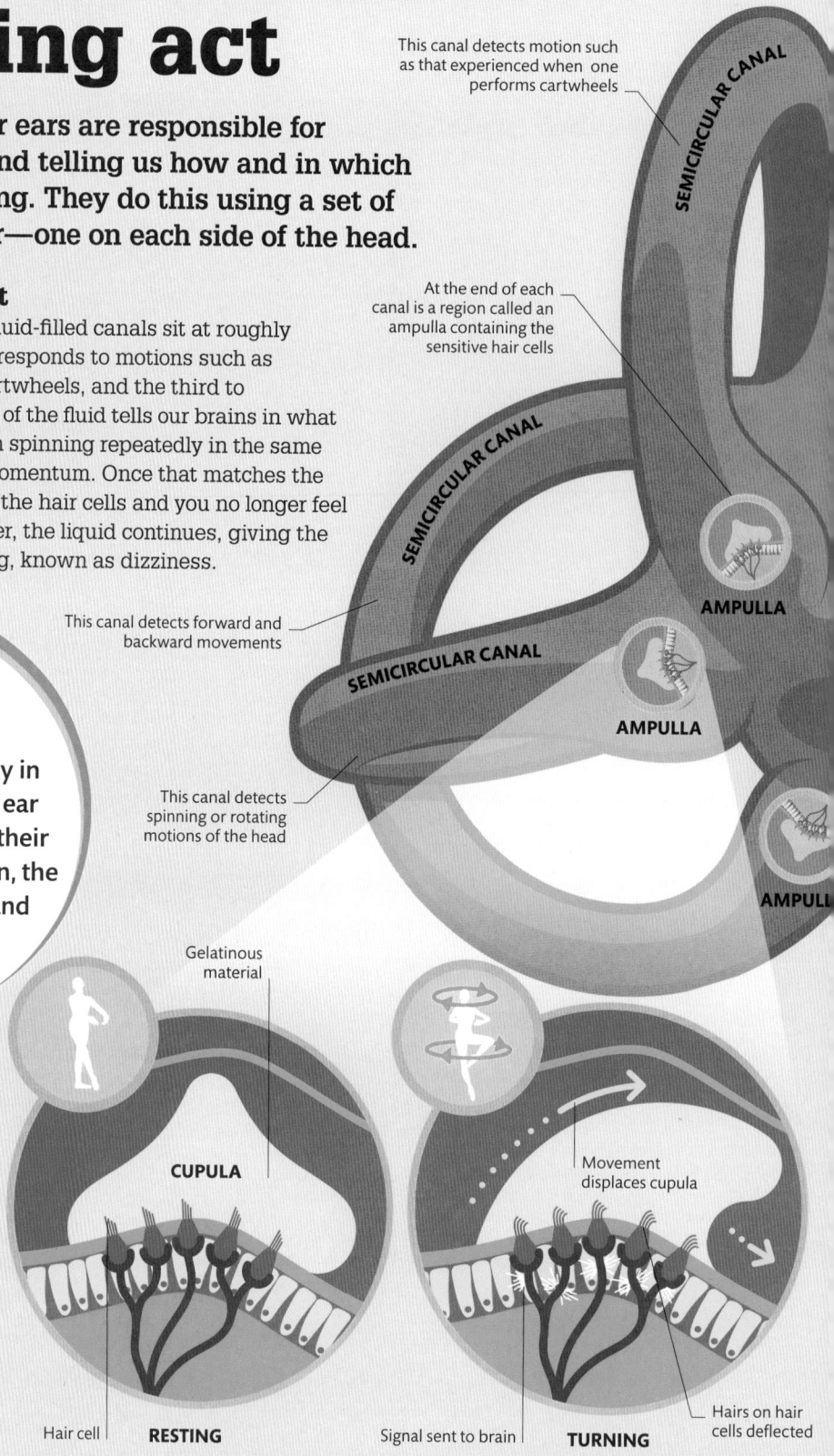

WHY DOES ALCOHOL MAKE YOUR HEAD SPIN?

Alcohol builds up quickly in the cupulas of the inner ear and makes them float in their canals. When you lie down, the cupulas are disturbed and the brain thinks you are spinning.

Turning sense organs

When you move, the liquid inside the canals moves too, but because it has inertia, it takes a while to start moving. This movement displaces a gelatinous mass called the cupula, disturbing the hair cells inside it and sending signals to the brain. When the cupula is bent in one direction, the nerves increase their rate of firing. If it is bent in the other direction, firing is inhibited—this tells the brain the direction of the motion.

Steady gaze

Your brain constantly adjusts the tiny movements your muscles make to keep you balanced. Inputs from the eyes and muscles combine with those from your inner ear to determine which way up you are.

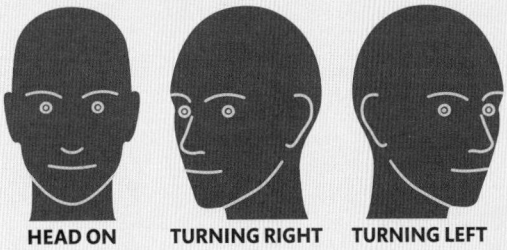

Correction reflex
Our eyes automatically correct for head movements, keeping the image on our retina stationary. Without this reflex, we would be unable to read, as the words would jump about every time our head moved.

BALLET DANCERS' BRAINS ADAPT TO SUPPRESS THE SENSATION OF DIZZINESS AFTER SPINNING

Gravity and acceleration

As well as turning motions, our inner ears sense straight-line acceleration—backward and forward, or up and down. We have two organs to sense acceleration—the utricle is sensitive to horizontal movements while the saccule detects vertical acceleration (such as the movement of an elevator). Both organs sense the direction of gravity relative to the head, such as when the head is tilted or level.

Utricle is sensitive to gravity and horizontal acceleration

UTRICLE

SACCULE

Saccule detects gravity and vertical acceleration

Heavy layer containing tiny stones

GELATINOUS LAYER

Cell fires

Hair cell

UPRIGHT

Heavy layer pulled by gravity

Hairs deflected

TILTED

Signal sent to the brain

Gravity sense organs
The hair cells in the utricle and saccule are within a gelatinous layer, topped with a structure containing tiny stones. Due to the weight of the structure, gravity moves it when the head is tilted, which in turn deflects the hairs. During acceleration, the stone-filled layer takes longer to start moving because of its greater mass. If there are no other cues, it can be hard to tell the difference between a head tilt and acceleration.

Hearing problems

Deafness and hearing problems are common but often treatable thanks to technological advances. Most people develop some form of hearing loss as they age due to damage to the components of the inner ear.

Causes of hearing problems

Deafness from birth is usually caused by genetic mutations that stop the ear from working properly. The hearing problems shown here can occur as a result of injury or illness throughout life.

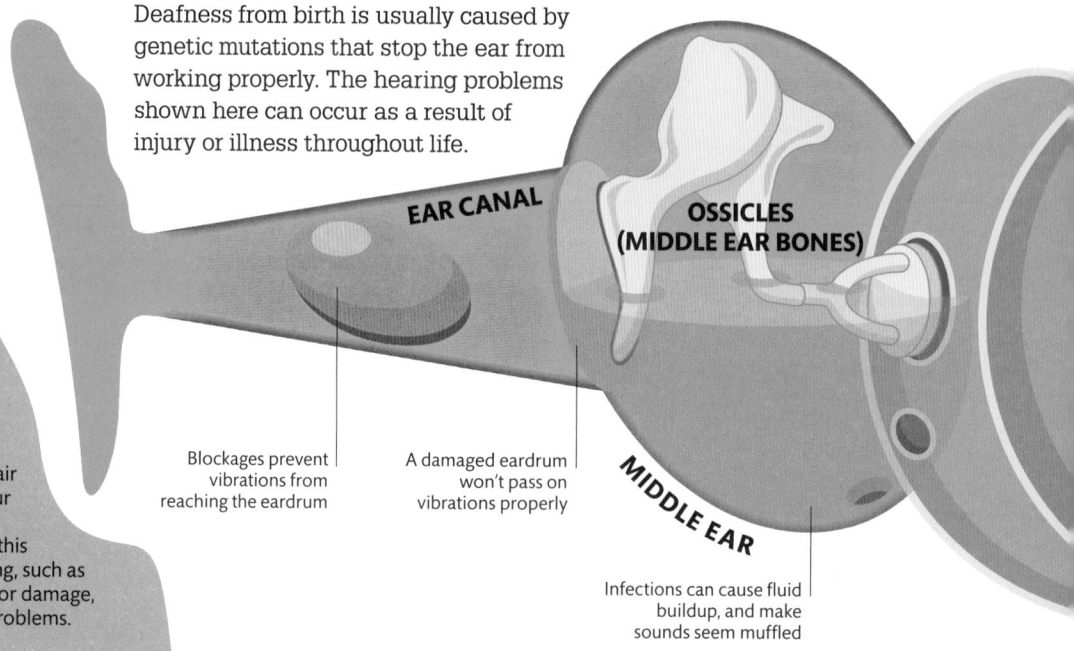

Blocked pathways
The ear converts sound waves in the air into nerve signals our brain can interpret. Anything that stops this process from working, such as a physical blockage or damage, can cause hearing problems.

How loud is too loud?

The decibel sound scale is logarithmic, and every 6 dB increase in volume doubles the sound energy. Loud noises can damage hair cells and above a certain level of damage the cells can't repair themselves, and die. If enough hair cells die, you can lose the ability to detect certain frequencies.

Causing damage
Any noise level above 85 dB can cause damage, depending on how long you are exposed to it.

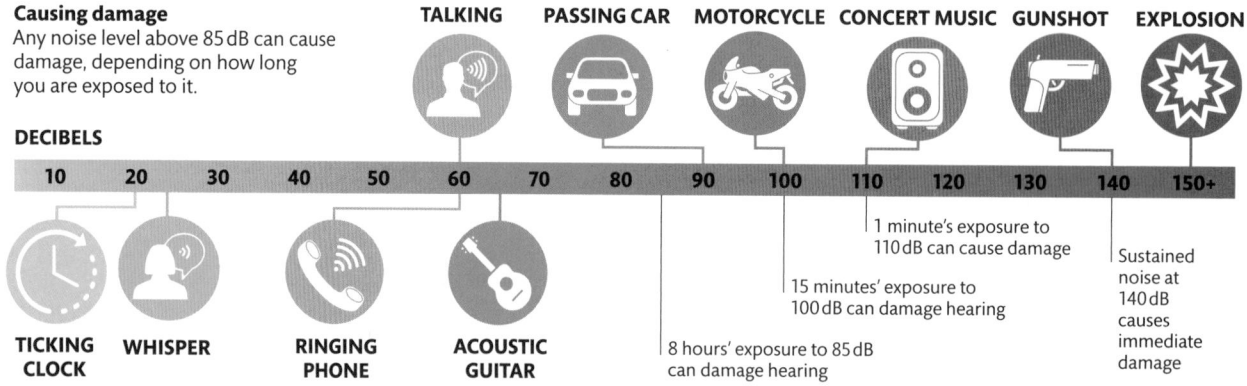

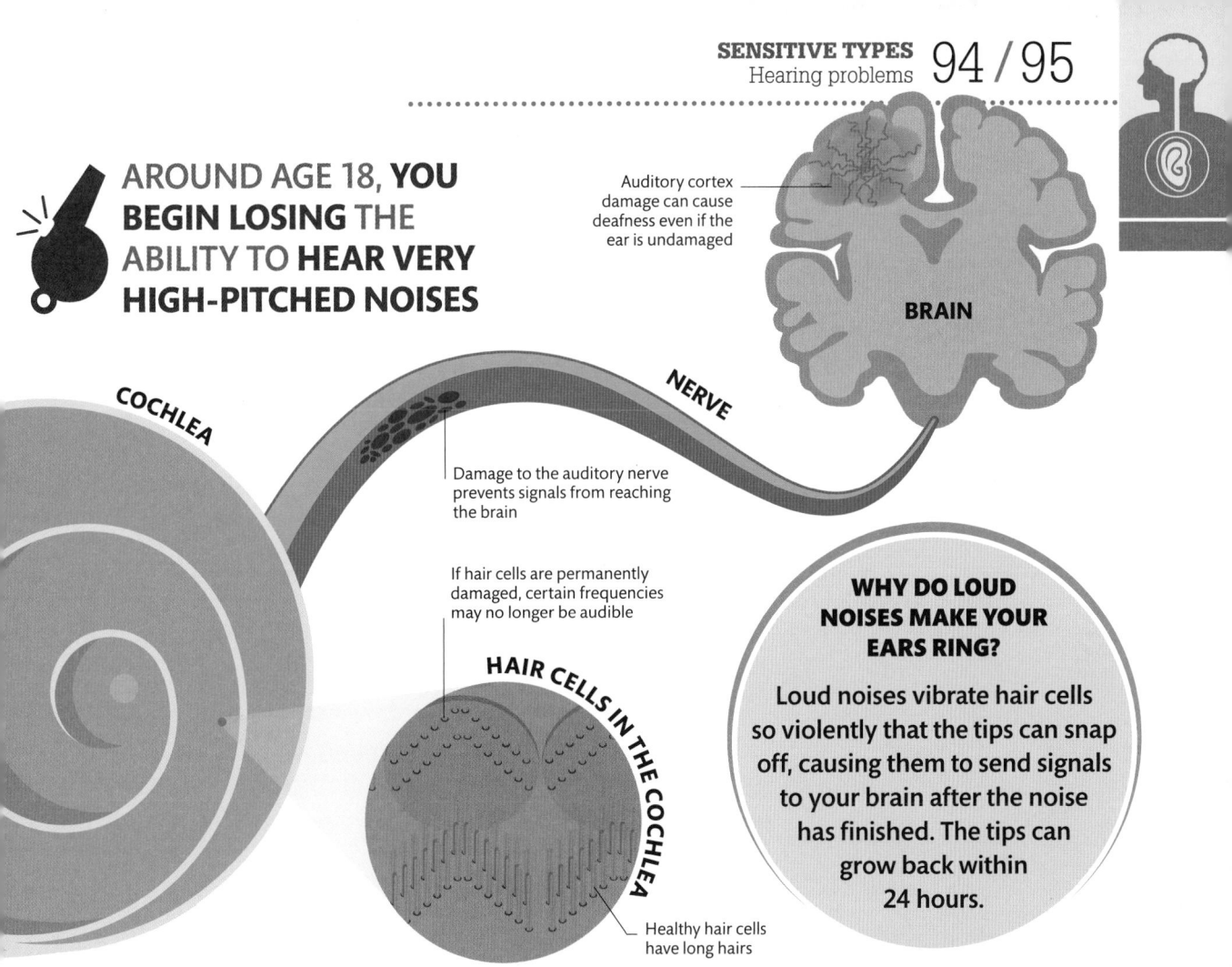

Cochlear implants

Normal hearing aids simply amplify sounds and cannot help people with damaged or missing hair cells. Cochlear implants replace the function of the hair cells, converting sound vibrations into nerve signals that the brain learns to interpret. Increased current through the electrodes within the cochlea produces a louder sound, while the position of the activated electrodes determines pitch.

How they work

External microphones detect sounds and send them to the processor. Signals then travel to the internal receiver via the transmitter, before passing as electric current to the electrode array inside the cochlea. Stimulated nerve endings send signals to the brain, and sounds are heard.

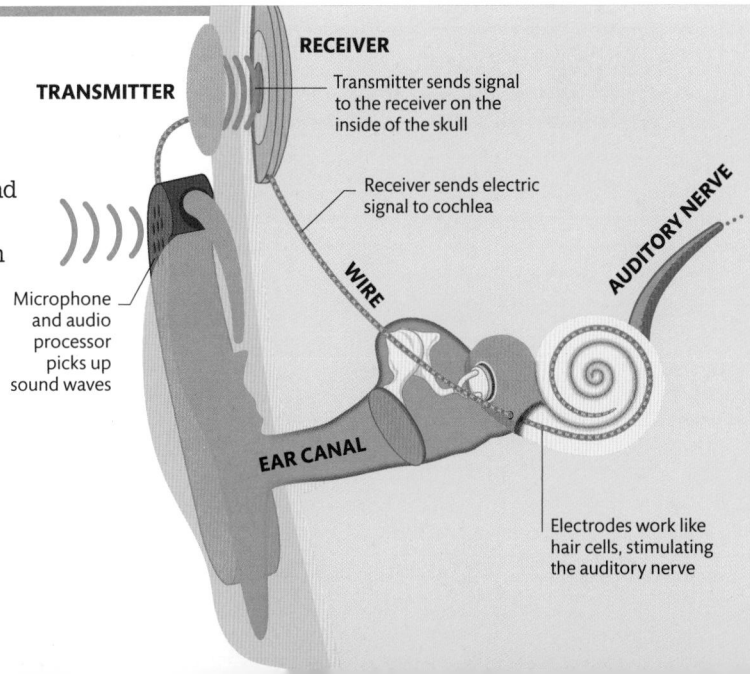

Catching a scent

Particles in the air are detected by sensory cells in your nose, and signals are sent to your brain so you can identify them as smells. These smells can invoke powerful emotions or memories because of physical links to your brain's emotional center.

Sense of smell

Anything that smells releases tiny particles, or scent molecules, into the air. When you inhale, these molecules pass into your nose, where the smell is detected by specialized nerve cells. Sniffing is an automatic response when catching a whiff—the more scent molecules you inhale, the easier it is to identify a smell. Our senses of smell and taste often work together when we are enjoying a meal, because scent molecules are released by the food we eat, which then pass into the rear of the nasal cavity.

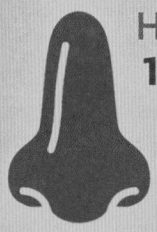

HUMANS HAVE AROUND **12 MILLION RECEPTOR CELLS** AND THEY CAN DETECT **10,000 DIFFERENT ODORS!**

2 Nose hair
At the entrance to the nose, hairs catch large particles of dust and debris, but admit the scent molecules, which are millions of times smaller.

1 Types of smell
Aromatic objects such as freshly baked bread, spoiled cheese, and burning things, release scent molecules. The type of molecules determines what you smell as well as the smell's intensity, since we are much more sensitive to some scent molecules than others.

LOSS OF SMELL

A complete lack of smell is called anosmia. Some people are born with anosmia, while others develop the condition after an infection or head injury. These instances can cause a severage of the nerve fibers, reducing the number of nervous signals they pass to the brain. Those with anosmia have reduced appetites and are more likely to suffer from depression—this is probably because of smells' links with the brain's emotional center. The sense can recover on its own or after drug treatment or surgery. For others, smell training, which probably leads to the regeneration of olfactory receptor cells, can help.

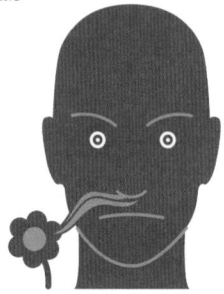

WHY DO WE HAVE NOSE BLEEDS?

Nasal membranes that line your nasal cavity are thin and filled with tiny blood vessels. These blood vessels can burst very easily to cause nose bleeds—either by breathing dry air which crusts and breaks the thin membrane or even by blowing your nose too hard.

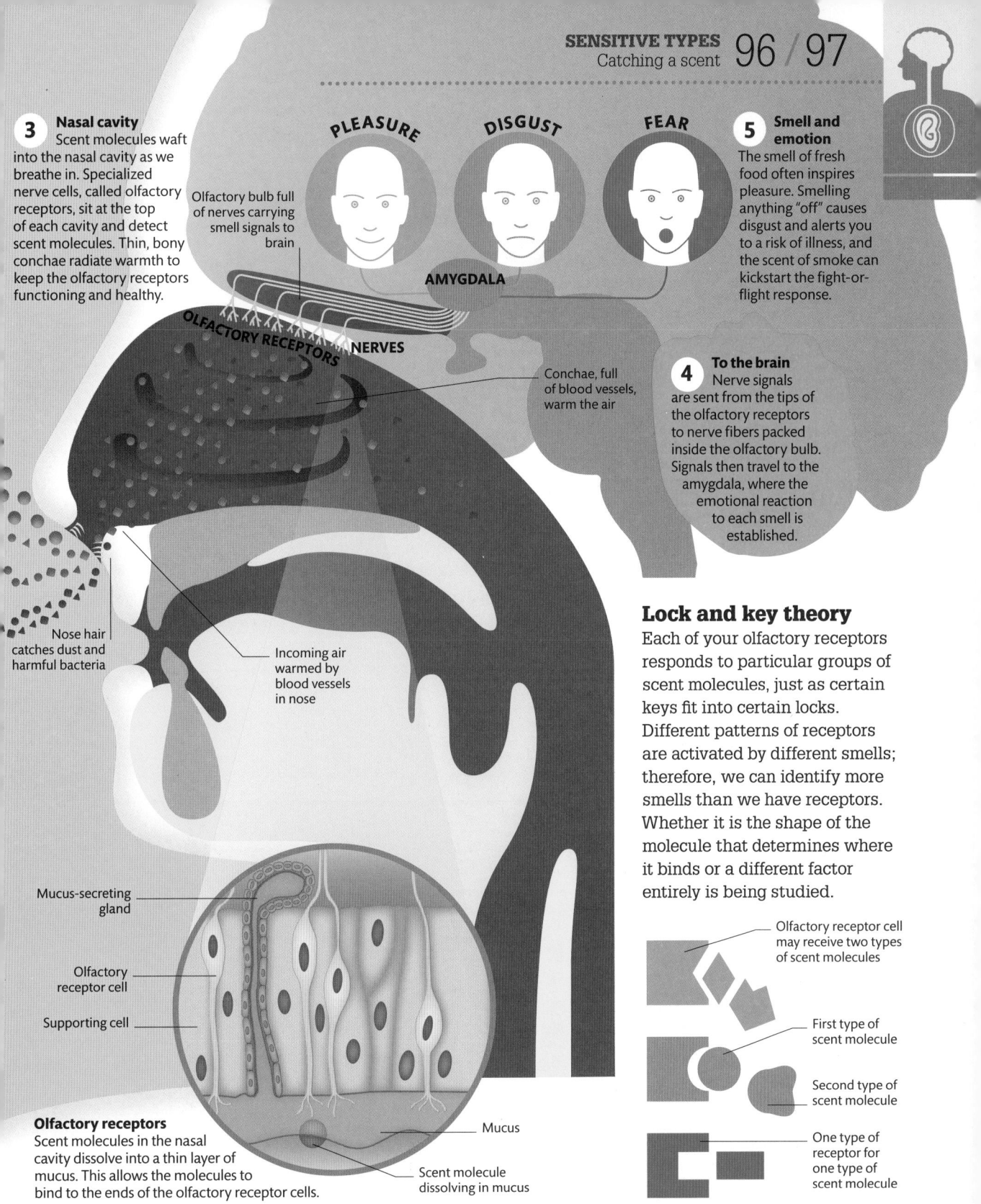

3 Nasal cavity
Scent molecules waft into the nasal cavity as we breathe in. Specialized nerve cells, called olfactory receptors, sit at the top of each cavity and detect scent molecules. Thin, bony conchae radiate warmth to keep the olfactory receptors functioning and healthy.

4 To the brain
Nerve signals are sent from the tips of the olfactory receptors to nerve fibers packed inside the olfactory bulb. Signals then travel to the amygdala, where the emotional reaction to each smell is established.

5 Smell and emotion
The smell of fresh food often inspires pleasure. Smelling anything "off" causes disgust and alerts you to a risk of illness, and the scent of smoke can kickstart the fight-or-flight response.

Lock and key theory

Each of your olfactory receptors responds to particular groups of scent molecules, just as certain keys fit into certain locks. Different patterns of receptors are activated by different smells; therefore, we can identify more smells than we have receptors. Whether it is the shape of the molecule that determines where it binds or a different factor entirely is being studied.

Olfactory receptors
Scent molecules in the nasal cavity dissolve into a thin layer of mucus. This allows the molecules to bind to the ends of the olfactory receptor cells.

On the tip of the tongue

Your tongue has thousands of chemical receptors, which detect some key chemical ingredients in your food and interpret them as one of five major taste sensations. However, not everyone's tongue is the same, which helps explain food preferences.

WHY DON'T CHILDREN LIKE COFFEE?

Children's dislike for bitter tastes may have evolved to protect us against poison. As we mature, we learn through experience to enjoy bitter tastes such as coffee.

Taste receptors

Our tongues are covered in tiny bumps (papillae), which contain taste receptors for chemicals that give us the five basic tastes—sour, bitter, salty, sweet, and umami (savory). Each receptor deals only with one taste, and there are receptors for all five tastes all over the tongue's surface. The flavor of food is a more complex sensation, of taste mixed with smell, that is detected when molecules travel up the back of the throat into the nose. This is why things taste bland when your nose is blocked.

Taste bud

A taste bud begins with a pore on the tongue papilla's surface. The pore lets in particles of food or drink, which contact taste receptor cells. The cells send signals to the brain when certain tastes are detected. Taste buds are also found on the insides of your mouth.

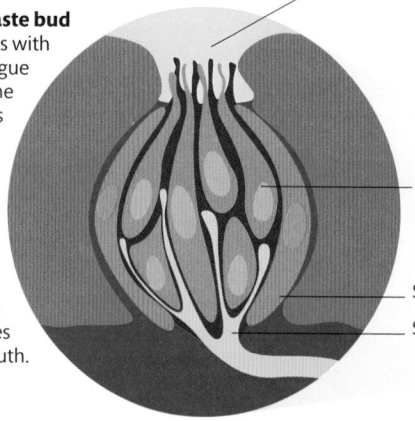

A papilla—a visible bump on the tongue that may contain taste buds sensitive to sour, bitter, salty, sweet, or umami tastes

SOUR

BITTER

SALTY

UMAMI

SWEET

SUPERTASTERS

Some people have many more taste buds than others. These supertasters can detect bitter substances that other people can't and generally dislike green vegetables and fatty foods. Supertasters are thought to make up 25 percent of the population.

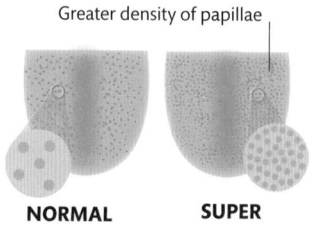

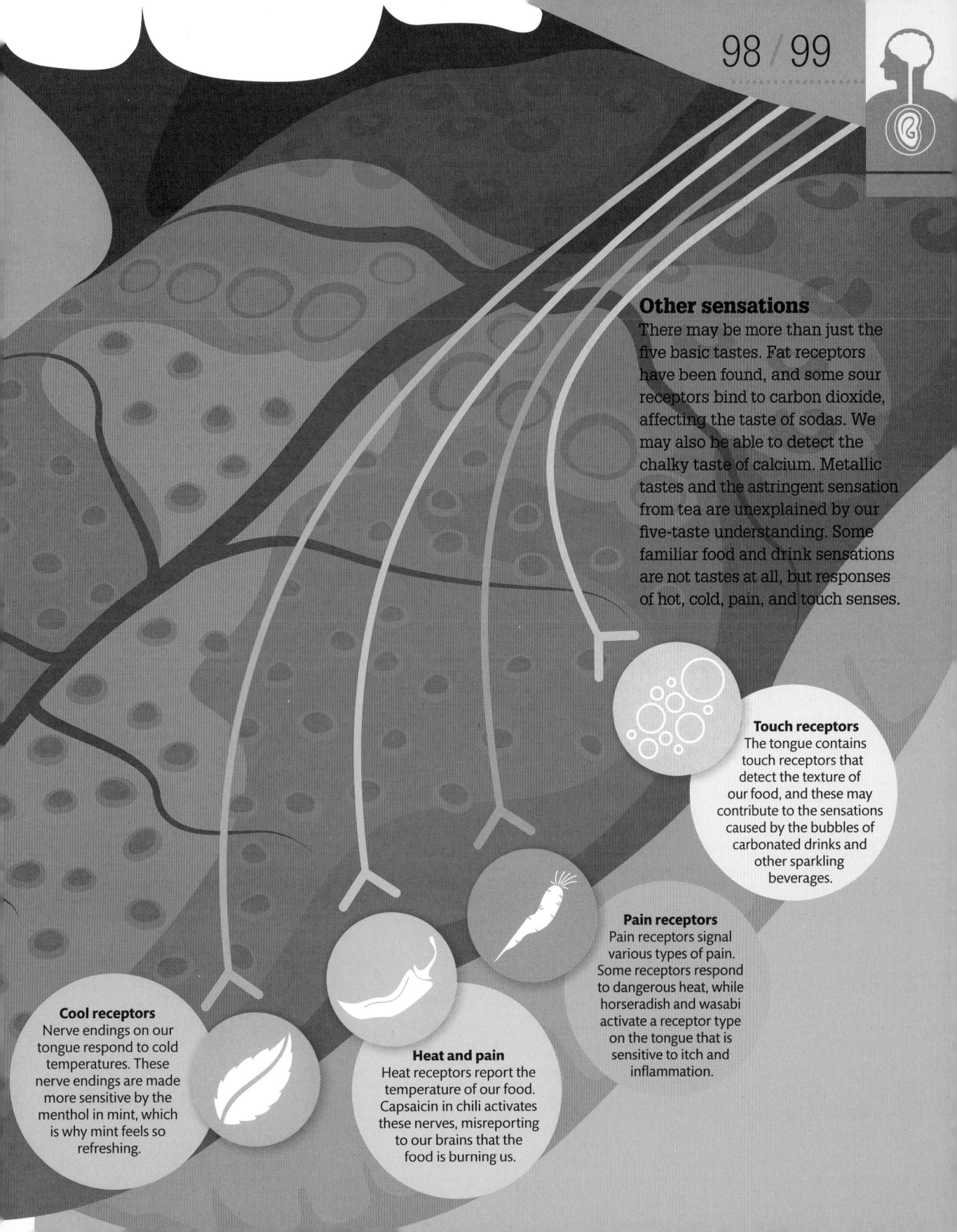

Other sensations

There may be more than just the five basic tastes. Fat receptors have been found, and some sour receptors bind to carbon dioxide, affecting the taste of sodas. We may also be able to detect the chalky taste of calcium. Metallic tastes and the astringent sensation from tea are unexplained by our five-taste understanding. Some familiar food and drink sensations are not tastes at all, but responses of hot, cold, pain, and touch senses.

Touch receptors
The tongue contains touch receptors that detect the texture of our food, and these may contribute to the sensations caused by the bubbles of carbonated drinks and other sparkling beverages.

Pain receptors
Pain receptors signal various types of pain. Some receptors respond to dangerous heat, while horseradish and wasabi activate a receptor type on the tongue that is sensitive to itch and inflammation.

Heat and pain
Heat receptors report the temperature of our food. Capsaicin in chili activates these nerves, misreporting to our brains that the food is burning us.

Cool receptors
Nerve endings on our tongue respond to cold temperatures. These nerve endings are made more sensitive by the menthol in mint, which is why mint feels so refreshing.

MIRROR BOX THERAPY

Many amputees suffer from "phantom limb" pain. The brain interprets the lack of sensory input from the missing limb as a sensation that the muscles are clenched and cramping. By tricking the brain into "seeing" the phantom limb with a mirror box, movements in the retained limb can often relieve the pain.

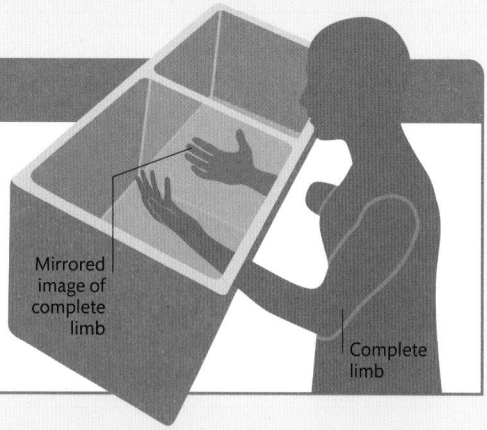

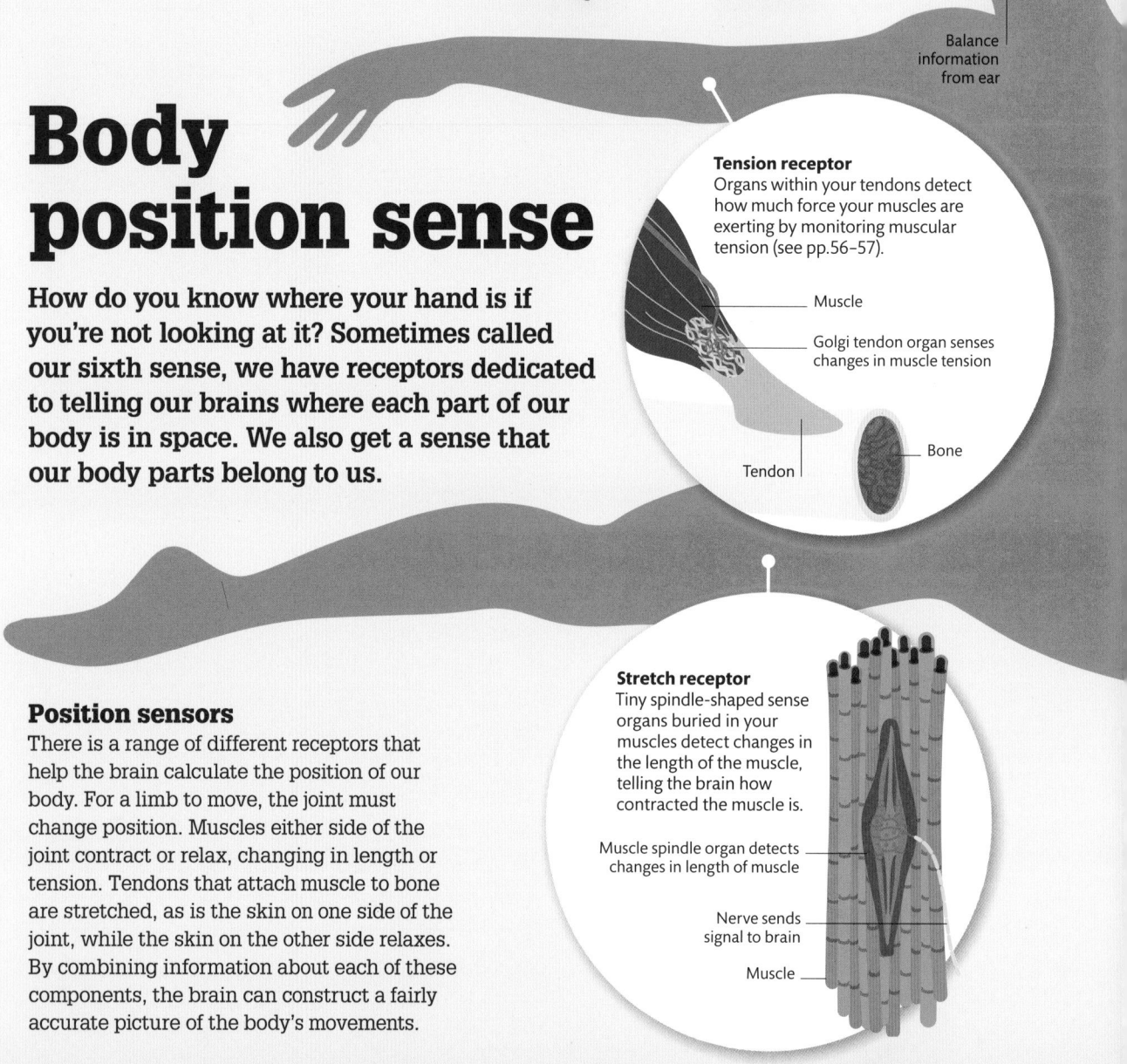

Body position sense

How do you know where your hand is if you're not looking at it? Sometimes called our sixth sense, we have receptors dedicated to telling our brains where each part of our body is in space. We also get a sense that our body parts belong to us.

Position sensors

There is a range of different receptors that help the brain calculate the position of our body. For a limb to move, the joint must change position. Muscles either side of the joint contract or relax, changing in length or tension. Tendons that attach muscle to bone are stretched, as is the skin on one side of the joint, while the skin on the other side relaxes. By combining information about each of these components, the brain can construct a fairly accurate picture of the body's movements.

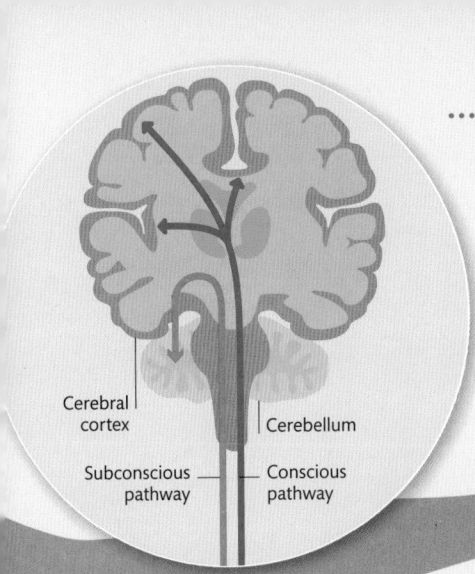

Integrator

The brain combines information from the sensors located in and around the muscles as well as your other senses to interpret how your body is positioned. The conscious element of this is controlled by the cerebral cortex and allows you to run, dance, or catch. The cerebellum, at the base of the brain, is in charge of the unconscious elements that keep you upright without your thinking about it.

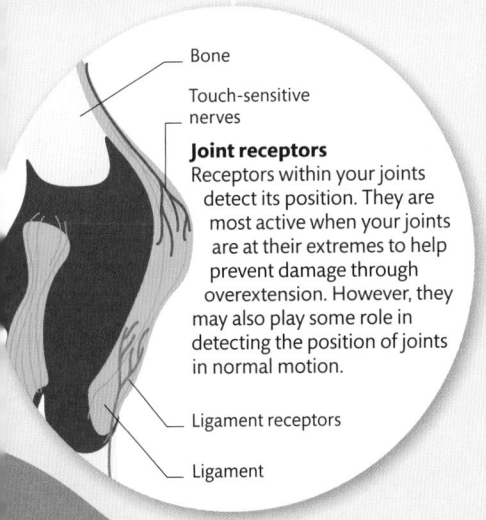

Joint receptors
Receptors within your joints detect its position. They are most active when your joints are at their extremes to help prevent damage through overextension. However, they may also play some role in detecting the position of joints in normal motion.

BODY OWNERSHIP SENSE

Your sense that your body is your own is more complicated and flexible than it seems. The rubber hand illusion shown here creates the feeling that a fake hand belongs to you. A similar technique can invoke out-of-body experiences, using a virtual reality headset. This flexibility allows us to cope if we lose a limb, or to include tools and prosthetics in what we think of as part of our body.

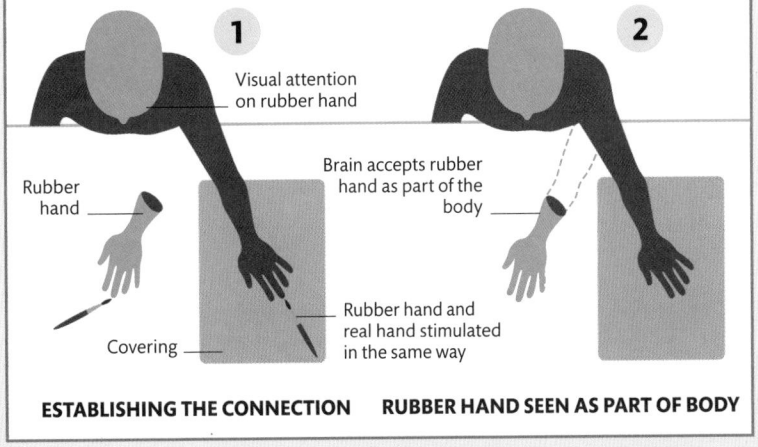

Skin stretch
Special receptors in the skin (see p.75) can detect stretch. This helps us determine the movements of our limbs, particularly changes in the angle of a joint, which causes the skin on one side to stretch while the skin on the opposite side is slackened.

BODY POSITION SENSORS IN THE JAW MUSCLES AND TONGUE HELP YOU **FORM THE RIGHT SOUNDS** WHEN YOU SPEAK

Integrated senses

Your brain makes sense of the world around you by combining information from all your senses. But, surprisingly, sometimes one sense can actually change how you experience another.

How senses can interact

Everything you experience is interpreted by your senses. When you see and pick up an item, you feel its shape and texture. You look for where sounds or smells are coming from and "eat with your eyes" before tasting your food. Your brain performs complex processing to integrate this information correctly. Sometimes, this combination of information can cause multisensory illusions. If information from different senses seems to conflict, the brain favors one sense over another, and depending on the situation, this can be helpful or misleading.

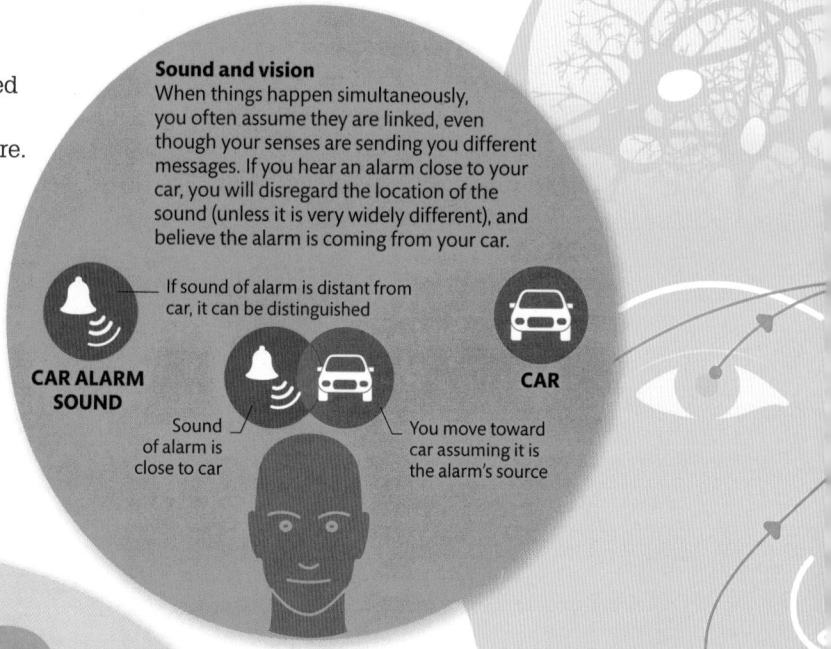

Sound and vision
When things happen simultaneously, you often assume they are linked, even though your senses are sending you different messages. If you hear an alarm close to your car, you will disregard the location of the sound (unless it is very widely different), and believe the alarm is coming from your car.

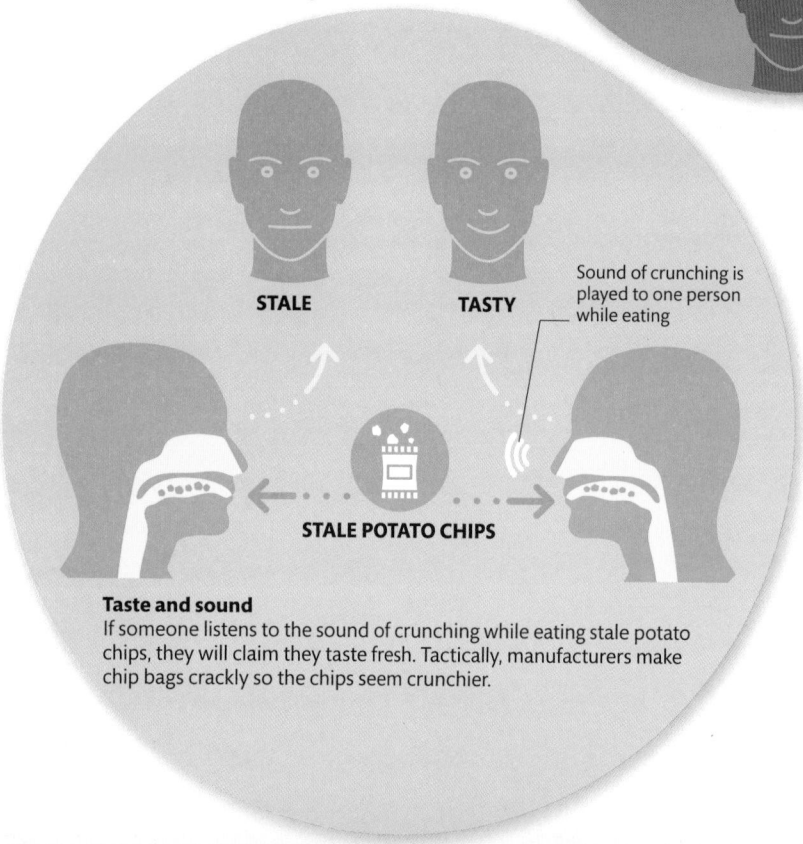

Taste and sound
If someone listens to the sound of crunching while eating stale potato chips, they will claim they taste fresh. Tactically, manufacturers make chip bags crackly so the chips seem crunchier.

IN NOISY ENVIRONMENTS YOU **LIP-READ**, USING **WHAT YOU SEE** TO **INTERPRET MUFFLED SPEECH**

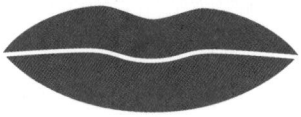

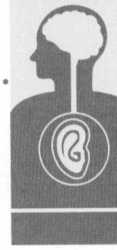

SOUNDS AND SHAPES

When shown these shapes and asked to name one Bouba and the other Kiki, most people call the spiky shape Kiki because of its spiky sound, while deciding the softer Bouba fits the rounded shape. This pairing holds across a wide variety of cultures and languages, indicating a link between the senses of sound and sight.

Smell and taste
Taste is a simple sense, made up of crude sensations such as "sweet" or "salty". Most of what you think of as flavor is actually what you are smelling. Smell can also influence the crude sense of taste itself. Smelling vanilla can make food or drink taste sweeter, but only in parts of the world where vanilla is a common flavor for sweet foods.

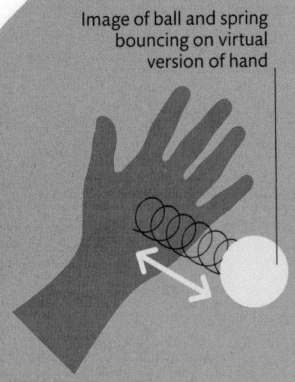

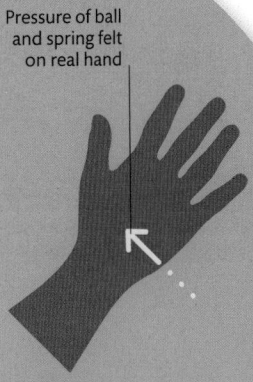

Touch and vision
When gamers pick up objects in virtual reality, visual cues give them physical sensations, even though their touch sense gives them no such information. What your eyes can see can actually influence what you feel.

Using your voice

Talking is achieved by a complex yet flexible network of nerve pathways in the brain and physical coordination of the body. Tone and inflection influences how words are spoken, which can add numerous meanings to even the simplest of sentences.

1 Thought process
Firstly, you must decide what words you want to say. This activates a network of regions in the left hemisphere of your brain, including Broca's area, drawing on your memory store of words.

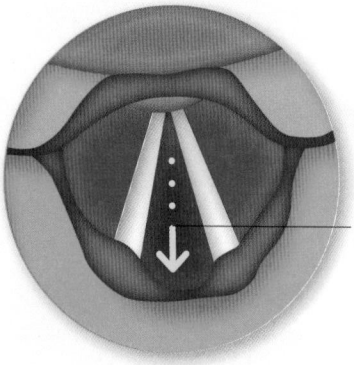

2 Breathing in
Your lungs provide the constant stream of air that you need in order to speak. When inhaling, vocal cords open to allow air to pass through, and then air pressure begins to build in the lungs.

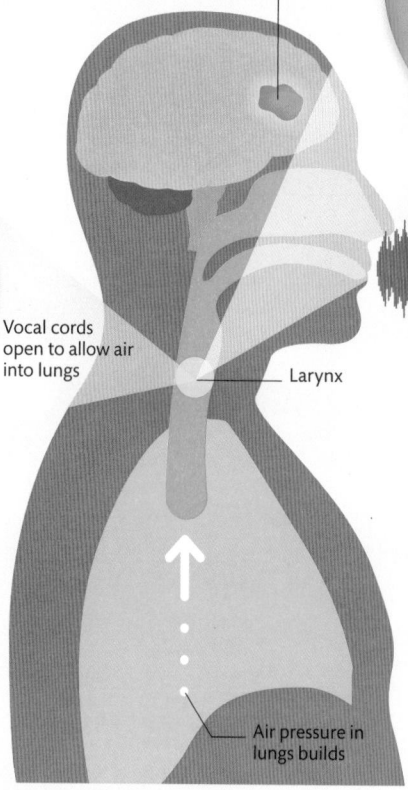

3 Producing sound
As you exhale, your vocal cords vibrate as the air stream passes them, making sound. Vibration speed dictates your voice's pitch, and this is controlled by muscles in the larynx. If you want to shout, you need a stronger airstream.

4 Articulation
Your nose, throat, and mouth act as resonators, while lip and tongue movements introduce specific sounds, altering the buzzing produced by the vocal cords into recognizable speech.

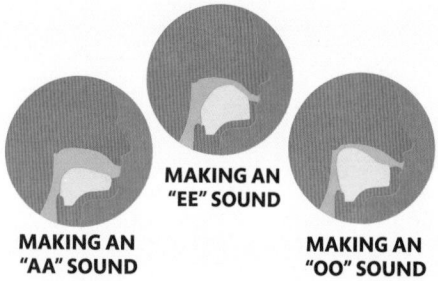

Making different sounds
Your tongue moves to mould sounds created by your vocal cords, aided by the teeth and lips. Changing the shape of your tongue and mouth produces vowels like aa, or ee, and the lips interrupt air flow to produce consonants, such as p and b.

How do you talk?

The brain, lungs, mouth, and nose all play vital roles when producing speech, but the voicebox, or the larynx, is the most important. Located in your throat above your windpipe, it contains two sheets of membrane that stretch across the inside. These are the vocal cords, and they are the structures that produce the sound you craft into speech.

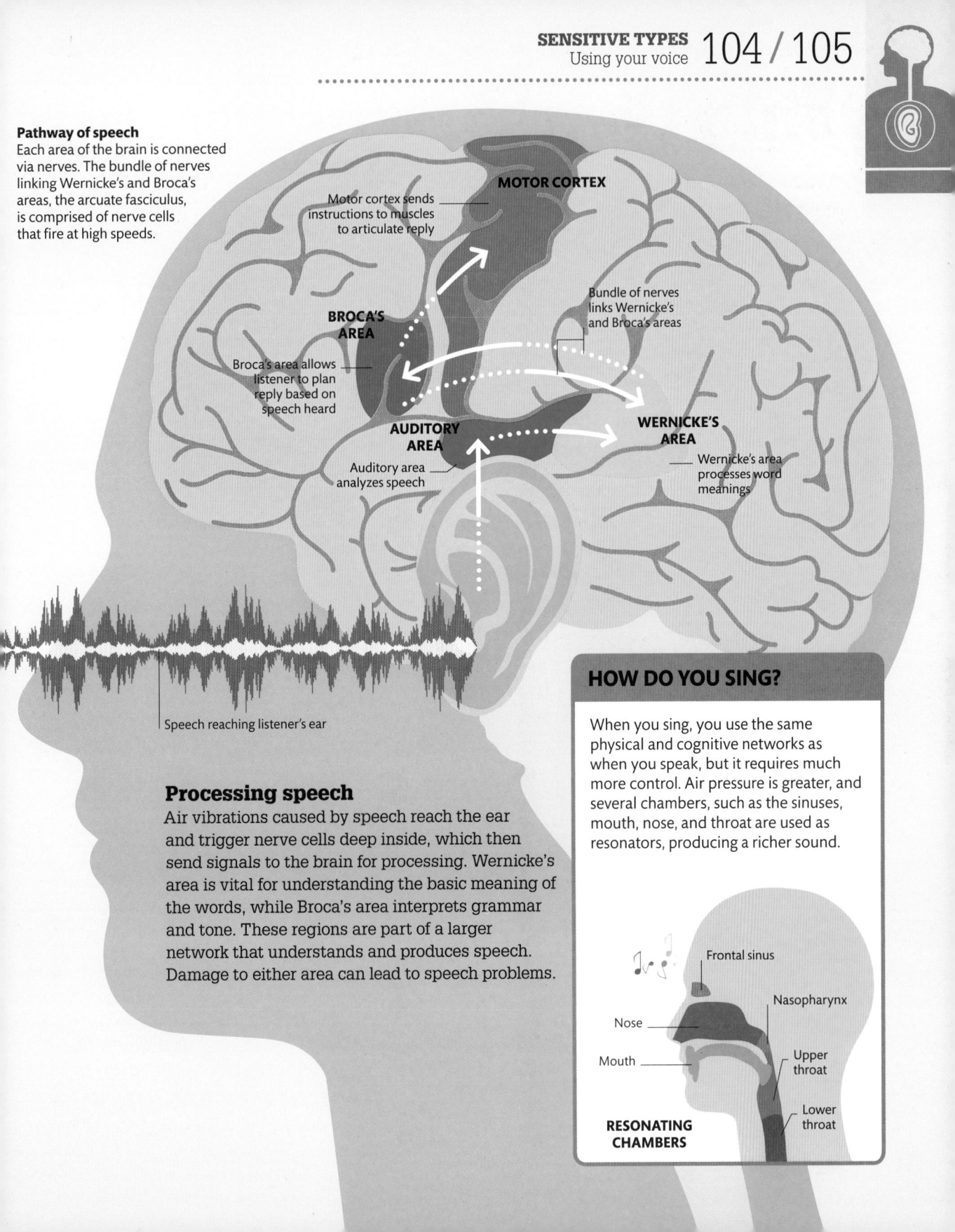

Pathway of speech
Each area of the brain is connected via nerves. The bundle of nerves linking Wernicke's and Broca's areas, the arcuate fasciculus, is comprised of nerve cells that fire at high speeds.

Processing speech

Air vibrations caused by speech reach the ear and trigger nerve cells deep inside, which then send signals to the brain for processing. Wernicke's area is vital for understanding the basic meaning of the words, while Broca's area interprets grammar and tone. These regions are part of a larger network that understands and produces speech. Damage to either area can lead to speech problems.

HOW DO YOU SING?

When you sing, you use the same physical and cognitive networks as when you speak, but it requires much more control. Air pressure is greater, and several chambers, such as the sinuses, mouth, nose, and throat are used as resonators, producing a richer sound.

Reading faces

We are a social species, so recognizing and understanding faces is vital for our survival. This means we have evolved to be very good at noticing them—even sometimes seeing them where they don't really exist, like on a piece of burned toast!

Importance of understanding faces

From birth, babies are fascinated by faces, and show a preference for looking at them above everything else. As you age, you not only quickly become an expert in recognizing faces, but also reading expressions. This allows you to identify those who would help or harm you. Individual faces can stay in your memory for a remarkable length of time, even if you haven't seen the person in years.

Facial expression cues
When recognizing a face, you look at the ratio between the eyes, nose, and mouth. Movements of these can help you detect emotions; for example, raised eyebrows and an open mouth would signal surprise. These signals are interpreted by your eyes and nerve signals are sent to the fusiform face area in your brain to be processed.

Fusiform face area
This area of the brain, named the fusiform face area, is activated when you look at faces. It is thought that this area of the brain is specialized in facial recognition. However, it also becomes active when you are looking at objects with which you are familiar—if you were a pianist, it may become active when you see a keyboard. Whether it is face-specific is still being studied.

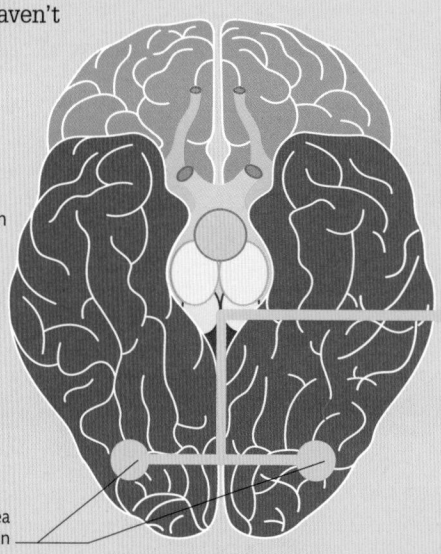

UNDERSIDE OF BRAIN

RECOGNIZING FACES

Humans tend to spot faces in random patterns and places—from cars to cheese, clothes washers and pieces of wood. This is because it was important for our ancestors to interpret the faces of others in order to thrive in a complex social hierarchy.

Expression muscles

Your face contains muscles that pull your skin and change the shape of your eyes and the position of your lips, making your face highly expressive. The ability to read these expressions on other faces allows you to judge other people's moods, intentions, and meanings. Faces tell us when to ask for a favor, when to leave a person alone, or when to offer comfort. Picking up even the subtlest cues, such as the furrowing of the brow or the curling of the lip, can mean the difference between interpreting a frown or a smirk correctly.

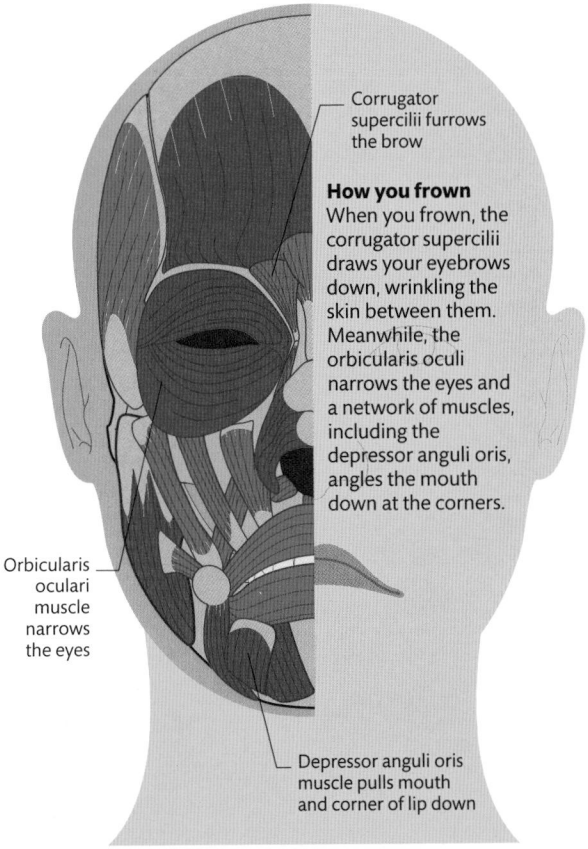

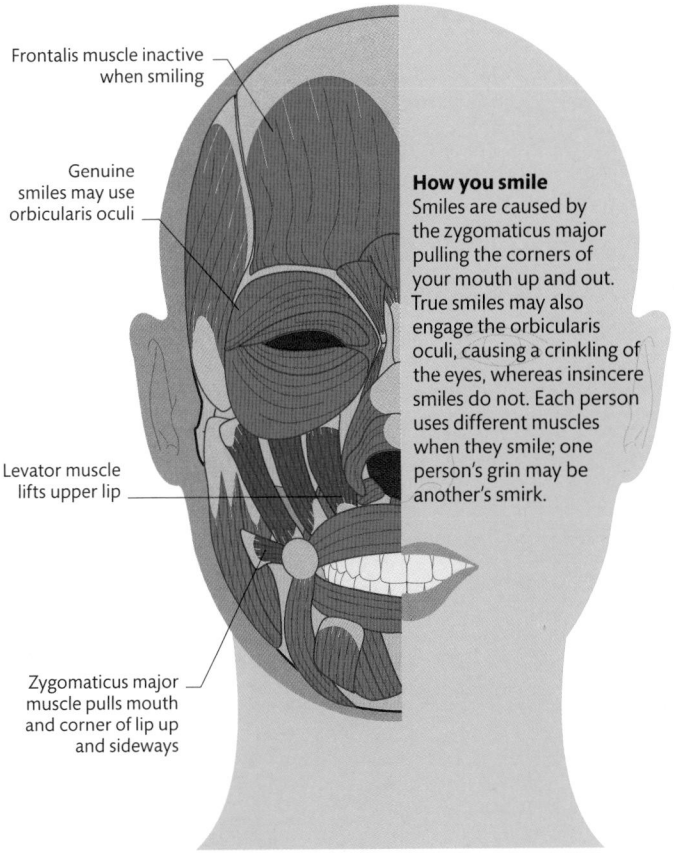

GAZE AND EYE CONTACT

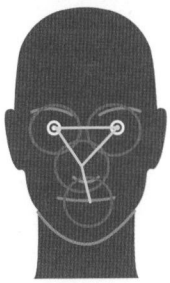

TYPICAL GAZE

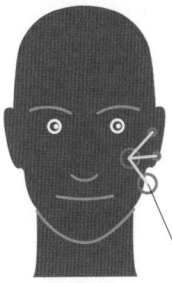

THOSE WITH AUTISM

People with autism (see p.246) usually don't focus on the eyes and mouth when looking at faces. They find socializing confusing and difficult, and may miss vital social cues when communicating. Babies may even exhibit this averted gaze, and they may go on to develop the condition, so it could be used as an early warning sign for autism.

People with autism show different patterns of looking behavior

PEOPLE BORN BLIND PRODUCE THE **SAME EXPRESSIONS AS SIGHTED PEOPLE** WHEN EMOTIONS ARE PROVOKED

What you don't say

You communicate using more than just your words. Facial expressions, tone of voice, and hand gestures can speak volumes, and noticing these signals is vital for understanding what someone really means.

INVADING SOMEONE'S PERSONAL SPACE CAN INSPIRE FEAR, AROUSAL, OR DISCOMFORT

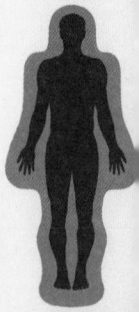

Nonverbal communication

When you are talking to someone, you are subconsciously picking up on subtle signals from the other person's voice, face, and body. Interpreting these signals correctly is most important when what is said could be ambiguous. Most of these signals allow you to gauge the mood of a person or group so you act appropriately in social situations. For example, in a meeting at work, assessing the body language and moods of your colleagues can be advantageous to you if you are waiting for the right time to pitch a big idea.

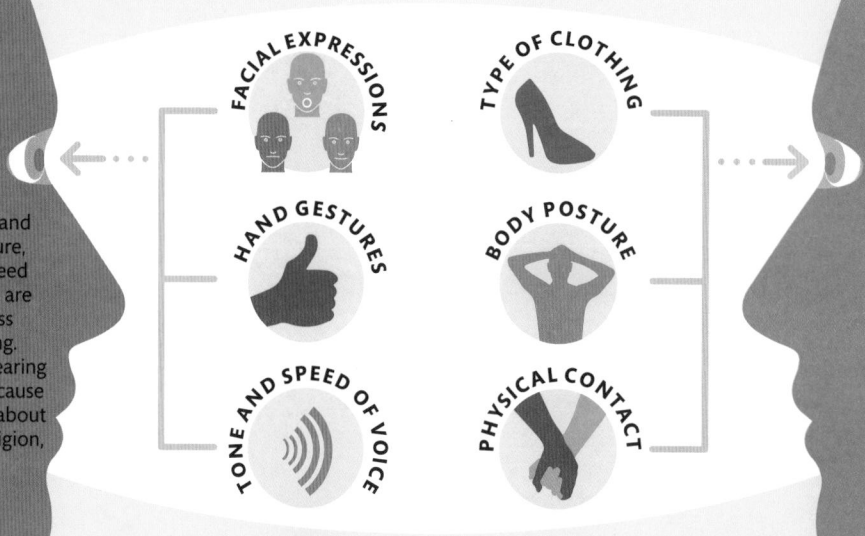

Types of signals
Facial expressions, hand gestures, body posture, and the tone and speed of somebody's voice are all signals you process when communicating. What someone is wearing is also important because it can provide clues about their personality, religion, or culture. Physical contact can add emotional weight to what is said.

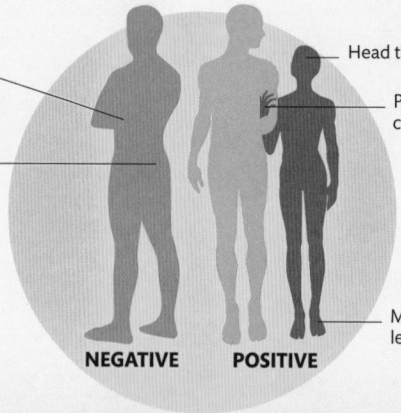

Body language
The way your body moves as you speak can often be just as telling as what you say. Holding eye contact, mirroring the facial expressions and posture of others, and physical contact, are generally interpreted as positive signals. Folded arms, hunched shoulders, and positioning yourself away from others can produce negative vibes.

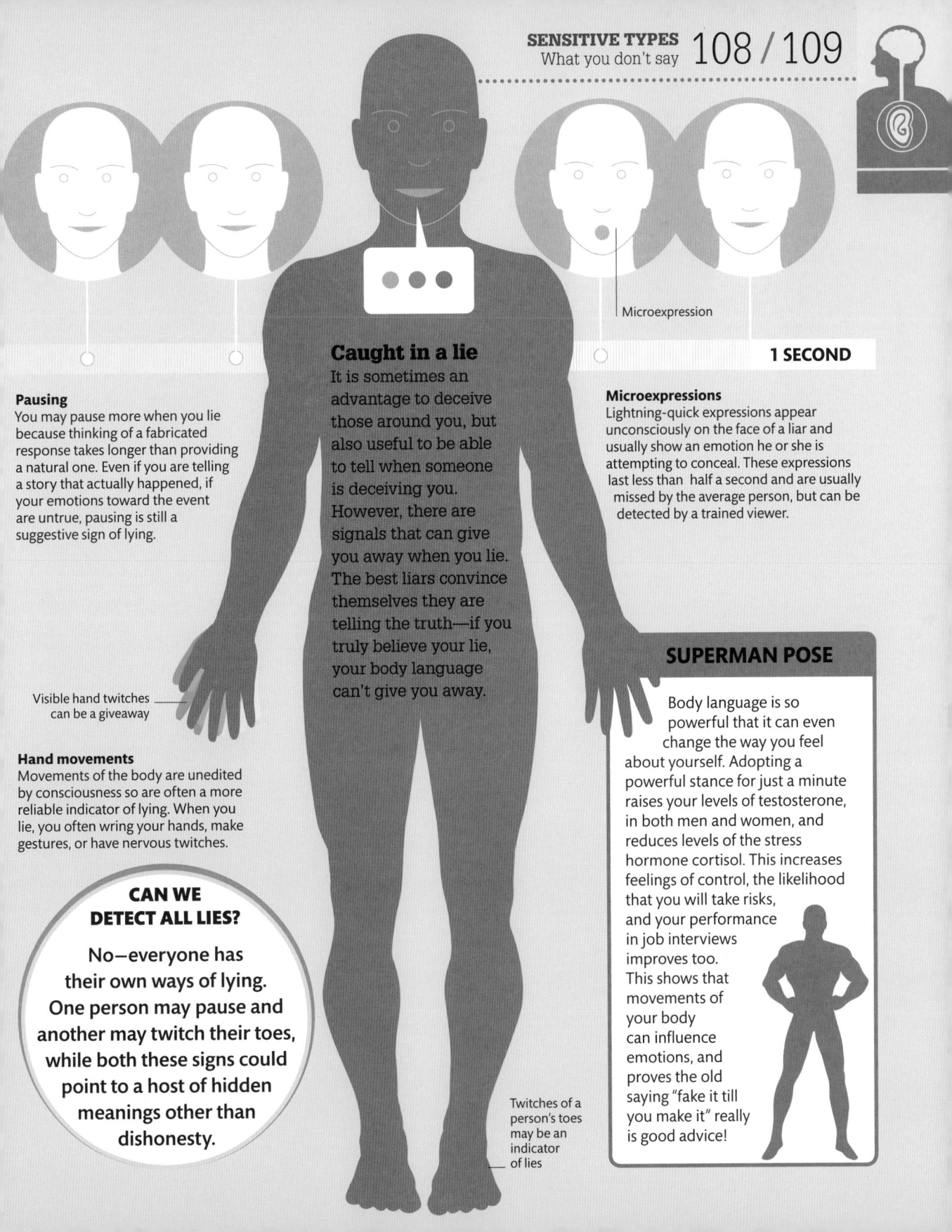

Caught in a lie

It is sometimes an advantage to deceive those around you, but also useful to be able to tell when someone is deceiving you. However, there are signals that can give you away when you lie. The best liars convince themselves they are telling the truth—if you truly believe your lie, your body language can't give you away.

Pausing
You may pause more when you lie because thinking of a fabricated response takes longer than providing a natural one. Even if you are telling a story that actually happened, if your emotions toward the event are untrue, pausing is still a suggestive sign of lying.

Microexpressions
Lightning-quick expressions appear unconsciously on the face of a liar and usually show an emotion he or she is attempting to conceal. These expressions last less than half a second and are usually missed by the average person, but can be detected by a trained viewer.

Hand movements
Movements of the body are unedited by consciousness so are often a more reliable indicator of lying. When you lie, you often wring your hands, make gestures, or have nervous twitches.

CAN WE DETECT ALL LIES?

No—everyone has their own ways of lying. One person may pause and another may twitch their toes, while both these signs could point to a host of hidden meanings other than dishonesty.

SUPERMAN POSE

Body language is so powerful that it can even change the way you feel about yourself. Adopting a powerful stance for just a minute raises your levels of testosterone, in both men and women, and reduces levels of the stress hormone cortisol. This increases feelings of control, the likelihood that you will take risks, and your performance in job interviews improves too. This shows that movements of your body can influence emotions, and proves the old saying "fake it till you make it" really is good advice!

THE HEART OF THE MATTER

Filling your lungs

Your lungs act like a giant pair of bellows, drawing air in and letting it out to extract oxygen and expel waste carbon dioxide. You breathe around 12 times per minute at rest and 20 times per minute or more during exercise; which all adds up to roughly 8.5 million breaths per year.

Controlling breaths

Your breathing rate speeds up or slows down due to signals from chemical receptors in the blood vessels. These receptors provide a feedback loop between the blood vessels, brain, and diaphragm.

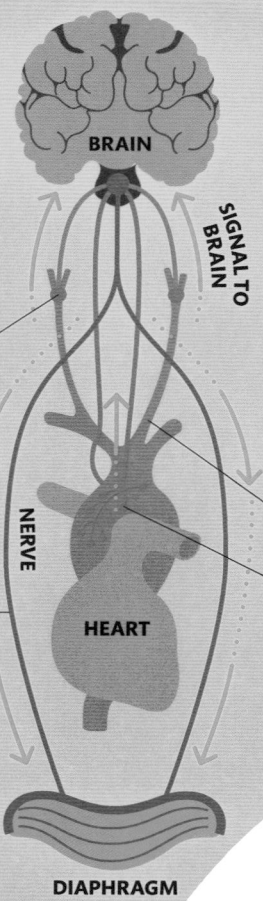

Feedback system
Chemical receptors detect changes in oxygen, carbon dioxide, and acidity levels in the blood. This information is sent to the brain, which controls the diaphragm's movements, increasing or decreasing rate and depth of breathing to keep blood levels constant.

Drawing breath

Air drawn in through the nose or mouth passes down the trachea, or windpipe, which channels air into the left or right bronchus, and then into smaller and smaller air passages called bronchioles. Between the trachea and the ends of the bronchioles, your airway divides 23 times.

1 Breathing in
Air is warmed and moistened as it passes through the nose or mouth. Nasal hairs filter out dust particles that could irritate the trachea or lungs and cause a coughing fit.

Air breathed in

NASAL CAVITY

TONGUE

Air traveling through the throat

Air traveling down the trachea

TRACHEA

Bronchiole

Lining of right lung

LUNG

SIZE MATTERS

The surface area of all the tiny air sacs (alveoli) in the lungs measures an incredible 753 sq ft (70 sq m)—this is 40 times greater than the surface area of your skin! This maximizes the amount of oxygen you can absorb.

Mechanics of breathing

Chest muscles and the rib cage influence breathing, but the main powerhouse is the diaphragm. It is a large domed muscle that separates the chest from the lower organs. To breathe in, the diaphragm contracts and pulls down like a piston. At the same time, muscles between your ribs contract, lifting your ribs so your lungs expand and air rushes in. When your diaphragm and chest muscles relax, air is forced out.

2 Into the lungs
Air travels down each bronchus into ever-smaller passages, eventually ending in tiny air sacs called alveoli. The lungs are separated from the chest by a pleural cavity filled with pleural fluid. This thin layer of fluid acts as a sticky lubricant, letting your lungs slide over your chest wall and preventing them from pulling away as you breathe out.

ALL YOUR AIRWAYS LAID END TO END WOULD MEASURE 1,490 MILES (2,400 KM)

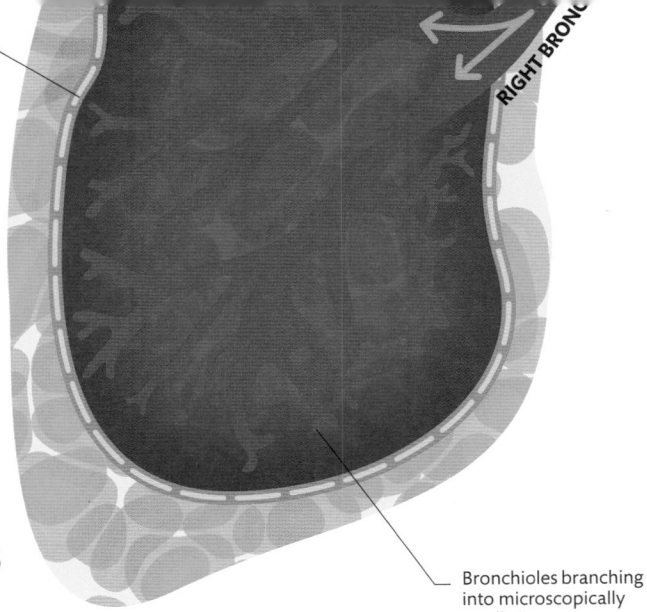

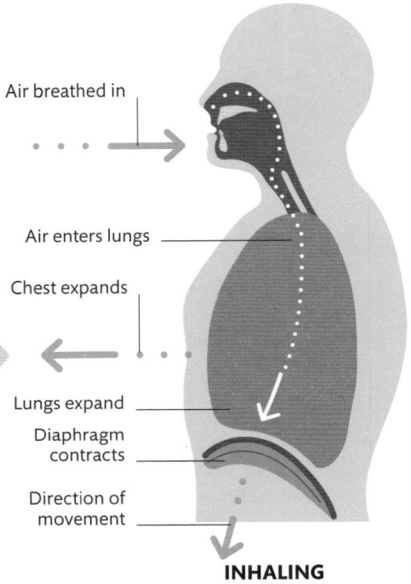

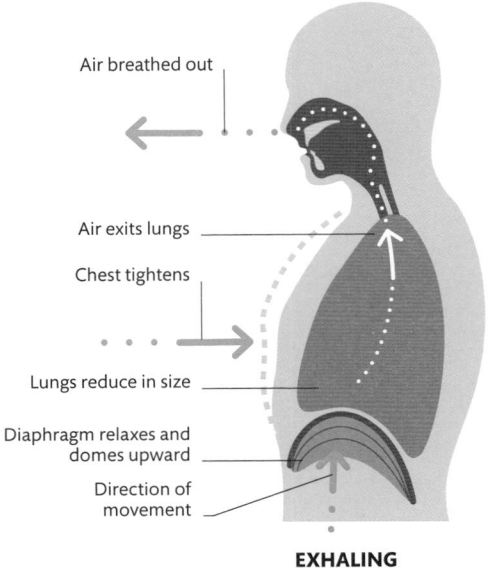

From air to blood

Every cell in your body needs oxygen and your lungs are highly adapted to extract this life-sustaining gas from the atmosphere. This extraction occurs from 300 million tiny air sacs called alveoli, which give your lungs a spongelike texture.

Deeper into the lungs

Inhaled air passes from the throat into the trachea to reach tiny branches called bronchioles. Mucus covers each bronchiole, which keeps them moist and traps inhaled particles. Each bronchiole is lined with thin strips of muscle. In people with asthma, a sudden constriction of these muscles narrows the airways, causing shortness of breath.

THE **AIR YOU BREATHE OUT** CONTAINS **16 PERCENT OXYGEN**, ENOUGH TO **RESUSCITATE SOMEONE!**

WHY CAN WE SEE OUR BREATH IN COLD AIR?

The air you breathe is warmed in your lungs, so when you exhale, water vapor in your breath condenses into clouds of water droplets.

Vein carries oxygen-rich blood to heart

Ring of stiff cartilage that stops bronchiole from collapsing

Alveolar sacs
The bronchioles lead into grapelike clusters of alveoli, each of which is wrapped in capillaries, the smallest type of blood vessel. In contrast to blood vessels in the rest of the body, it is the arteries that carry oxygen-poor blood to the capillaries.

Artery carries oxygen-poor blood from the heart to the lungs

LUNGS

VEIN

BRONCHIOLE

ARTERY

CLUSTER OF ALVEOLI

Capillaries wrap around every alveolus

Types of gas
Oxygen
Carbon dioxide

Exhaled air contains 100 times more carbon dioxide than inhaled air

Inhaled air contains 21 percent oxygen

One-cell-thick wall of alveolus

One-cell-thick wall of capillary

Blood going back to heart to be pumped around the body

ALVEOLUS

Blood plasma rich in carbon dioxide

Oxygen-poor red blood cell

Carbon dioxide entering air

CAPILLARY

Oxygenated red blood cell

Oxygen entering red blood cell

1 Carbon dioxide
Carbon dioxide diffuses from the blood plasma through the one-cell-thick walls of the capillary and alveolus. Blood can absorb oxygen and get rid of carbon dioxide simultaneously.

2 Oxygen
The oxygen we breathe diffuses from alveolar air into the blood. Here, it is captured by red blood cells, turning them, and the blood, bright red.

Gas exchange

Capillaries are in such close contact with alveoli that gases are able to cross over rapidly. Carbon dioxide leaves the blood in exchange for oxygen, and the newly oxygenated blood is distributed around the body by the heart. Since you do not exhale all your inhaled air in one breath, oxygen-poor and oxygen-rich air mixes in your lungs, which is why exhaled air contains oxygen.

HIGH ALTITUDE

At high altitudes, air is thinner and less oxygen is present. You may find yourself automatically taking deep breaths, since your body will detect lower amounts of oxygen in your bloodstream than it normally expects.

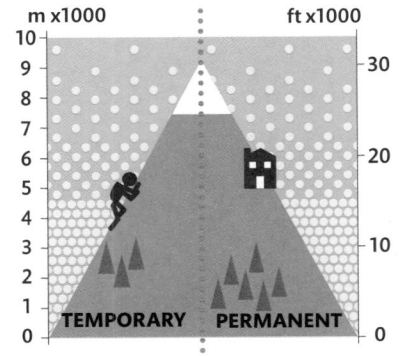

Acclimatizing
People who travel to high altitudes can adapt by producing more red blood cells to carry more oxygen in their circulation. Full adaptation takes around 40 days, but is not permanent.

Adapting
Those who live their entire lives at high altitudes may inherit larger lungs, wider chests, and more efficient oxygen-processing genes in order to cope permanently with the hardships.

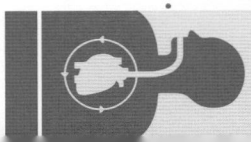

Why do we breathe?

The oxygen we breathe is vital for staying alive because we use it to create energy. Tiny capillaries, the smallest type of blood vessel, transport oxygen to the 50 trillion cells that make up your body. One person uses about 1,160 pints (550 liters) of oxygen per day.

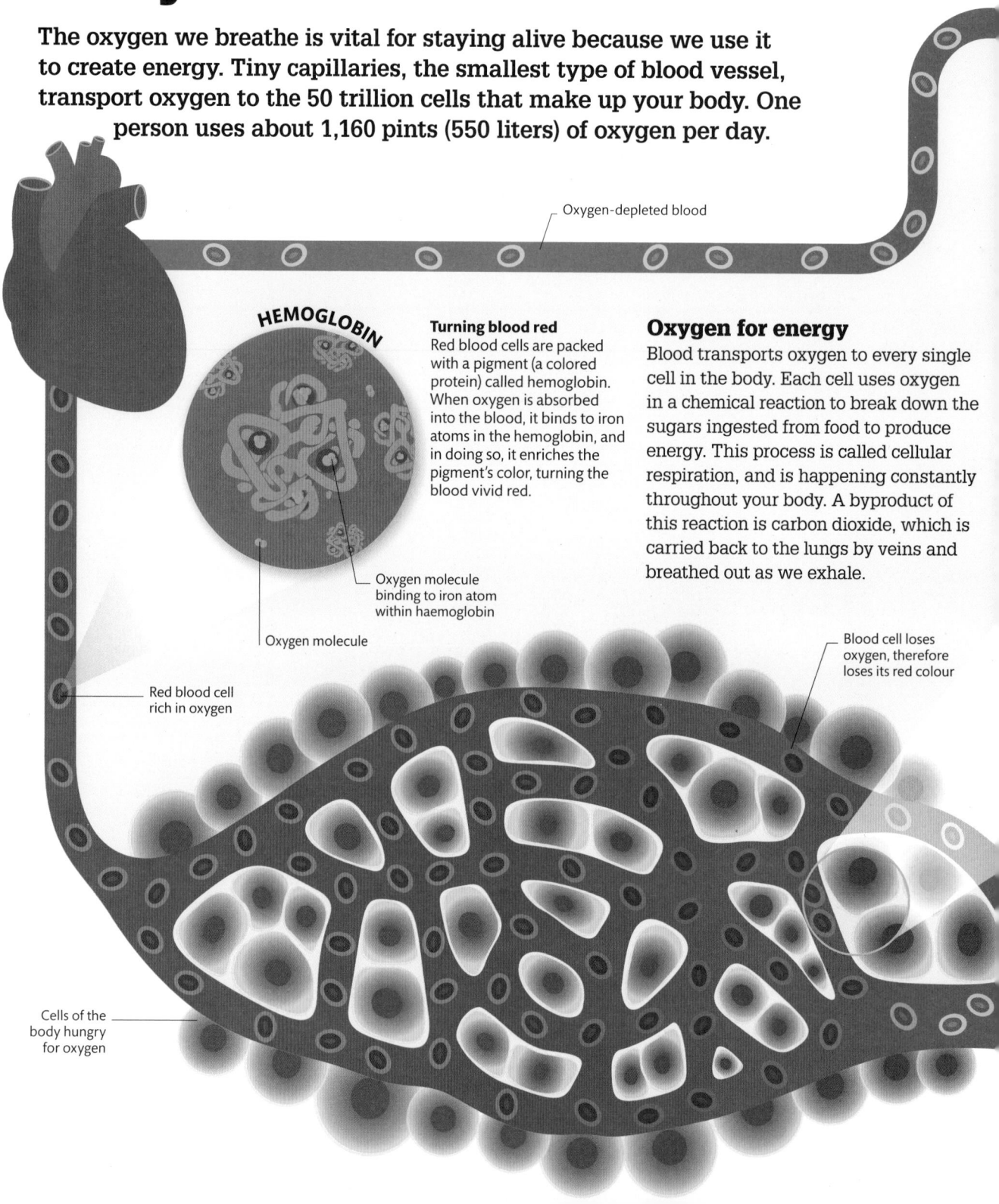

Turning blood red
Red blood cells are packed with a pigment (a colored protein) called hemoglobin. When oxygen is absorbed into the blood, it binds to iron atoms in the hemoglobin, and in doing so, it enriches the pigment's color, turning the blood vivid red.

Oxygen for energy

Blood transports oxygen to every single cell in the body. Each cell uses oxygen in a chemical reaction to break down the sugars ingested from food to produce energy. This process is called cellular respiration, and is happening constantly throughout your body. A byproduct of this reaction is carbon dioxide, which is carried back to the lungs by veins and breathed out as we exhale.

Gas exchange
Oxygen diffuses, or drifts, from where it is in high concentrations (in red blood cells) to where there is a low concentration (in body cells). Likewise, carbon dioxide diffuses from the body cells into the blood.

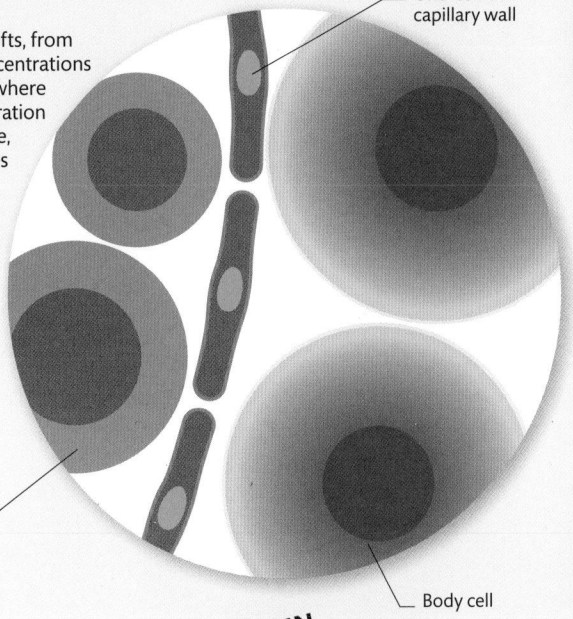

THIN CAPILLARIES

Capillaries connect tiny arteries (arterioles) with tiny veins (venules). The thin walls of capillaries allow the exchange of oxygen and carbon dioxide. They are thin enough to access all body tissues from bones to skin, yet only just wide enough for red blood cells. Red blood cells even have to change their shape to squeeze through some capillaries.

HUMAN HAIR
0.08MM

BLOOD CAPILLARY
0.008MM

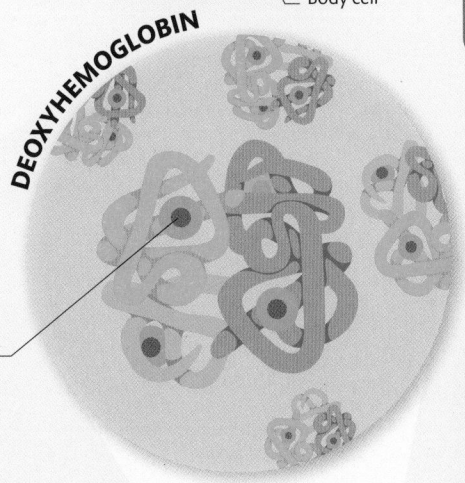

Blue blood?
When hemoglobin is carrying oxygen, it is called oxyhemoglobin. When it releases oxygen into your body tissues, it becomes deoxyhemoglobin, and turns a dark red color—the color of oxygen-depleted blood. The blood is not really blue, even though veins look blue beneath your skin.

IF YOU HOLD YOUR BREATH, THERE IS ENOUGH OXYGEN IN YOUR BLOOD TO STAY CONSCIOUS FOR SEVERAL MINUTES

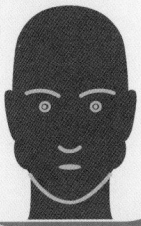

Red blood cell without oxygen

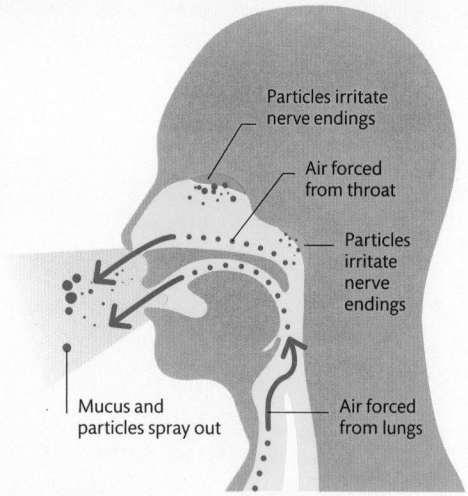

Sneezing

This reflex aims to remove irritants from the nasal cavities, and can be triggered by inhaled particles, infection, or allergies.

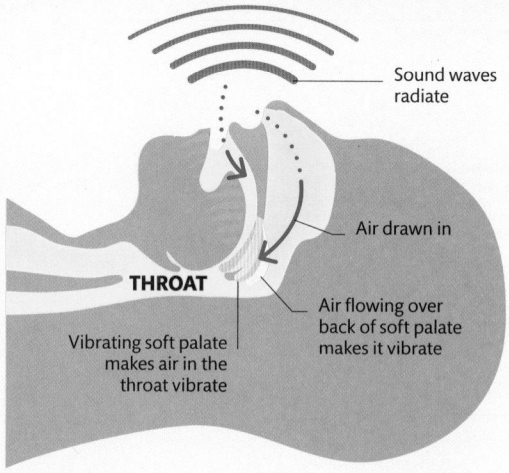

Snoring

A partial collapse of the upper airway during sleep will cause snoring. The tongue falls back and the soft palate vibrates as you breathe.

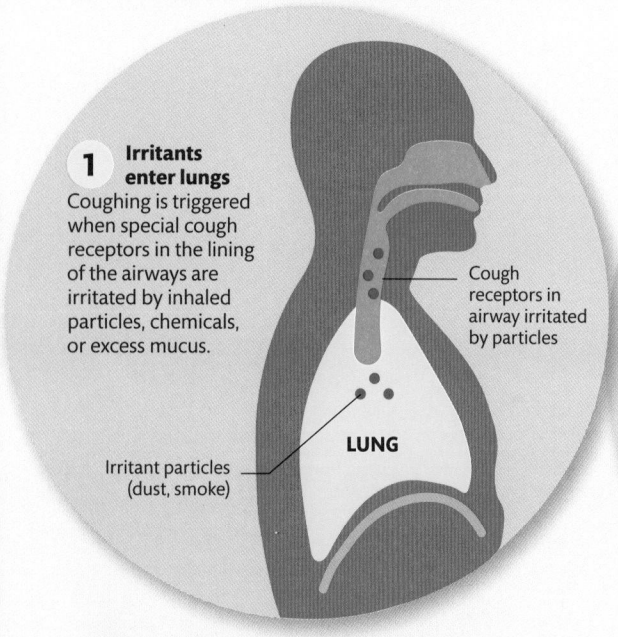

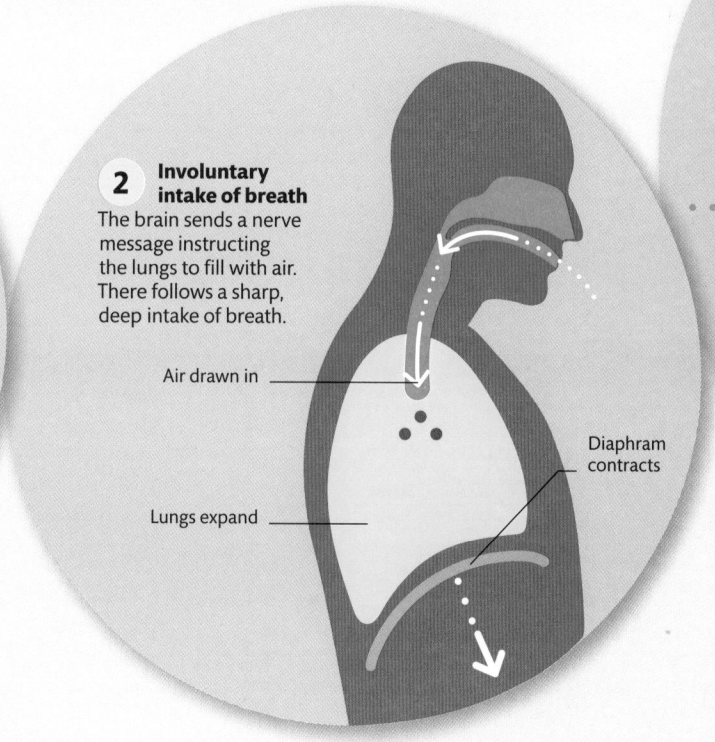

Coughs and sneezes

The respiratory system leaps into sudden action without our conscious control. Its reflex actions get rid of particles in the airways with coughs and sneezes. The functions of hiccups and yawns, however, are more mysterious.

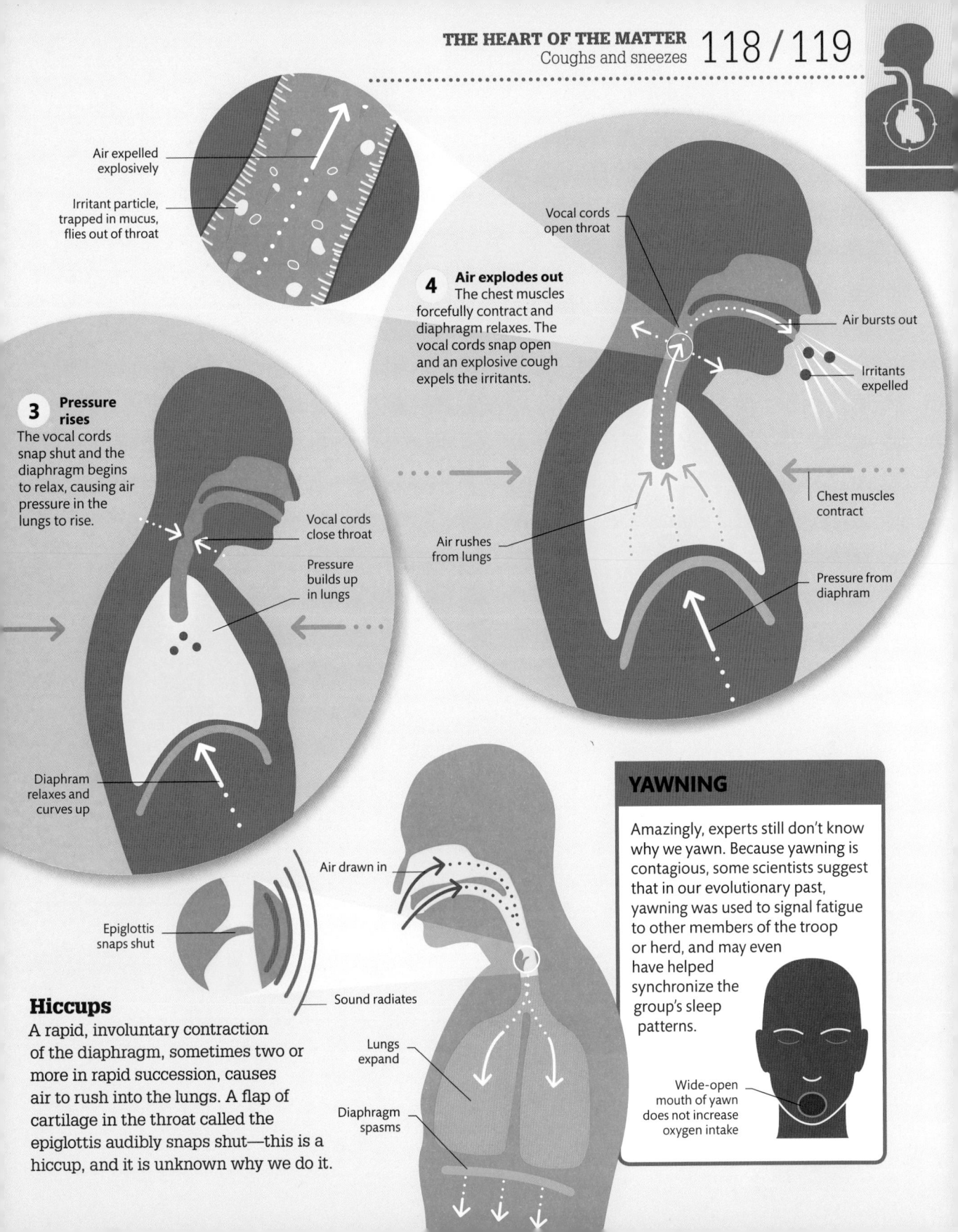

Hiccups

A rapid, involuntary contraction of the diaphragm, sometimes two or more in rapid succession, causes air to rush into the lungs. A flap of cartilage in the throat called the epiglottis audibly snaps shut—this is a hiccup, and it is unknown why we do it.

The many tasks of your blood

Your heart and blood vessels contain around 10½ pints (5 liters) of blood, which transports everything your cells need or produce, such as oxygen, hormones, vitamins, and wastes. Blood carries nutrients from food for processing and toxins for detoxification to the liver, and transports wastes and excess fluid to the kidneys, which expel it from the body.

What is blood made of?

Blood consists of a fluid called plasma, in which float billions of red and white blood cells, plus platelets—the cell fragments involved in blood clotting. Blood also contains wastes, nutrients, cholesterol, antibodies, and protein clotting factors that travel within the plasma. The body carefully controls blood temperature, acidity, and salt levels—if these vary too much, blood and body cells could not function properly.

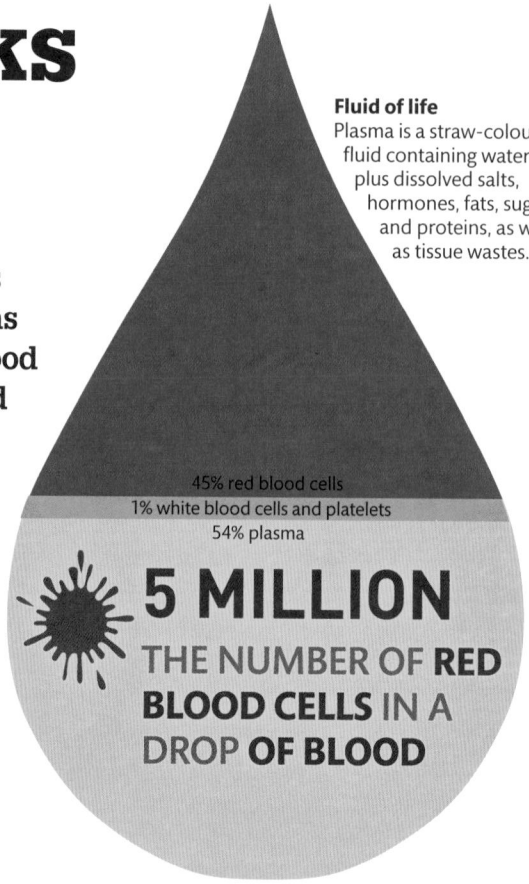

Fluid of life
Plasma is a straw-coloured fluid containing water plus dissolved salts, hormones, fats, sugars, and proteins, as well as tissue wastes.

Oxygen transport

Most oxygen is carried within the red blood cells. A small amount of oxygen is also dissolved in plasma. After a red blood cell collects oxygen from the lungs, it takes around a minute to complete one circuit around the body. During this circuit, oxygen diffuses into the tissues and carbon dioxide is absorbed into the blood. Oxygen-depleted blood cells are then taken back to the lungs, where the blood releases carbon dioxide and the cycle starts again.

WHERE IS BLOOD MADE?

Strangely, blood is actually manufactured in bone marrow in your flat bones (such as the ribs, sternum, and shoulder blades)—millions of blood cells are produced every single second!

Double circulation
Oxygen-depleted blood is pumped from the right side of the heart to the lungs. Blood rich in oxygen from the lungs is pumped from the left side of the heart out to the body.

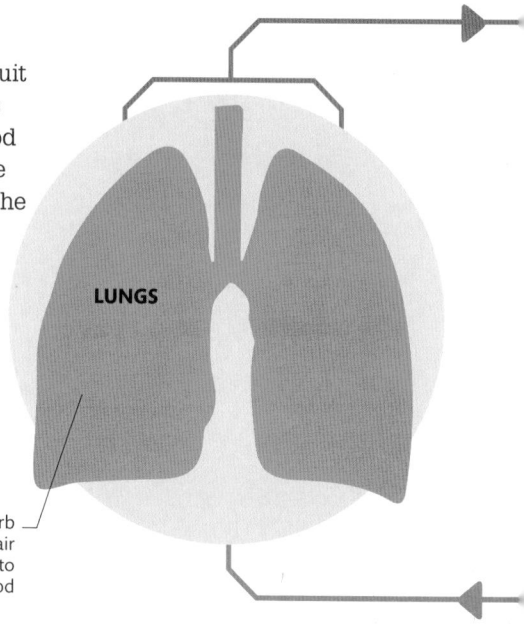

What the body needs

All the living cells throughout the body need various things to help them function properly. Blood carries these vital supplies, such as oxygen, salts, fuel (in the form of glucose or fats), and protein building blocks—amino acids—for growth and repair. Blood also carries hormones, such as epinephrine, which are chemicals that affect the behavior of cells.

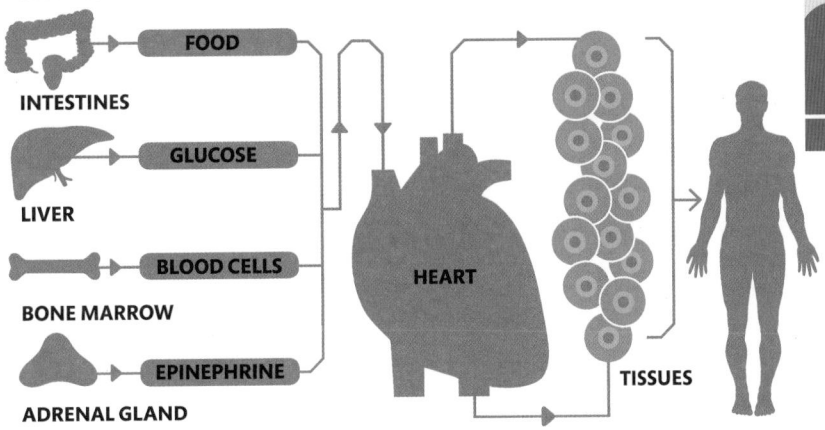

What the body doesn't need

Wastes, such as lactic acid, are produced as by-products of normal cell function. Blood quickly carries the wastes away to prevent imbalances. Some wastes may be transported to the kidneys, to be expelled in urine, or can be carried to the liver to be converted back into something that the cells need.

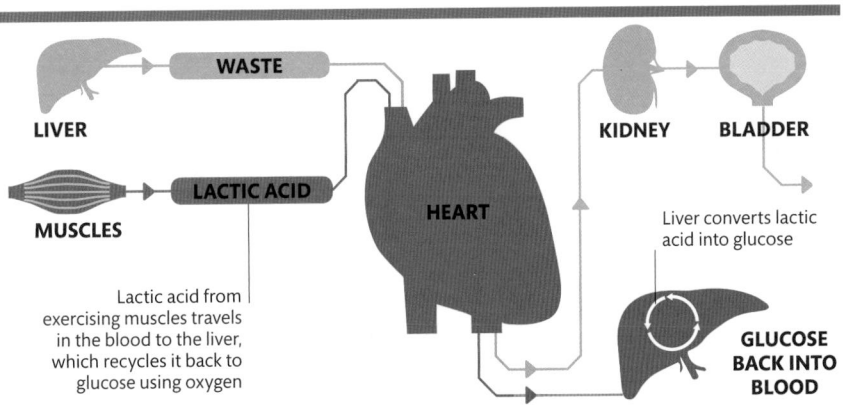

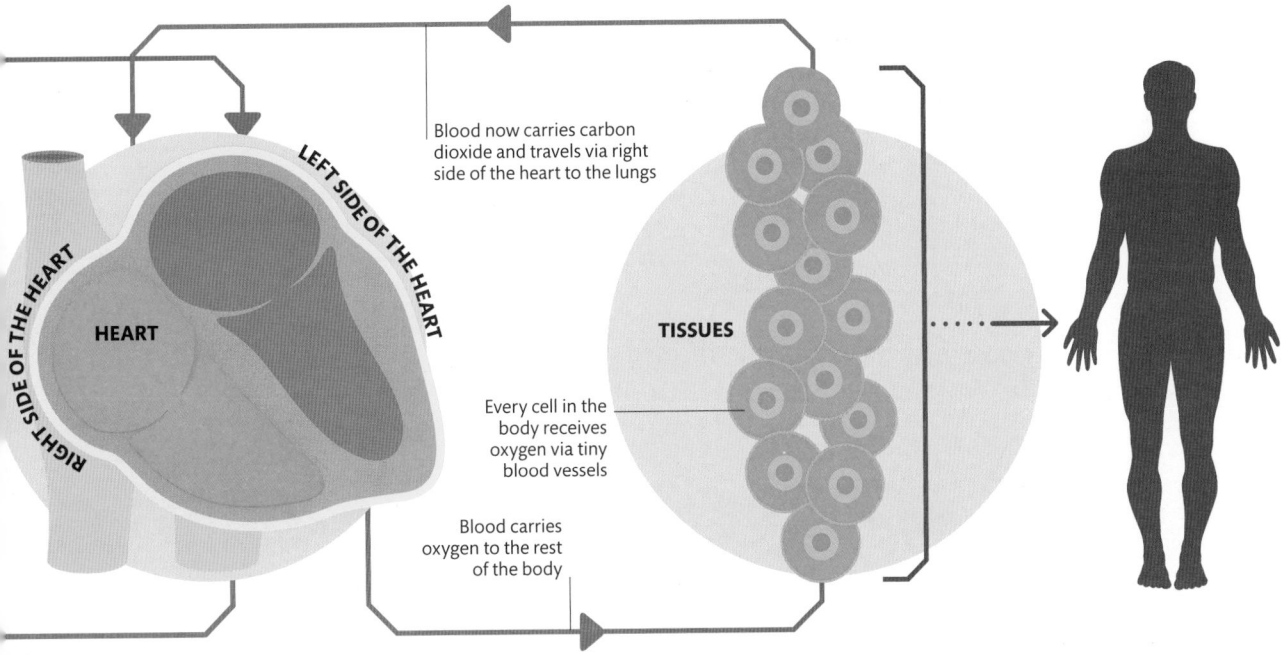

How the heart beats

The heart is a fist-sized muscular organ that contracts and relaxes around 70 times a minute. This keeps blood flowing around the lungs and body, transporting life-giving oxygen and nutrients.

Heart cycle

Your heart is a muscular pump that is divided into two halves, left and right. Each side of the heart is further divided into two chambers—an upper atrium and a lower ventricle. Valves prevent backflow so that blood keeps traveling in the correct direction. A patch of muscle acts as a natural pacemaker, generating an electric signal that makes the heart muscle cycle between contraction and relaxation. The rhythmic squeezing of the heart pumps blood from its right side to the lungs and from the left side to the rest of the body.

ECG recording

Electric impulses within the heart can be recorded by electrodes to produce an electrocardiogram (ECG). Each heartbeat produces a characteristic trace on the ECG display. Its shape is made up of five phases—P, Q, R, S, and T, each of which is a sign of a particular stage of the heartbeat cycle.

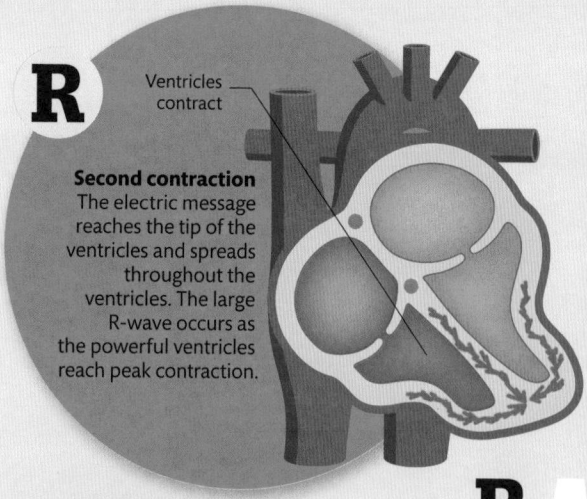

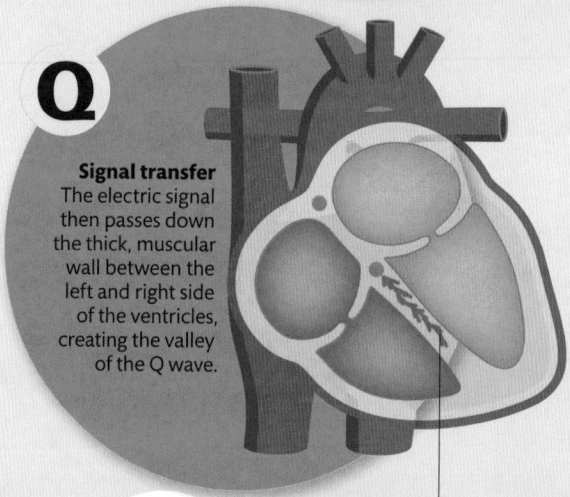

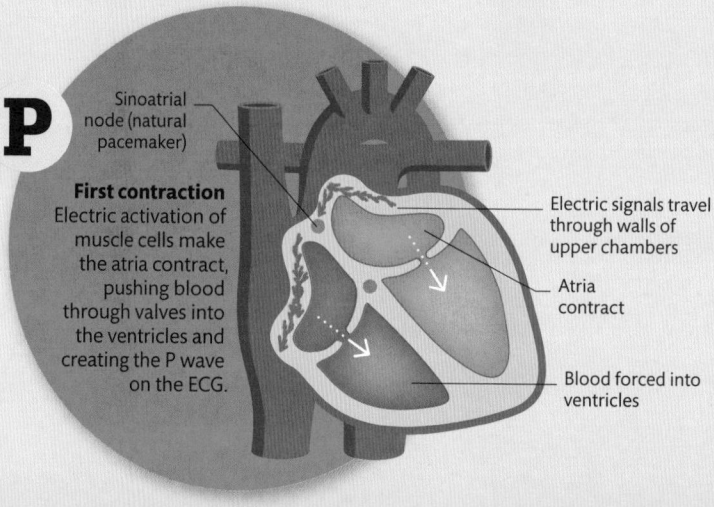

R

P

Q

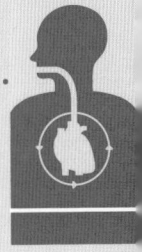

WHAT CREATES THE SOUND OF THE HEARTBEAT?

The heart has four valves, and the opening and closing, in pairs, of these valves produces the familiar lub-dub noises of the beating heart.

How electric signals travel

The heart's pacemaker, the sinoatrial node, is a region of muscle in the upper right atrium. It starts a regular electric impulse that is conducted throughout the heart by specialized nerve fibers. Heart muscle cells are adept at spreading electric messages rapidly, so the heart muscle contracts in an orderly sequence, first the two atria followed by the two ventricles.

Oxygen-rich blood from the lungs is pumped to the rest of the body

Atrium relaxed

S

Electricity travels back
The S wave and flat ST segment occur as the ventricles are contracting and emptying of blood. The atrial muscle cells have recharged, ready for the next contraction.

Electricity travels back up towards atria

Blood from heart's right side is pumped to the lungs

Ventricles still contracted

Natural pacemaker

Specialized cells
Natural pacemaker cells in the heart are "leaky" and allow a flow of ions (charged particles) in and out. This generates a regular electric impulse that causes the heart to beat. Heart (cardiac) muscle cells have branched fibers that let electric messages spread quickly to the neighboring muscle cells.

Electric current

Cardiac muscle cell

T

T

Heart recharges
The final T wave of the ECG trace occurs as the ventricular muscle cells recharge, or repolarize. The heart rests as the muscle cells get ready for the next contraction.

HEART MUSCLE CELLS RECHARGE

S

WITH **EACH BEAT,** EACH LOWER CHAMBER **PUMPS** 2 ⅓ FL OZ (70 ML) **OF BLOOD–** NEARLY ⅕ OF A **BLOOD DONATION BAG**

How blood travels

Blood travels through arteries, capillaries, and veins. Arteries have muscular, elastic walls to even out surges in pressure as the heart pumps. Veins have thinner walls and can distend to help lower blood pressure. If blood pressure rises too high, damage increases the risk of a heart attack or stroke.

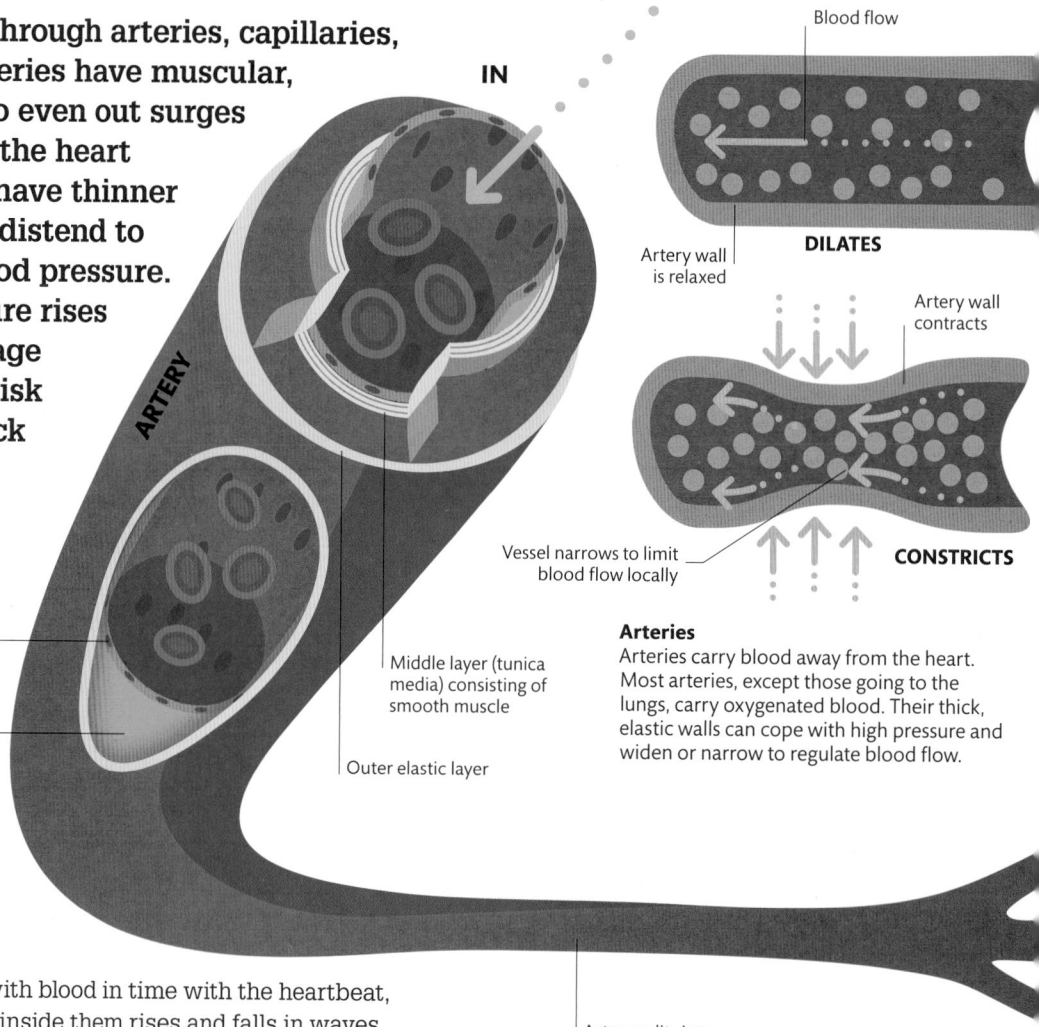

Arteries
Arteries carry blood away from the heart. Most arteries, except those going to the lungs, carry oxygenated blood. Their thick, elastic walls can cope with high pressure and widen or narrow to regulate blood flow.

Blood pressure

The arteries pulse with blood in time with the heartbeat, and so the pressure inside them rises and falls in waves. Arterial pressure is greatest just after the heart contracts (systolic blood pressure) and is lowest when the heart rests between beats (diastolic blood pressure). Pressure is much lower in the capillaries as they are so numerous they spread the force widely. Once blood reaches the veins, its pressure is minimal.

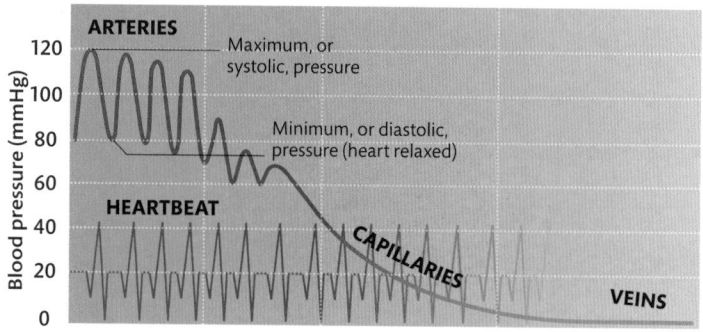

Ranges of pressure
Blood pressure is measured in millimeters of mercury (mmHg) and typical blood pressure varies rhythmically between 120 and 80 mmHg. Although the pressure is lower in both capillaries and veins, blood pressure never reaches 0 mmHg.

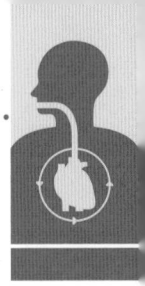

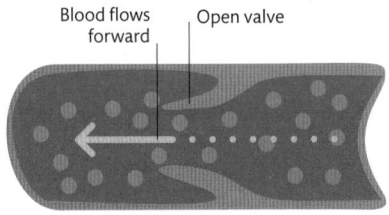

VALVE OPEN

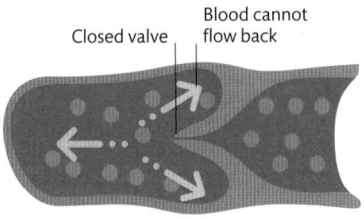

VALVE CLOSED

Veins
Veins carry blood back to the heart. Pressure in them is very low (5–8 mmHg) and the long veins in the legs have a one-way valve system to prevent backflow due to gravity.

Route through the body

Blood pulses away from the heart in large arteries, which divide to form smaller arterioles. From the arterioles, blood enters a network of capillaries. In lung capillaries, blood collects oxygen and releases carbon dioxide gas. In body capillaries, blood releases oxygen and collects carbon dioxide. Blood then flows into venules, which join up to form veins returning blood to the heart.

OUT

VEIN

Layer of smooth muscle

Outer elastic layer

Valve

Tunica intima

Capillaries
Capillaries form an extensive network that branches finely through body tissues. The entrance to some capillaries is protected by muscle rings (sphincters), which can shut down that part of the network.

CAPILLARIES

Small venules join up to form a larger vein

Small venules

Measuring blood pressure
To measure your blood pressure, a nurse inflates a cuff around your arm until the pressure is high enough to stop arterial blood flow. Pressure is then slowly released until blood can just squirt past the cuff, producing a distinct sound that pinpoints systolic blood pressure. As cuff pressure continues to fall, sounds suddenly stop at the point where blood flow is no longer constricted, which pinpoints diastolic blood pressure.

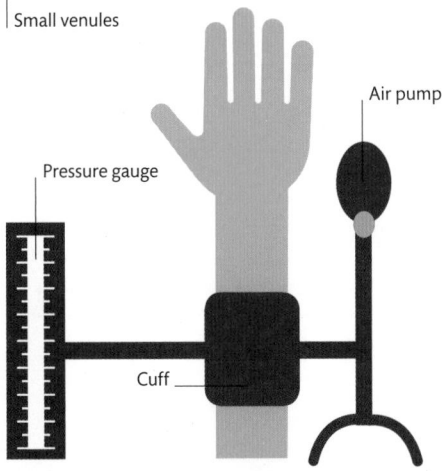

WHY IS HIGH BLOOD PRESSURE SO HARMFUL?
High blood pressure damages artery linings. This can trigger a buildup of cholesterol-laden plaque, which hastens hardening and furring up of the arteries.

Broken blood vessels

Blood vessels permeate the tissues of the body. Their thin walls allow oxygen and nutrients to pass but are easily damaged. Repair systems allow blood to clot so that any damage is quickly fixed, but sometimes unwanted clotting causes a blockage.

WHY DO PEOPLE GET DEEP-VEIN THROMBOSIS ON LONG FLIGHTS?

Blood can clot by mistake inside a healthy vessel due to sluggish blood flow, especially when someone sits still for hours. Such a clot, or thrombosis, can block a vein.

Bruising

When a part of the body is knocked, tiny blood vessels may rupture and leak blood into surrounding tissues. Some people bruise more easily than others, especially the elderly. This is sometimes related to blood-clotting disorders or nutrient deficiency such as lack of vitamin K (needed to make clotting factors) or vitamin C (needed to make the protein collagen).

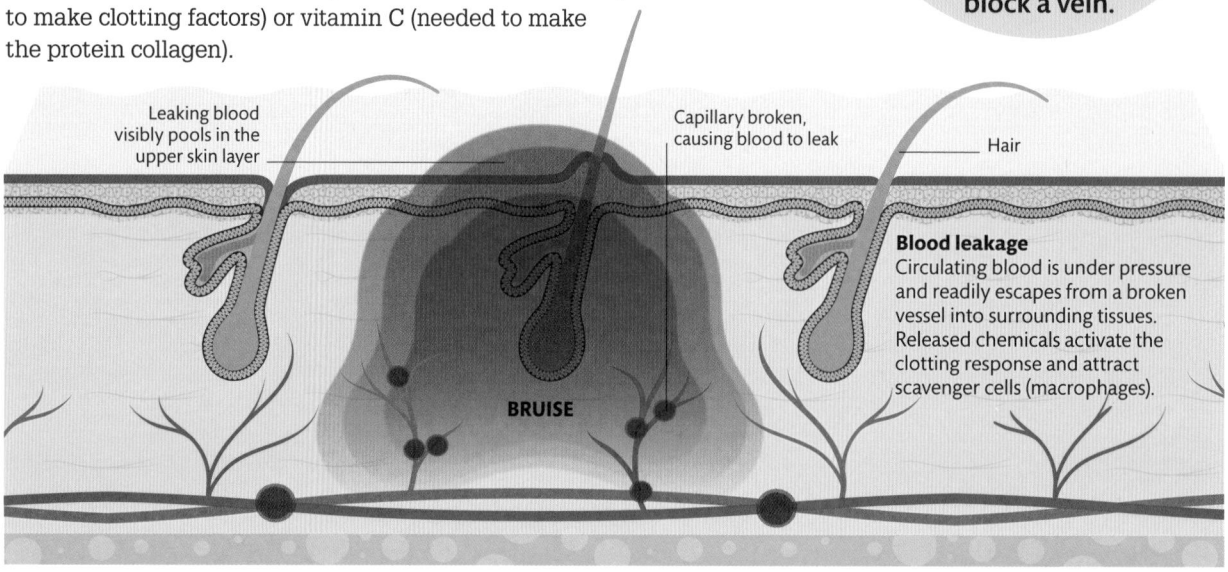

Blood leakage
Circulating blood is under pressure and readily escapes from a broken vessel into surrounding tissues. Released chemicals activate the clotting response and attract scavenger cells (macrophages).

Clotting

A damaged blood vessel must be sealed quickly to prevent blood loss. A complex sequence of reactions causes inactive proteins dissolved within the blood to activate and plug the damage. The blood vessel may constrict to slow blood flow and reduce blood loss from the circulation.

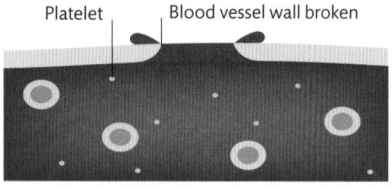

1 Initial opening
Exposure of proteins such as collagen in a broken blood vessel wall immediately attracts cell fragments called platelets.

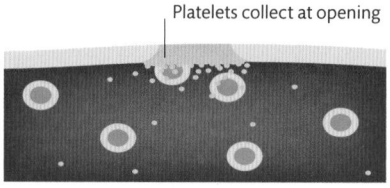

2 Forming a clot
Platelets stick together and release chemicals that trigger fibrin—a protein circulating in the blood—to form fibers.

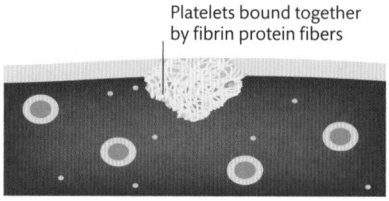

3 Holding the clot
A sticky web of fibrin fibers forms a net that binds platelets together. The web traps red blood cells to form a clot.

How bruises heal
Bruises start purple—the color of oxygen-poor blood cells seen under the skin. Scavenging macrophage cells recycle the spilled red blood cells as they clean up the area, converting the red blood pigments into first green, then yellow pigments.

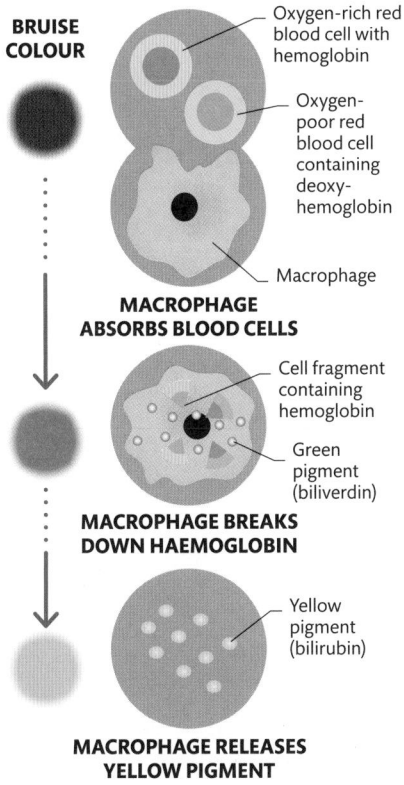

Varicose veins

Varicose veins are a price we pay for walking on two legs rather than four. Valves in the long leg veins let blood travel up against gravity. In surface veins, these valves can collapse, and blood pools, forming bulges. Varicose veins may be hereditary and may also result from increased pressure during pregnancy.

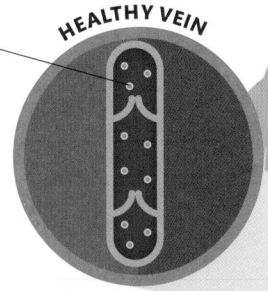

Healthy valves
A series of valves stops blood from flowing backward. This allows blood to flow up the length of the leg against the pull of gravity.

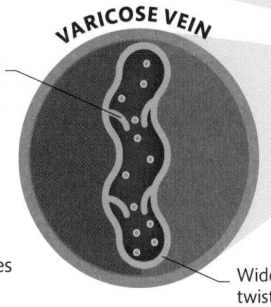

Pressure builds
When weak valves give way, gravity causes blood to fall backward and pool in the veins. Increased pressure causes the veins to become dilated and twisted.

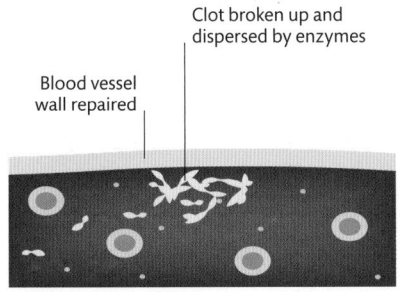

4 Clot dissolves
Cells that repair the wound also release enzymes that slowly break down the platelet/fibrin clot—a process called fibrinolysis.

Blocked blood vessels

Raised blood pressure or high glucose levels slowly damage artery walls. Platelets stick to injured areas to fix the damage. If blood cholesterol levels are also high, this seeps into affected areas, causing a buildup that narrows the artery and restricts blood flow. If arteries supplying heart muscle are affected, it can cause a heart attack. When blood flow to the brain is reduced, the memory is affected.

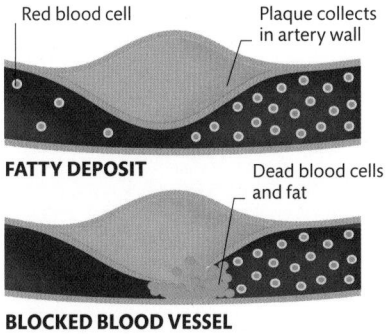

Limiting blood flow
Fatty deposits may collect in damaged areas in arteries to form plaques. These deposits cause the arteries to narrow and stiffen, restricting blood flow.

Heart problems

The heart is a vital organ—if it stops pumping blood, cells will not receive the oxygen and nutrients they need. Without oxygen or glucose, brain cells cannot function and you lose consciousness.

Vulnerable vessels

Heart muscle needs more oxygen than any other muscle in the body and the heart has its own coronary arteries to supply its needs because it cannot absorb oxygen from the blood in its chambers. The left and right coronary arteries are relatively narrow and prone to hardening and furring up (narrowing) – a potentially life-threatening process known as atherosclerosis.

IS LAUGHTER REALLY THE BEST MEDICINE?

It may very well be true– laughter can increase your blood flow and relax your blood vessel walls.

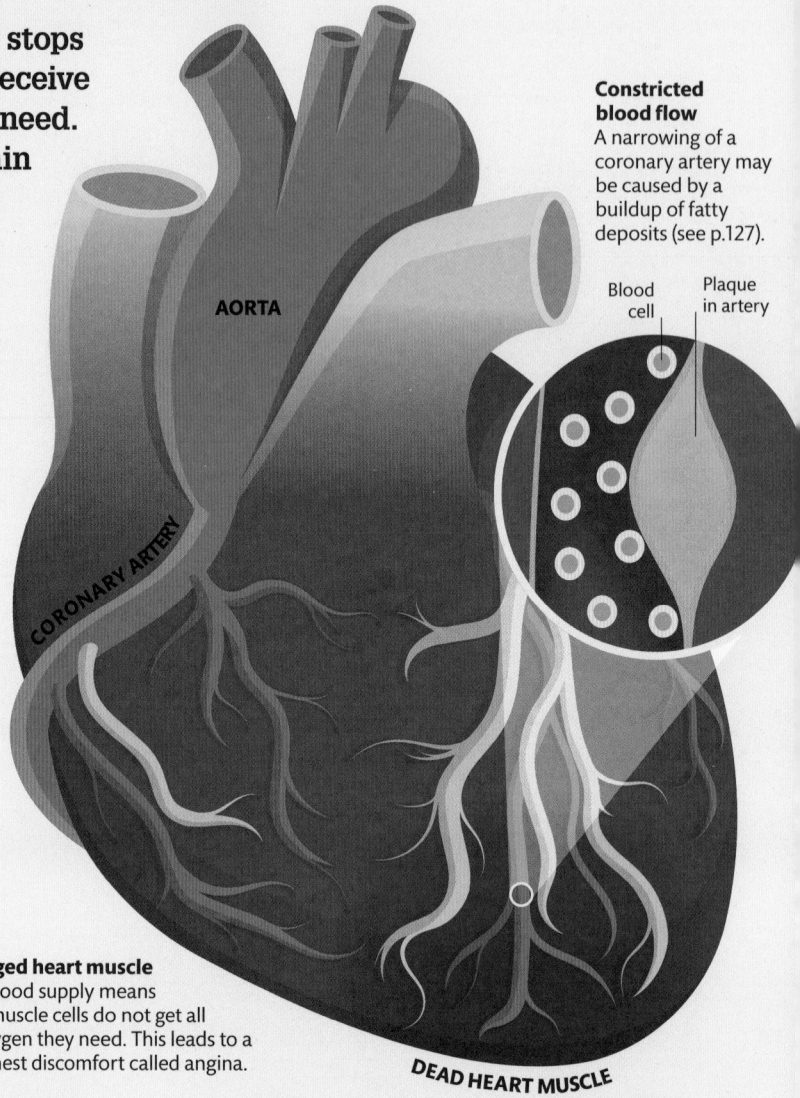

Constricted blood flow
A narrowing of a coronary artery may be caused by a buildup of fatty deposits (see p.127).

Damaged heart muscle
Poor blood supply means heart muscle cells do not get all the oxygen they need. This leads to a tight chest discomfort called angina.

Decreasing oxygen supply
The heart has specialized cardiac muscle cells whose branched fibers spread electrical messages quickly. Characteristic changes on an ECG (electrocardiogram) help doctors diagnose whether chest pain is due to poor blood supply (angina) or muscle cell death (heart attack).

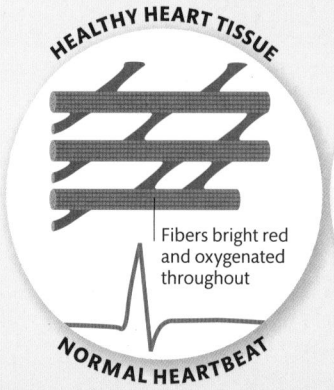

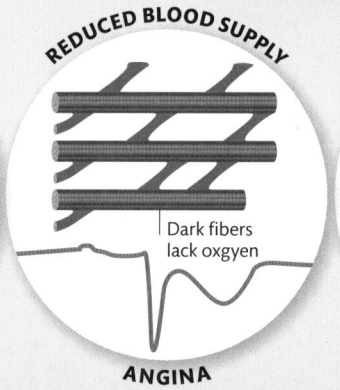

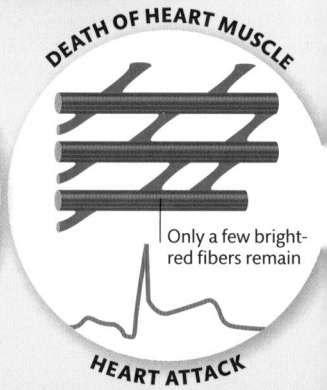

Heart rhythm problems

If the heart is beating too fast, too slowly, or irregularly, medics say that it has arrhythmia, or abnormal heart rhythm. Most arrhythmias are harmless, such as premature extra beats that feel like a flutter or skipped heartbeat. Atrial fibrillation is the most common type of serious arrhythmia, in which the two upper chambers of the heart (atria) beat irregularly and fast. This can cause dizziness, shortness of breath, and fatigue, and also increases the risk of suffering a stroke. Some arrhythmias can be treated with drugs. Some need defibrillation to reset and normalize electrical activity.

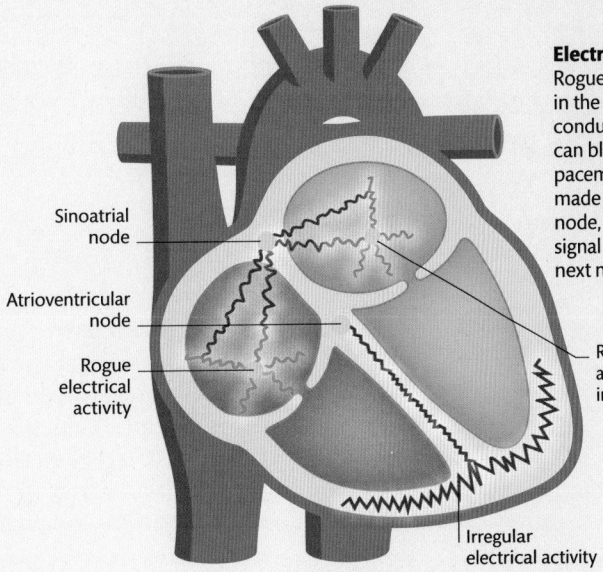

Electrical activity
Rogue activity arising in the electrical conduction system can block the normal pacemaker impulses made by the sinoatrial node, stopping the signal reaching the next node.

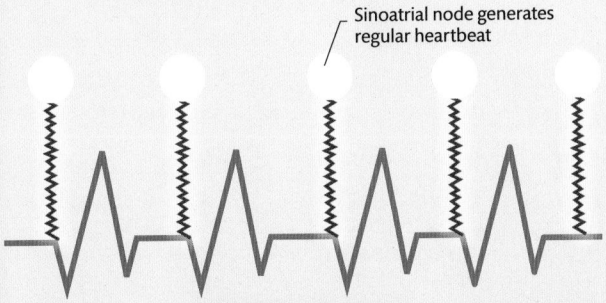

NORMAL HEART BEAT

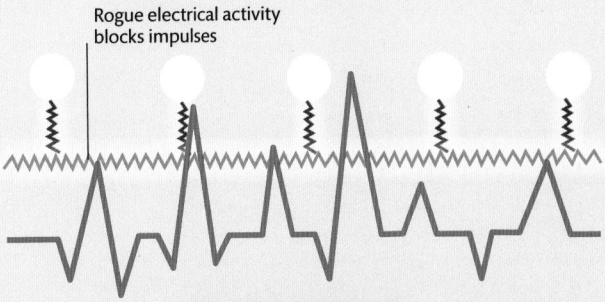

IRREGULAR HEARTBEAT

Electrical interference
The coordinated beating of the heart relies on a clear signal reaching the ventricles from the sinoatrial node. If rogue electrical activity gets in the way, the heart's rhythm of contraction is disturbed and can become erratic.

THE HUMAN **HEART BEATS** MORE THAN **36 MILLION** TIMES A YEAR—ABOUT **2.8 BILLION TIMES** IN AN AVERAGE LIFETIME

DEFIBRILLATION

Some life-threatening arrhythmias can be treated by defibrillation. A burst of electricity is delivered to the chest in an attempt to re-establish normal heart electrical activity and contraction. Defibrillation only works if a "shockable" rhythm is present, such as ventricular fibrillation. It cannot restart the heart if no electrical activity is detected (asystole). Cardiopulmonary resuscitation can trigger electrical activity so that defibrillation can be tried.

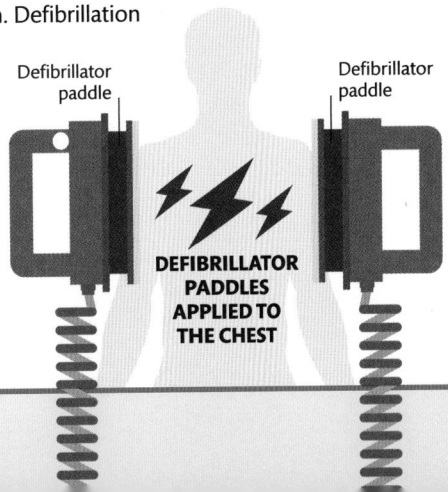

Exercising and its limits

When you go for a jog or a sprint, extra blood is pumped to your muscles, providing you with the vital ingredient to make energy—oxygen. Deep, regular breaths replenish your muscles with oxygen and set your pace.

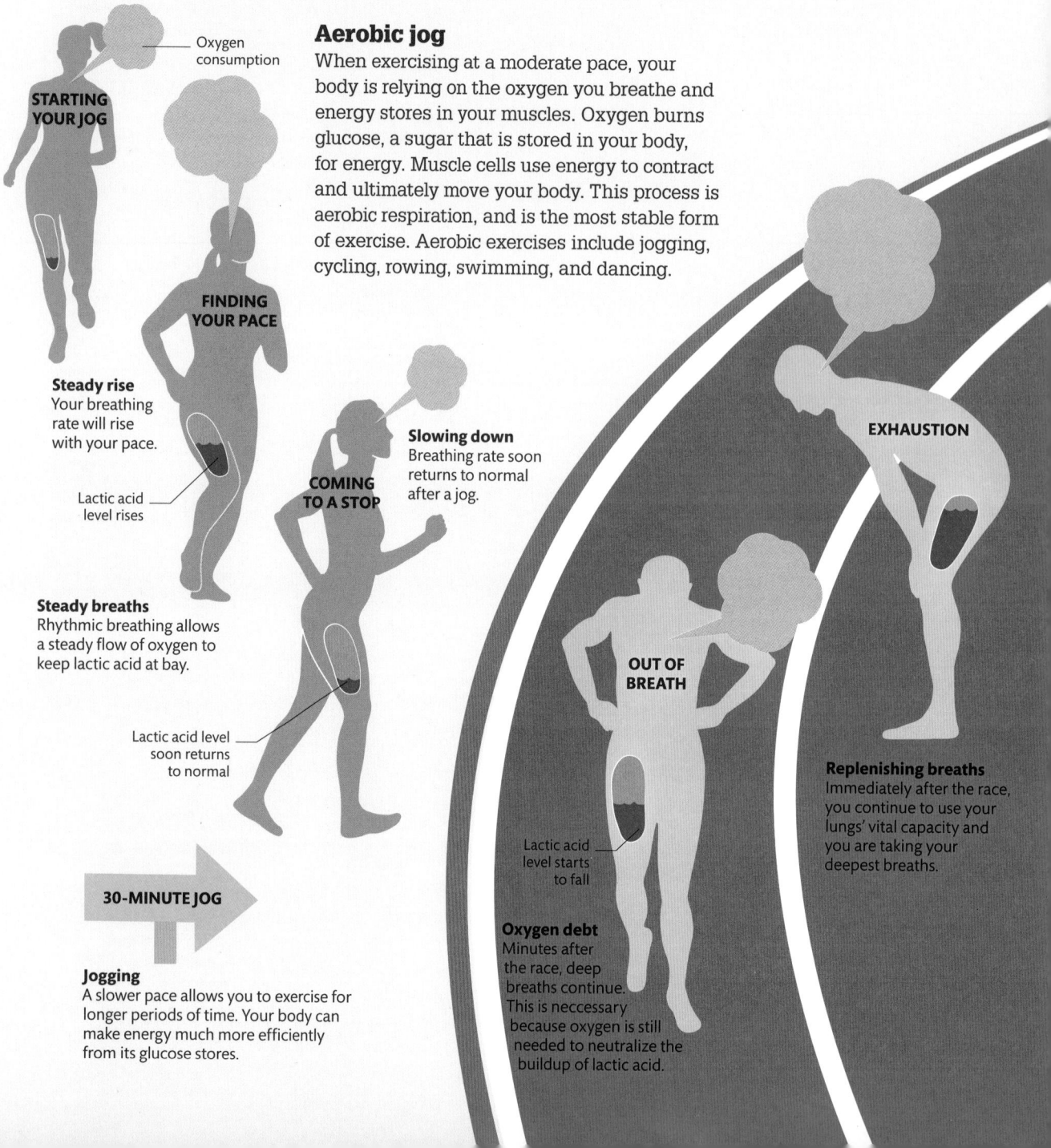

Aerobic jog

When exercising at a moderate pace, your body is relying on the oxygen you breathe and energy stores in your muscles. Oxygen burns glucose, a sugar that is stored in your body, for energy. Muscle cells use energy to contract and ultimately move your body. This process is aerobic respiration, and is the most stable form of exercise. Aerobic exercises include jogging, cycling, rowing, swimming, and dancing.

All systems go
Lactic acid builds up quickly in the muscles. Oxygen intake lags behind.

Crouch
You prepare to take deeper breaths.

ON YOUR MARKS

EXERTING YOURSELF

REACHING YOUR LIMIT

High level of lactic acid

30-SECOND SPRINT

Sprinting
Exerting yourself in a short space of time causes your body to make energy inefficiently, which releases a lot of lactic acid, causing the "burn."

Breaking point
You become dizzy and feel the "burn." Lactic acid will eventually reach a level where your muscles simply cannot contract. The breaths you take are as deep as possible to maximize the amount of oxygen you absorb.

Anaerobic sprint

During strenuous exercise, your body demands energy more quickly than you can provide oxygen to make it. Muscles can continue to break down glucose without oxygen in a process known as anaerobic respiration. It is great for short bursts of energy, but it generates excessive lactic acid in your muscles and is unsustainable. Now, oxygen is needed, not to help burn glucose, but to convert the build up of lactic acid into glucose—for future energy. This is known as paying the oxygen debt and leaves you out of breath for some time after an intense sprint.

Reaching your limit

A buildup of lactic acid in your body is the reason why you get tired during exercise. Lactic acid interferes with muscle contraction, which results in physical exhaustion. Oxygen is needed to get rid of lactic acid, which is why you breathe heavily after exercise. This buildup of lactic acid happens during both aerobic and anaerobic exercise, but it occurs quicker in the latter. Brain cells can only burn glucose for fuel and as exercising muscles deplete the body's available glucose supplies, mental fatigue also sets in.

EFFECT OF LACTIC ACID IN MUSCLES

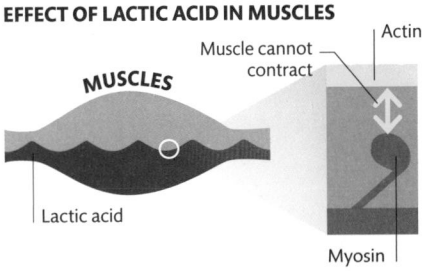

HYDRATION

Drinking water during exercise helps regulate body temperature through sweating and flushes away lactic acid. Water in blood plasma is sweated out, so your blood thickens and your heart works harder to pump blood around the body. This is called cardiac drift, and it's one reason why you can't respire aerobically and jog forever.

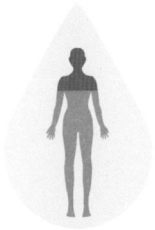

FULLY HYDRATED: 75%

LIMIT OF SAFE DEHYDRATION: 70%

Fitter and stronger

Exercise that makes your heart race and your lungs breathe hard and deep is called cardiovascular—it strengthens the heart and improves stamina. In contrast, exercise that forces you to contract muscles repetitively is called resistance training, and it can build and strengthen your muscles.

Cardiovascular exercise

When you perform cardiovascular exercise, such as jogging, swimming, bicycling, or brisk walking, you train your cardiovascular system. Your heart rate climbs, beating faster in order to pump more blood around your body, especially to the chest muscles that influence the depth of your breaths. As your body's demand for oxygen increases, your breathing rate and depth rises accordingly. Your blood is saturated with as much oxygen as possible to provide your body with the energy it needs.

Chest muscles
Muscles within the neck, chest wall, abdomen, and back coordinate to expand and reduce the size of your ribcage, so that the volume of air your lungs inhale and exhale increases.

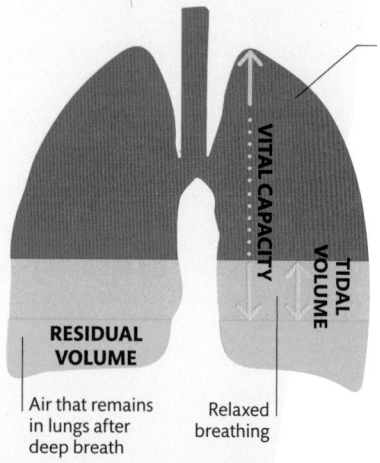

Lung capacity
Your tidal volume is the volume of air that flows into your lungs during a relaxed breath. If you try to breathe all the air out of your lungs, some air remains as your residual volume, and cannot be breathed out. Your vital capacity, the deepest breath you can take when training, is the rest of your lung volume excluding the residual volume.

Scalene muscles contract to raise upper ribs

Internal intercostal muscles contract and tilt ribs downward

Lung volume reduced as muscles contr and ribs tilt

COLLARBONE

LUNG

RIB

STERNUM

External intercostal muscles contract and tilt ribs upward

Rectus abdominis muscle pulls ribcage downward

Lung volume increases due to ribs tilting upward

External oblique muscles contract and shorten to pull ribs downward

INHALING

EXHALING

WHICH TYPE OF EXERCISE BURNS MORE FAT?

It depends on the individual, but a combination of both cardio and weight training will result in greater fat loss than just doing one or the other.

Resistance training

Weight training builds your muscle, but so does dancing, gymnastics, and yoga – they are all forms of resistance training. A repetition (rep) is one complete motion of exercise. A set is a group of consecutive reps that will contract a particular muscle, or multiple muscles, repeatedly. You can target muscles to grow by choosing to perform a selection of sets and reps over a period of time. The fewer reps you are able to do per set, the tougher your workout.

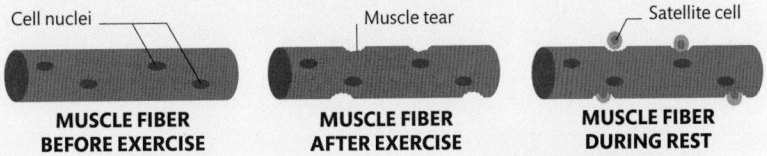

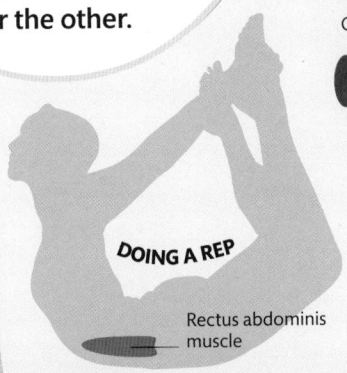

Bow pose
Yoga is a good way to grow muscle steadily. The bow pose forces the rectus abdominis muscle to contract and tear slightly. Repeating this as a "rep" will start the muscle growth process.

Muscle growth process
Exercise tears muscle fibers, which are then repaired by satellite cells. Although muscle fibers are single body cells, they have many nuclei, and they incorporate the satellite cells, along with their nuclei—growing as they do so. During a break from exercise, your muscle fibers shrink, but they retain the nuclei from the satellite cells and regain their size quickly after retraining.

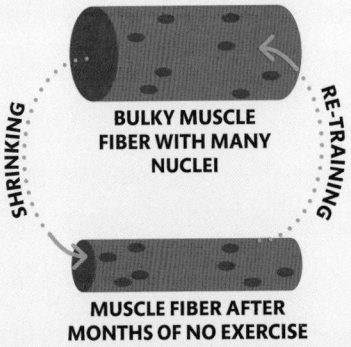

RATES OF EXERCISE

Exercise intensity can be expressed as the percentage of your maximum heart rate. When you go for a jog, you are working your heart at about 50 percent of its potential power. Athletes who have reached their peak fitness can work their heart at maximum strength—100 percent. A fitness instructor can give you a target heart rate to reach when training (which varies with age) while achieving your fitness goals.

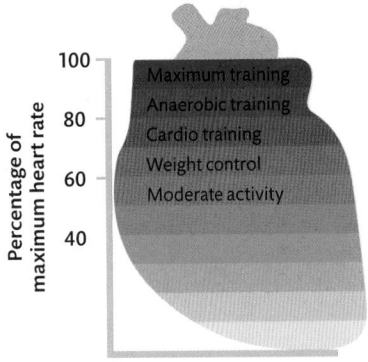

WHEN YOU SLEEP, HORMONES THAT STIMULATE MUSCLE GROWTH ARE RELEASED

Maximizing your fitness

While exercising is necessary to maintain health, regular training can improve your overall fitness. Your body will adapt to tough training regimes; muscles get thicker, breaths get deeper, and your state of mind is enhanced.

Positive results of regular exercise

If you exercise regularly, you will see widespread improvements across your body. Adults benefit from just 30 minutes of brisk exercise on most days, while children need at least 60 minutes of running around. Keeping yourself active is vital for improving your organs and muscles, and by exerting yourself in steady sessions your body systems will become more efficient and eventually will start to function at the best of their ability.

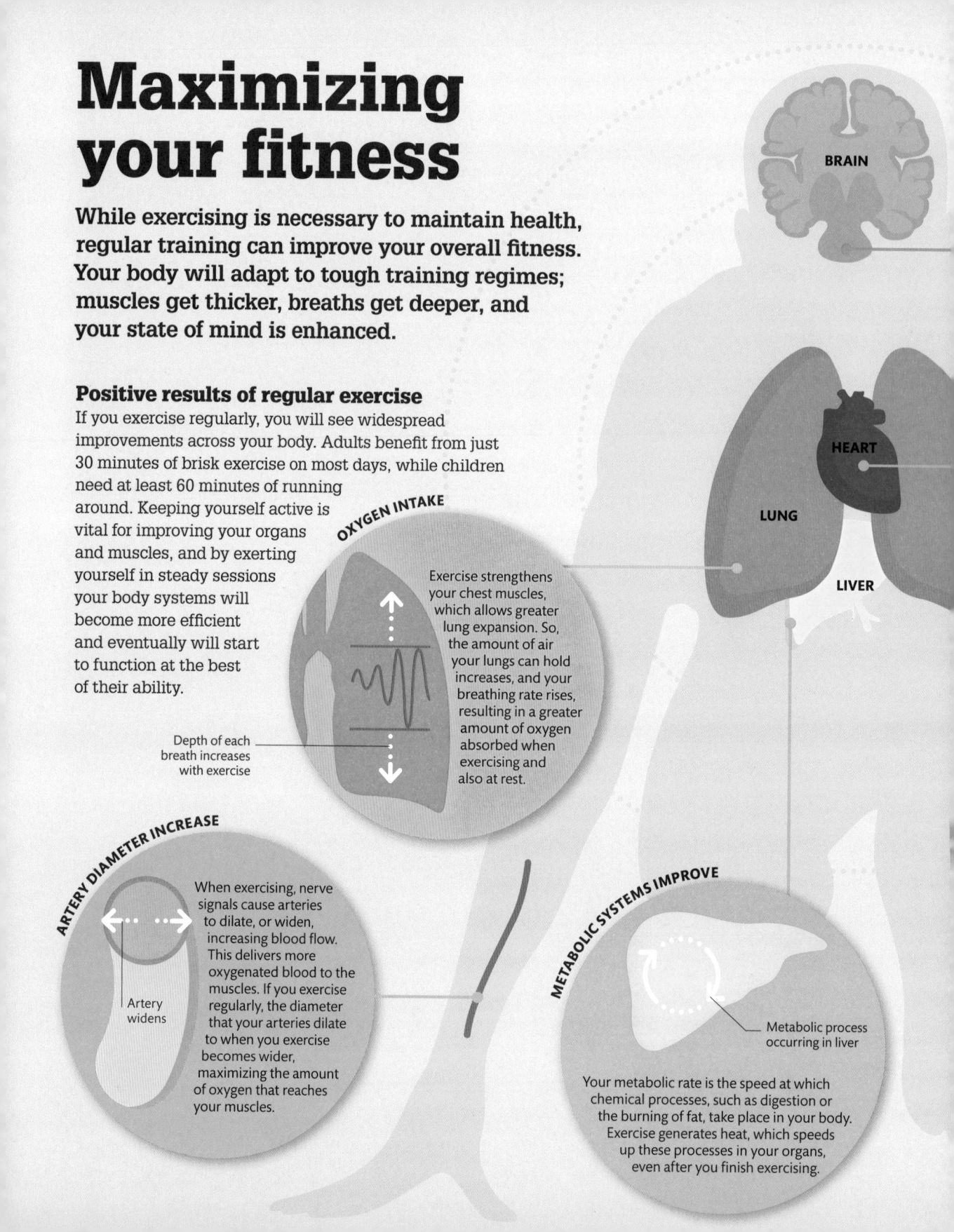

COGNITIVE IMPROVEMENT

Regular exercise increases the delivery of blood, oxygen, and nutrients to the brain. In turn, this stimulates new connections between brain cells, improving general mental abilities. Exercise also boosts the levels of neurotransmitters such as serotonin in the brain, raising your mood.

STRONGER CARDIAC MUSCLE

Cardiac muscle fibers grow in size, but not via satellite cells as is the case in muscles in the rest of your body. Instead, their existing fibers grow stronger. Your heart's contractions become stronger too, and it distributes blood more thoroughly around the body, lowering your resting heart rate.

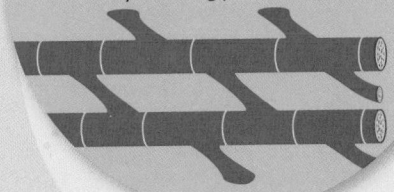

STRONGER MUSCLES

Having strong muscles increases your physical strength, strengthens your bones, improves posture, flexibility, and how much energy you burn during exercise and while at rest. Strong muscle is also more resilient to exercise-induced injury.

Reaching your maximum

During a training program, for most people, the effort you put in reaps great benefits at first, as your fitness increases from your untrained level. Further improvements become ever harder to achieve as you approach your own physiological limits, which depend on age, gender, and other genetic factors. You reach your maximum more quickly with a higher-intensity training program. The best athletes explore their limits, looking for opportunities to extend them.

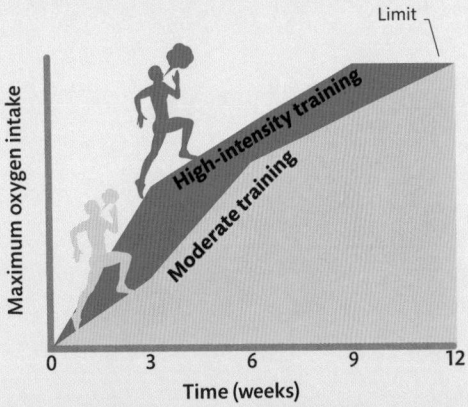

RESTING HEART RATES

Athletes have low heart rates at rest because training enhances the strength of their cardiac muscle. Compared to those who are untrained, athletes' heart contractions are stronger, and blood is distributed more efficiently with every heartbeat. A trained athlete may have a pulse rate as low as 30–40 beats per minute at rest.

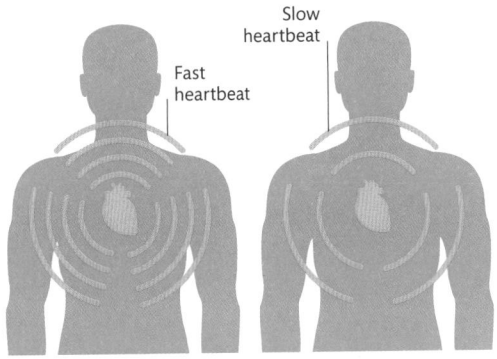

IN AND OUT

Feeding the body

Although the body can manufacture many vital chemicals, a lot of the materials we need must be acquired by eating. The energy needed to fuel the body is gained entirely through the food we consume. Once nutrients are absorbed into the bloodstream, they are then transported to different parts of the body, where they are put to innumerable tasks.

WHAT IF I DON'T GET WHAT I NEED?

Your body systems will start to fail and you may be afflicted with deficiency diseases. For example, if you do not have enough minerals in your diet, your bones will not grow properly.

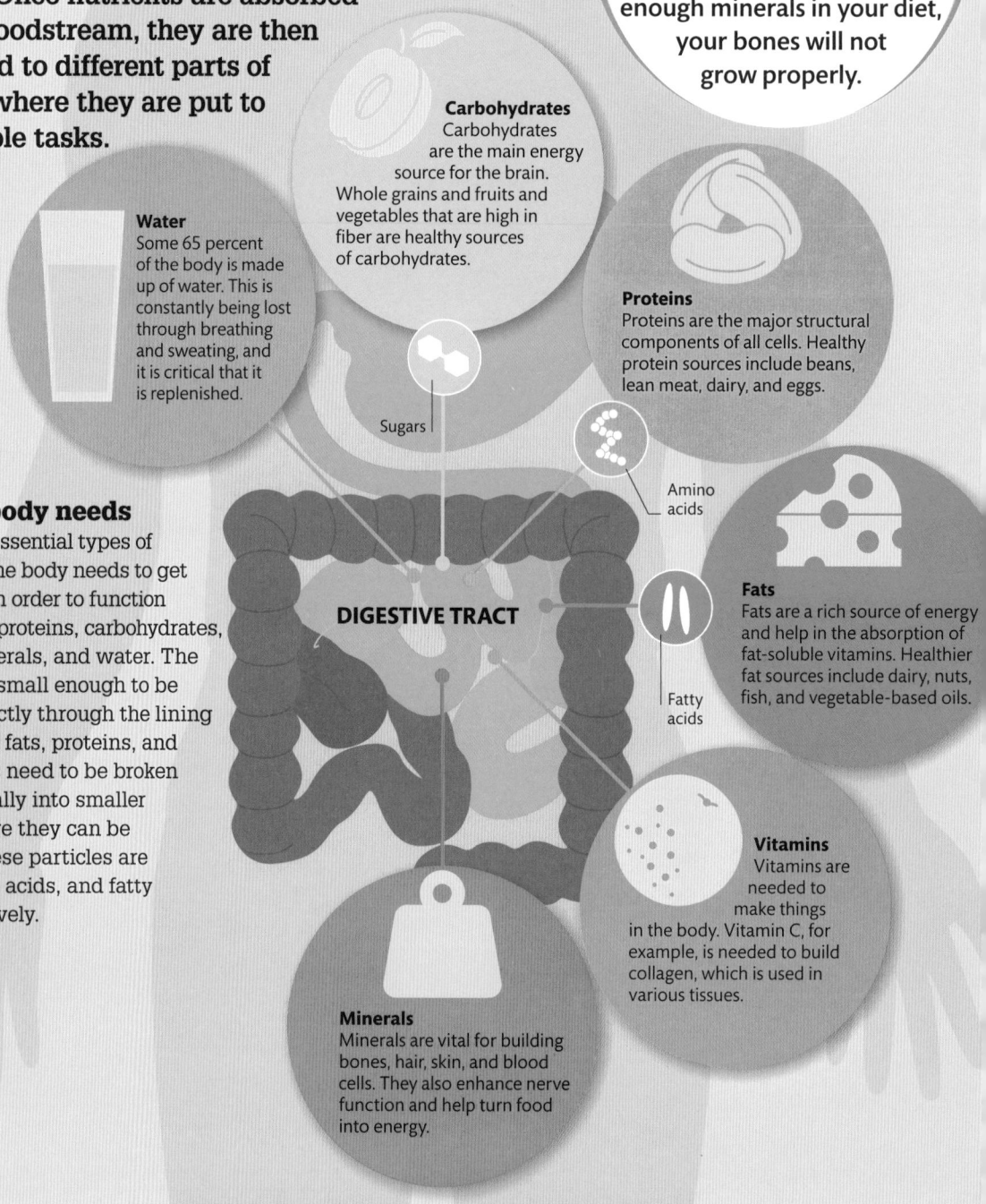

What the body needs

There are six essential types of nutrient that the body needs to get from the diet in order to function properly: fats, proteins, carbohydrates, vitamins, minerals, and water. The last three are small enough to be absorbed directly through the lining of the gut, but fats, proteins, and carbohydrates need to be broken down chemically into smaller particles before they can be absorbed. These particles are sugars, amino acids, and fatty acids respectively.

Building an eye

Every tissue in our body is built and maintained by the nutrients we absorb from our food. The tissues of the human eye, for example, are built from amino acids and fatty acids, and fueled by sugars. The membranes and spaces are filled with fluids, and vitamins and minerals are needed to convert light into an electrical impulse—the basis of vision itself.

THE LIVER CAN STORE UP TO 2 YEARS' WORTH OF VITAMIN A

Cell membranes
All the cells of the eye (and the rest of the body) are surrounded by membranes that are built using fatty acids and proteins.

Energy
The eyes are an extension of the brain, and just like the brain, they need the sugars we get from carbohydrates for energy.

The food of sight
Like all organs of the body, the eye utilizes all six of the essential nutrients. These give it structure and enable it to send visual information to the brain.

Fluids
The eye is filled with fluid, which maintains the pressure in the eye and provides nutrients and moisture to the inner eye tissues. This fluid is 98 percent water.

Tissue structures
Eyelashes are made up of the protein keratin, which is built from amino acids. Other tissues of the eye are made of the protein collagen.

Vision
Vitamin A is bound to proteins in the eye known as visual pigments. When light hits the cells, the vitamin A changes shape, sending an electrical impulse to the brain.

Red blood cells
The tissues of the eye are oxygenated by the red blood cells, which need the protein hemoglobin and the mineral iron in order to carry oxygen.

How does eating work?

Eating is the process of breaking down food into molecules that are small enough to be absorbed into the bloodstream. For the food, this involves a 30 ft (9 m) journey through a series of organs known collectively as the gut, or the gastro-intestinal tract.

The journey of food

Food begins as a (usually) appetizing meal, and ends with us taking trips to the toilet. Between these stages, the food has done its job—released its nutrients in a four-stage process involving the mouth, the stomach, the small intestine, and the large intestine. The liver and pancreas also play roles, as do the hormones leptin and ghrelin. On average, it takes 48 hours for food to pass through the body.

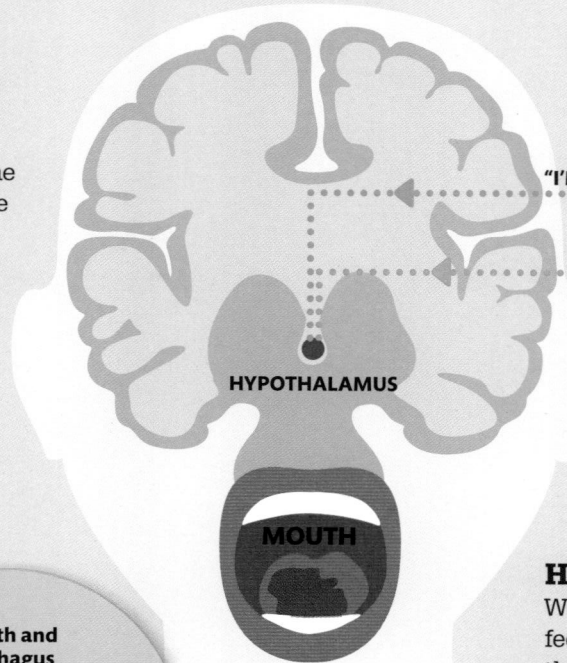

Hunger
A few hours after eating, the hormone ghrelin is secreted by the stomach. This sends a signal to the brain, which readies the gut for food.

Ghrelin's signal makes us feel hungry

Leptin's signal makes us feel full

Satisfaction
When we have eaten enough, the hormone leptin is released by our fat tissues. This signals the brain to put the gut back on "standby" mode.

Hunger and satisfaction

We eat when we feel hungry, and stop when we feel full. However, we are not responsible for these feelings. When we are low on nutrients, the hormone ghrelin is released by the stomach, making us feel hungry—and when we are full, the hormone leptin is released by our fat tissues, inhibiting our appetites.

Nutrient absorption
Some nutrients take longer to be absorbed than others, but most are absorbed in the small intestine.

- Vitamins
- Sugars
- Amino acids
- Minerals
- Fatty acids
- Water
- Blood flow

1 Mouth and esophagus
Stage one starts with the mechanical breakdown of food by chewing. This mixes the food with saliva, which begins to digest it chemically. The food is then swallowed, which drops it into the esophagus (see p.142).

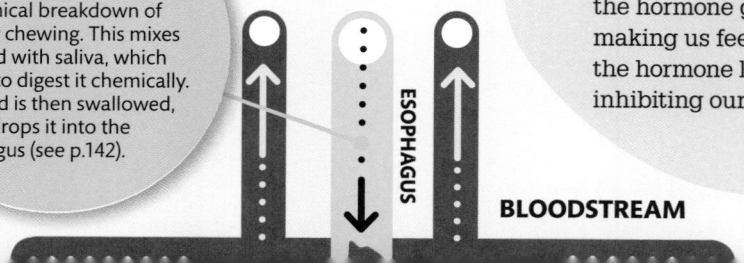

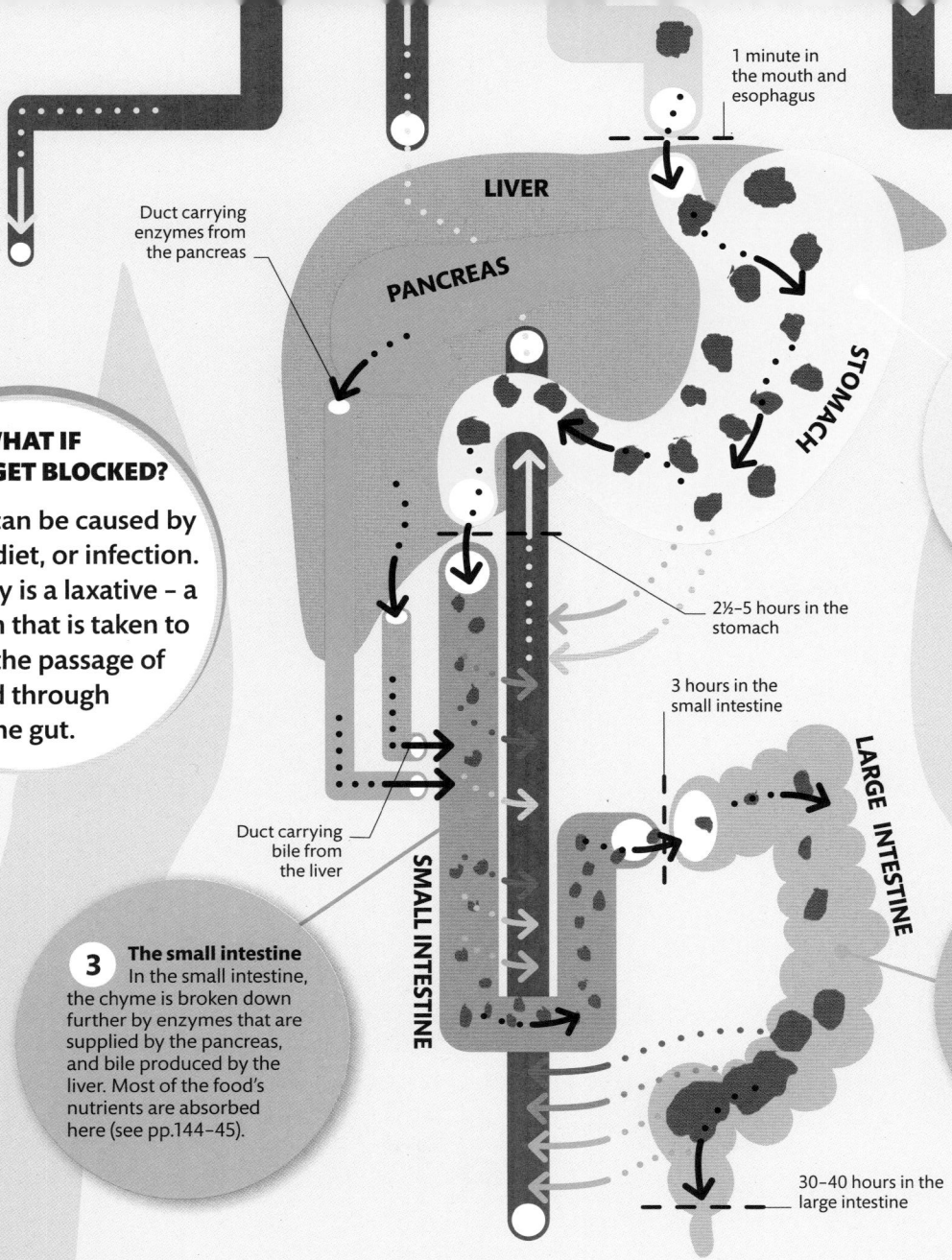

WHAT IF THINGS GET BLOCKED?

Blockages can be caused by stress, bad diet, or infection. One remedy is a laxative – a medication that is taken to smooth the passage of food through the gut.

2 The stomach
Muscular contractions in the esophagus propel the food into the stomach. Here it is doused in gastric juices, which turn it into a soupy mixture called chyme (see p.143).

3 The small intestine
In the small intestine, the chyme is broken down further by enzymes that are supplied by the pancreas, and bile produced by the liver. Most of the food's nutrients are absorbed here (see pp.144–45).

4 The large intestine
Most of the water from the food is absorbed in this last section of the gut, along with a few final nutrients. At the same time, the indigestible parts of the food are pressed into feces and stored for removal (see pp.146–47).

A mouth to feed

The long and convoluted journey taken by food through the body begins with a brief stay in the mouth and an acid bath in the stomach. The goal of this first stage of digestion is to turn food into chyme—a soup of nutrients that is then moved on to the small intestine for processing.

Heading south

The route from mouth to stomach is a vertical one, via a connecting tube called the esophagus. The food is propelled by gravity and by muscular contractions in the esophagus known as peristaltic waves.

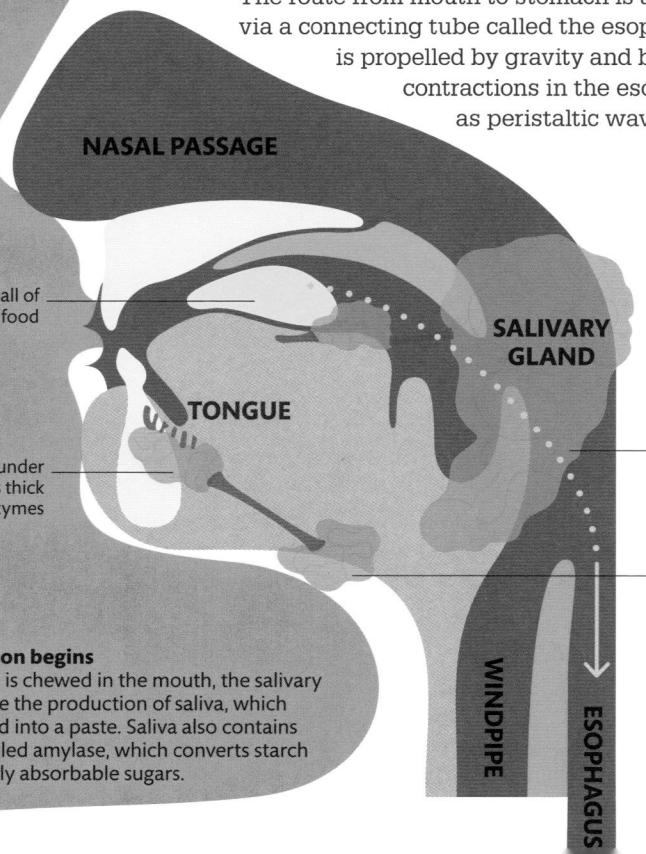

Chewing creates a ball of saliva-saturated food

The salivary gland under the tongue produces thick saliva containing enzymes

The salivary glands in the cheeks produce watery saliva

Another salivary gland under the jaw releases saliva at the base of the tongue

1 Digestion begins
As food is chewed in the mouth, the salivary glands increase the production of saliva, which helps turn food into a paste. Saliva also contains an enzyme called amylase, which converts starch into more easily absorbable sugars.

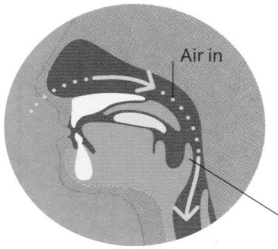

Chewing
When food is in the mouth, the epiglottis stands up to keep the windpipe open. This allows us to breathe through our noses while chewing.

Epiglottis up

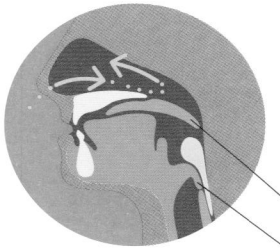

Swallowing
When we swallow, the epiglottis folds down, closing off the windpipe. At the same time, the soft palate rises to block off the nasal cavity.

Soft palate up

Epiglottis down

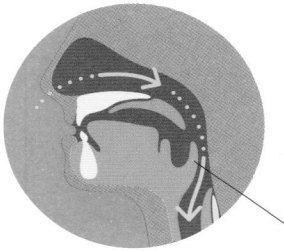

Ready to chew again
When food has entered the esophagus, the epiglottis and soft palate return to their former position. This enables us to breathe and chew again.

Epiglottis up

How to avoid choking

Since we both eat and breathe through our mouths, it is vital that our windpipes can be closed off when we swallow. Luckily, our bodies have a pair of built-in safety devices—a small flap of cartilage in the throat called the epiglottis, and a piece of flexible tissue in the roof of the mouth called the soft palate.

WHY DO WE GET INDIGESTION?

Indigestion, or "heartburn," is the inflammation of the stomach by the stomach's own acidic juices. It is commonly caused by overeating, stress, or drinking too much alcohol.

2 Into the stomach
Food enters the stomach via a ring of muscle. For several hours the food is then churned by three different muscles in the stomach. In a violent procedure that we are barely aware of, this mixes the food with the gastric juices that are secreted by glands in the stomach lining.

3 Gastric juices
The stomach's juices include the extraordinarily corrosive hydrocholoric acid, which kills bacteria, and pepsin, an enzyme that converts protein into smaller molecules called peptides. Also released is gastric lipase, an enzyme that begins the process of breaking down fat, and mucus. Mucus forms a slimy layer that protects the stomach from its own digestive juices.

4 Moving on
After being churned in the stomach for 3–4 hours, all the food has been turned into chyme. This chemical mixture is then squirted through another ring of muscle at the base of the stomach into the neck of the small intestine. Once here, digestion begins in earnest.

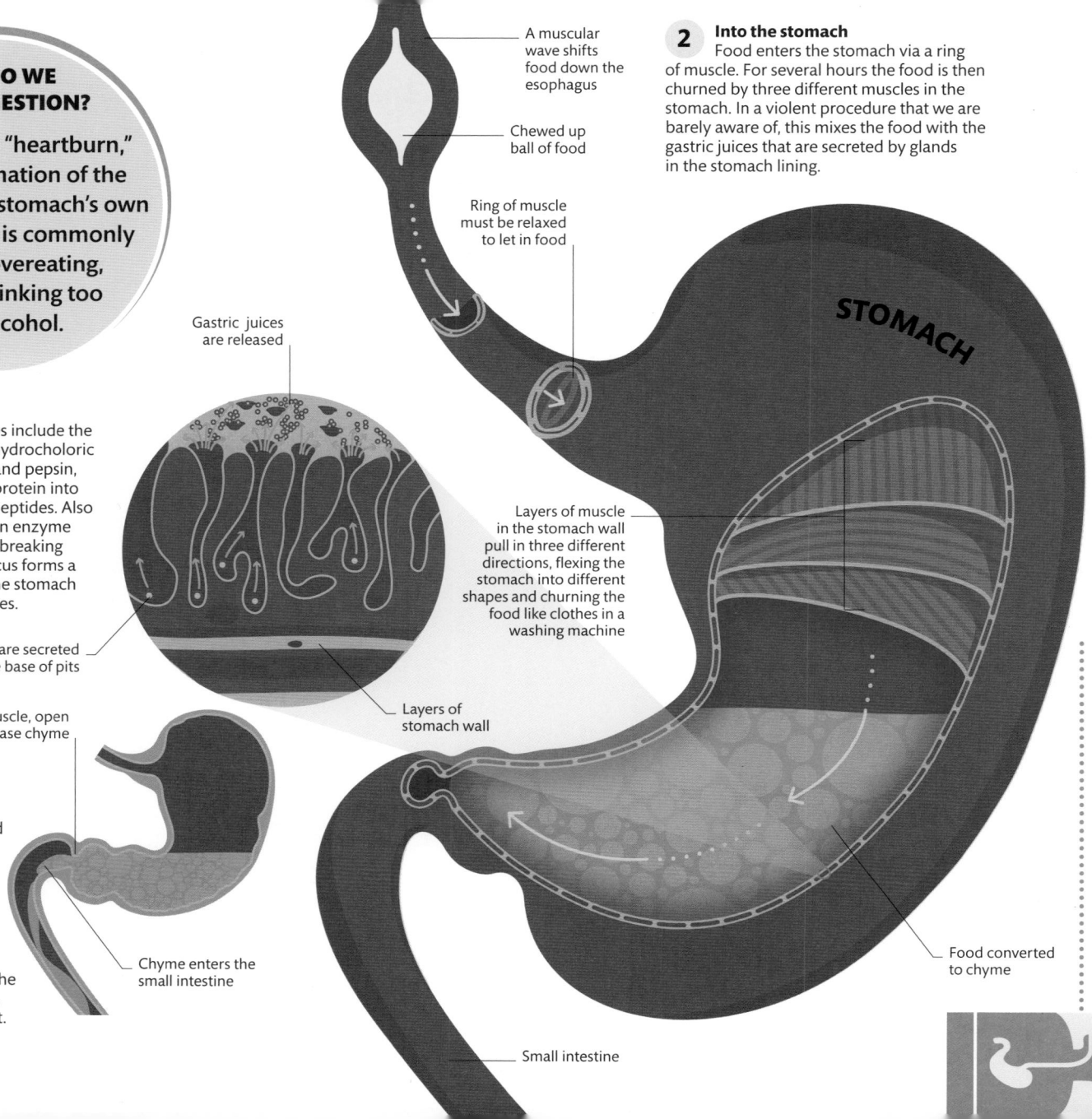

Gut reaction

Once food has been turned into chyme in the stomach, it is squirted into the small intestine. Here, in a frenzy of chemical activity, it is broken down further and finally absorbed by the blood. Each day, around 24 pints (11.5 liters) of food, liquids, and digestive juices pass through the small intestine.

Organs in concert

To help it digest, the small intestine gets help from three other organs: the pancreas, which makes enzymes; the liver, which makes bile; and the bile-storing organ, the gallbladder.

1 Bile factory
One of the liver's many jobs is to produce bile—a bitter liquid that turns fats into more readily digestible fatty droplets. Once produced, bile is stored in the gallbladder.

2 Bile store
When food leaves the stomach bile leaves the gallbladder and heads for the small intestine. There it mixes with incoming enzymes from the pancreas.

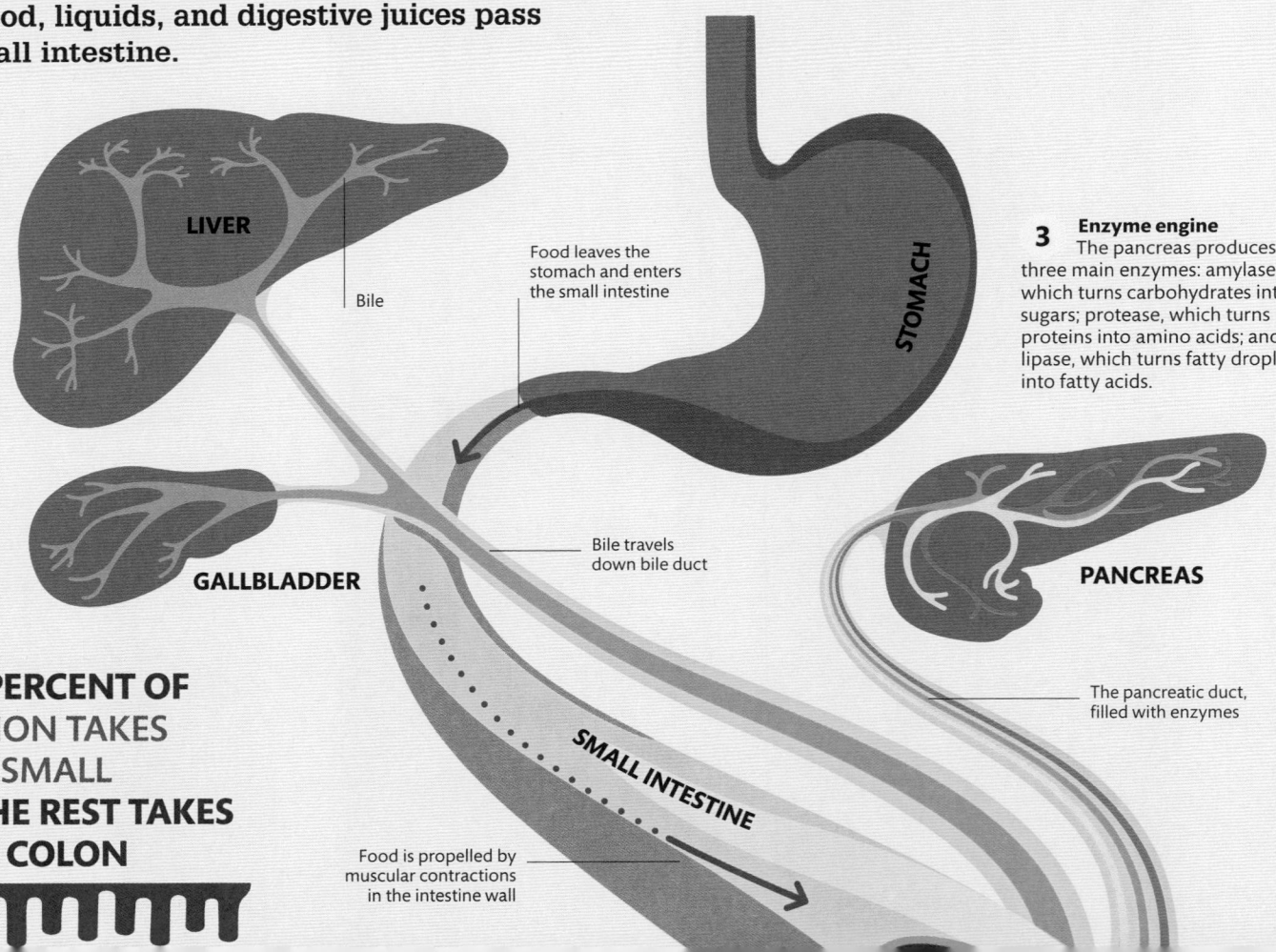

3 Enzyme engine
The pancreas produces three main enzymes: amylase, which turns carbohydrates into sugars; protease, which turns proteins into amino acids; and lipase, which turns fatty droplets into fatty acids.

AROUND 95 PERCENT OF ALL ABSORPTION TAKES PLACE IN THE SMALL INTESTINE—THE REST TAKES PLACE IN THE COLON

CHEWING THE FAT

Fats are particularly hard to digest. Even after being drenched in hydrochloric acid in the stomach, they are still not fit for enzyme consumption. This is where bile comes in. In a process called emulsion, bile turns fats into fatty droplets, which are then small enough for enzymes to attack.

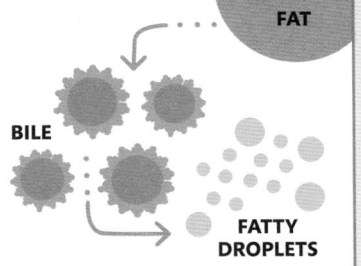

Opening of ducts carrying digestive juices

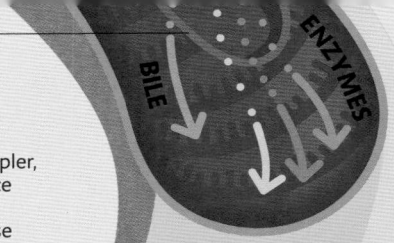

4 Absorption begins
For 3–5 hours bile and enzymes work together, reducing nutrients to simpler, absorbable forms. Absorption takes place in the intestine wall, which is lined by thousands of fingerlike projections. These projections, called villi, greatly increase the surface area of the intestine and thereby its capacity for absorbing nutrients.

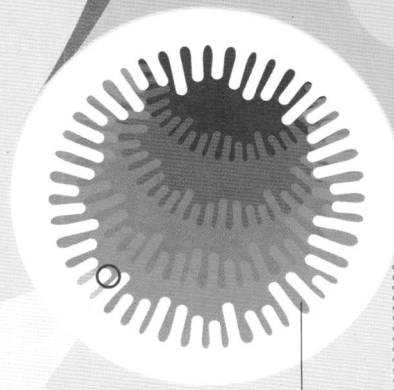

Thousands of villi line the intestine wall

Lipase digests fatty droplets, producing fatty acids

Protease digests protein, producing amino acids

Amylase digests carbohydrates, producing sugars

FATTY DROPLET

PROTEIN

CARBOHYDRATE

Fatty acids

Amino acids

Sugars

5 Into the blood
Villi absorb nutrients and channel them into the blood. Once there, they are taken to the liver and distributed around the body. Meanwhile, the remaining chyme enters the final part of the gut (see pp.146–47). Although not shown here, fat digestion has another complication; on entering the villi, the fatty acids take a trip through the lymph system before finally entering the bloodstream.

Dissolved fatty acids

Dissolved amino acids

Dissolved sugars

BLOODSTREAM

Up, down, and out

The final stage of digestion takes place in the large intestine—a 4 ft- (2.5 m-) long tube that frames the small intestine. Here, bacteria set to work fermenting carbohydrates, releasing nutrients that are vital for human health. At the same time, fecal matter is compacted, stored, and ejected.

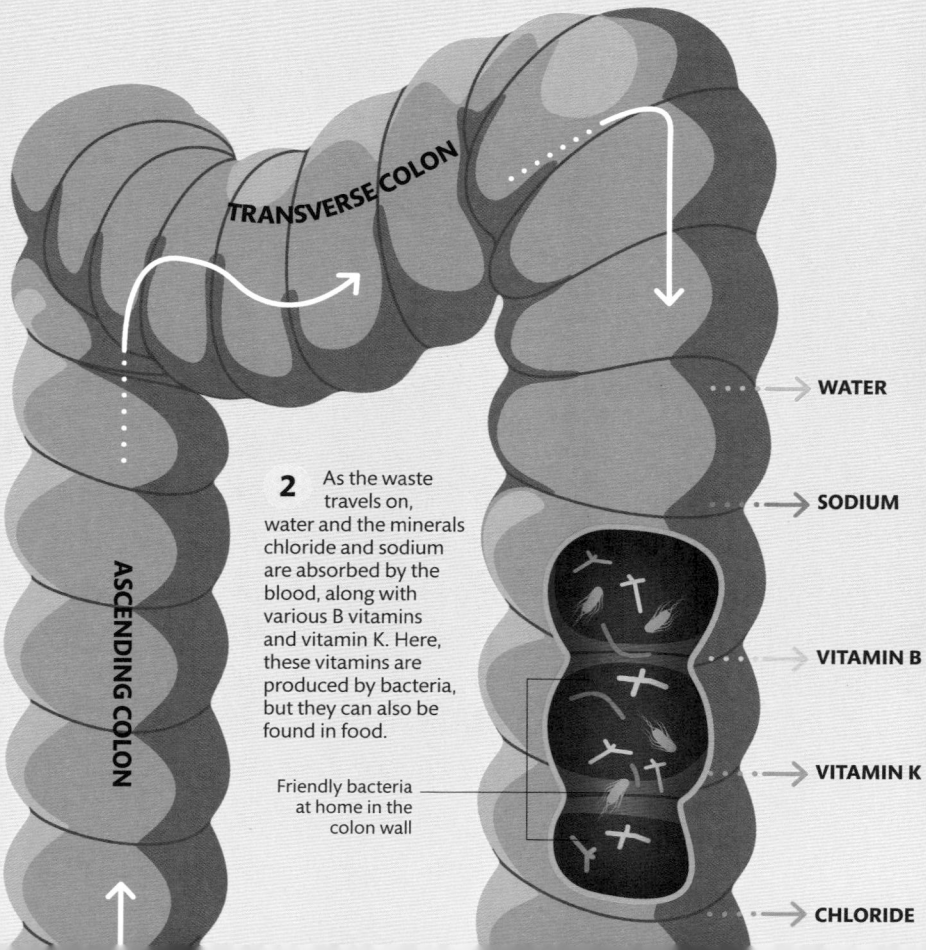

2 As the waste travels on, water and the minerals chloride and sodium are absorbed by the blood, along with various B vitamins and vitamin K. Here, these vitamins are produced by bacteria, but they can also be found in food.

Friendly bacteria at home in the colon wall

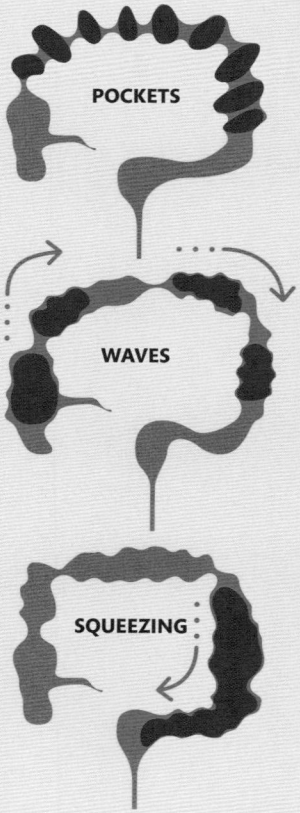

Forming pockets
Every 30 minutes or so, bands of muscles in the colon squeeze to form pockets in which fecal material is churned and mixed. It hardly moves forward at all.

Muscular waves
With similar movements to those in the esophagus and the small intestine, muscles squeeze feces in waves up from the cecum and along the colon.

Mass movements
Stimulated some three times per day by food entering the stomach, slow-moving waves of muscular squeezing shift feces along the colon into the rectum.

Keeping regular

Waste takes 19–36 hours to move through the large intestine, allowing time for water to be absorbed. If the feces are rich in fiber, their bulk carries them promptly through the system.

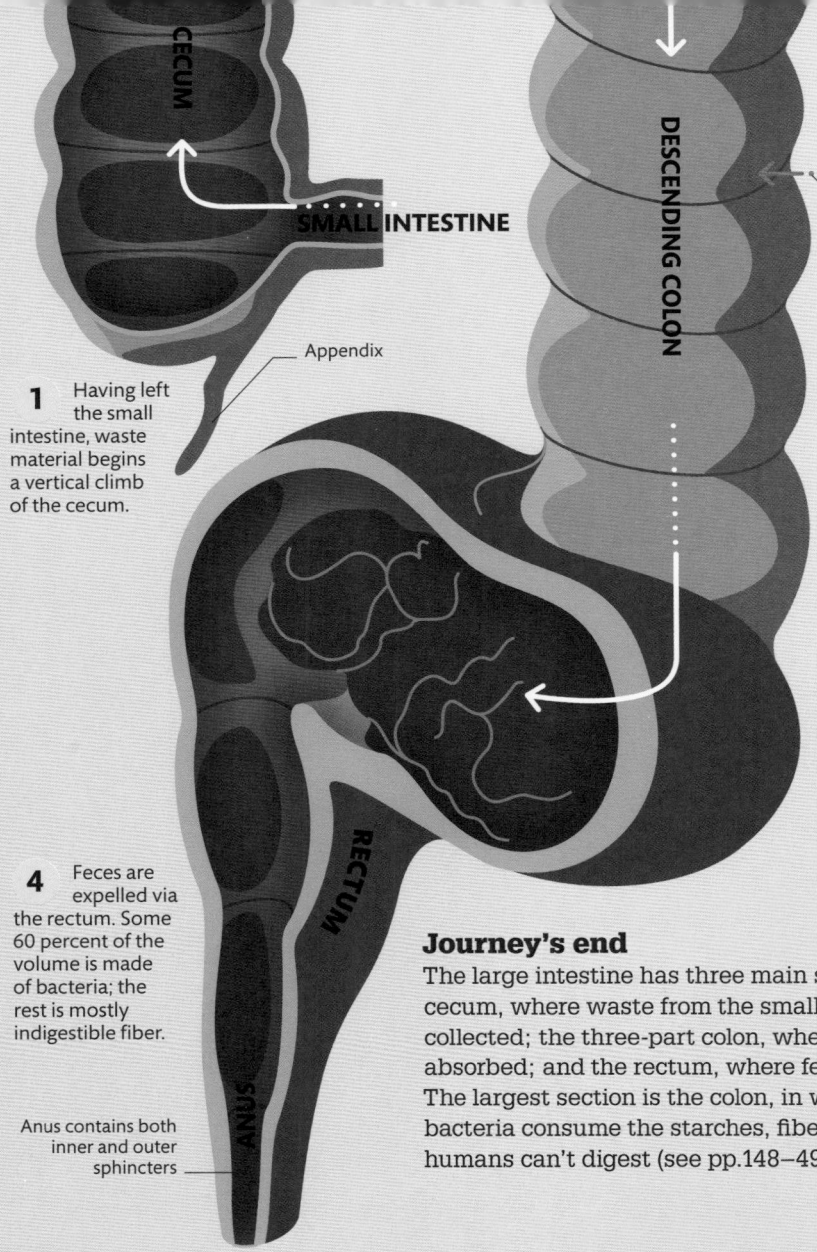

1 Having left the small intestine, waste material begins a vertical climb of the cecum.

Potassium and bicarbonate are absorbed by the colon to replace the sodium absorbed by the bloodstream

3 Feces are compacted into the lower colon. They are kept moist by mucus secreted from the colon walls.

4 Feces are expelled via the rectum. Some 60 percent of the volume is made of bacteria; the rest is mostly indigestible fiber.

Journey's end

The large intestine has three main sections: the cecum, where waste from the small intestine is collected; the three-part colon, where nutrients are absorbed; and the rectum, where feces are expelled. The largest section is the colon, in which colonies of bacteria consume the starches, fiber, and sugars that humans can't digest (see pp.148–49).

WHY DO WE HAVE AN APPENDIX?

The appendix is possibly the remnant of an organ that helped our ancestors digest foliage thousands of years ago. Today, however, it plays no obvious role, except perhaps as a safe refuge for gut bacteria.

WHEN NATURE CALLS

When feces enter the rectum, stretch receptors trigger a "need to go" reflex by sending impulses to the spinal cord. Motor signals from the spine then tell the internal anal sphincter to relax. At the same time, sensory messages to the brain make a person aware of the need to defecate, and the person makes a conscious decision to relax the external anal sphincter. On a healthy diet, this happens between three times a day and once every three days.

Bacterial breakdown

Over 100 trillion beneficial bacteria, viruses, and fungi live in the digestive tract. Known collectively as gut microbes, they provide us with nutrients, help us digest, and help defend us against harmful microbes (see pp.172–73).

Swallowing microbes

We receive our first microbes at birth, and more enter our bodies every day of our lives. They enter through the nose and mouth and travel to the stomach, where conditions are too acidic for many to take up permanent residence. The small intestine is likewise too acidic, but many microbes survive just long enough to move into the colon, where they play a vital role in digestion.

90 PERCENT OF ALL THE CELLS IN OUR BODIES ARE BACTERIAL RATHER THAN HUMAN

Lactobacilli are common stomach bacteria that are used in probiotic medical treatments. They fight off other bacteria that cause diarrhea

STOMACH

Helicobacter pylori is a foe, causing ulcers as it burrows into the stomach lining

CHYME

70 percent of all gut microbes live in the large intestine

LARGE INTESTINE

Harmful bacterium invading the small intestine

Wall of friendly bacteria

SMALL INTESTINE

Substances released by friendly bacteria to ward off invaders

ANTIBIOTICS

Antibiotics destroy or slow down the growth of bacteria, but they aren't able to discriminate between harmful and friendly bacteria. As a consequence, the friendly microbes in the gut suffer when we take antibiotics. The diversity of gut bacteria starts to decrease as soon as the antibiotic course starts and reaches a minimum about 11 days later. The populations soon bounce back after treatment, but overuse of antibiotics can cause them permanent damage.

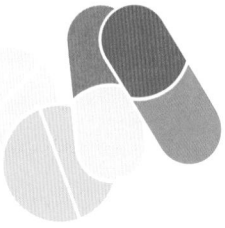

In it together

Although many of the bacteria that get into our bodies are harmful, most of them protect us against microscopic enemies. They do this both by taking up space (lining the intestine walls), and releasing substances that kill harmful bacteria.

Digesting what we can't

The microbes in the colon use the carbohydrates we can't digest for energy. They ferment fibre such as cellulose, which help us absorb dietary minerals such as calcium and iron, are used to produce vitamins, and have other benefits in the body. The microbes themselves also secrete essential vitamins, such as vitamin K.

WHAT'S THAT SMELL?

Fermentation by gut microbes produces a number of different gases, including hydrogen, carbon dioxide, methane, and hydrogen sulfide. In large amounts, these can cause bloating and flatulence. The most gas-producing foods include beans, corn, and broccoli—but onions, milk, and artificial sweeteners are major offenders too.

CORN BROCCOLI

LARGE INTESTINE

Gases produced by fermentation

Bacteria digesting carbohydrates

Carbohydrates

Nutrients being absorbed by the large intestine

Wall of friendly bacteria

Vitamin K plays a vital role in blood clotting

ACETIC ACID

BUTYRIC ACID

PROPIONIC ACID

VITAMIN B

VITAMIN K

BLOODSTREAM

Acetic acid is vital for muscle health

Absorption into the bloodstream

Butyric acid produces energy for gut cells

Propionic acid helps tissues respond to insulin

Vitamin B helps us convert food into energy

WHAT ARE PROBIOTICS?

Probiotics are the opposite of antibiotics. They are live bacteria that are consumed—in yogurts or tablets—to fortify gut bacteria that have been damaged by antibiotics or disease.

Cleaning the blood

As blood travels through the body, it picks up a great deal of waste and excess nutrients. These would quickly reach life-threatening levels without the kidneys, whose job it is to flush them out of the system.

THE ENTIRE BLOOD STREAM IS FILTERED BY THE KIDNEYS 20–25 TIMES PER DAY

Waterworks

It takes 5 minutes for blood to pass through the kidneys. It enters waste-laden and leaves clean, having passed through countless microscopic filters that turn the waste into urine. The urine then flows to the bladder, at which point we feel the need to urinate. A major component of urine is urea—a waste product formed in the liver (see pp.156–57).

STONES IN THE BODY

So much waste passes through the kidneys that even the smallest amount of a mineral can build up and form a stone. These so-called "kidney stones" can pass out of the body without incident, but some can become large enough to block the ureter. The causes of kidney stones include obesity, bad diet, and not drinking enough water.

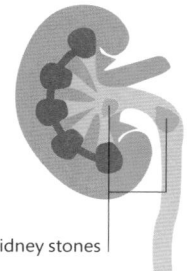

Each nephron is anchored to the middle part of the kidney, called the medulla

Waste in the form of urine is collected in the medulla

CORTEX

MEDULLA

1 Dirty blood in
Waste-laden blood enters the kidney via the renal artery. This artery branches out into a forest of capillaries that feed around a million microfilters known as nephrons. After being filtered, clean blood leaves the kidney via the renal vein.

Dirty blood flows in

Clean blood flows out

RENAL PELVIS

RENAL ARTERY

RENAL VEIN

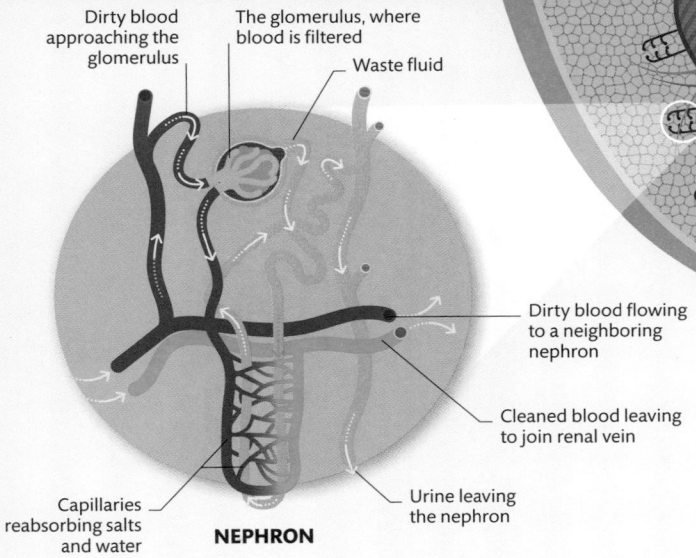

2 The filtration process
As the blood passes through a nephron, it is forced through a tiny filter called a glomerulus, which lets urea and other wastes pass, but keeps blood cells and valuable proteins in the bloodstream. On the far side, the waste fluid passes on a long loop through the kidney, where its composition of salts and water is fine-tuned, before it flows into urine-collecting ducts.

3 Collecting urine
The urine-collecting tubes of the medulla join together as they coverge on the renal pelvis. Here the urine flows past the renal artery and the renal vein and enters a tube called the ureter. The ureter connects the kidney to the bladder.

4 Waste disposal
Muscular contractions squeeze the urine along the ureter—which is why our bladders fill even when we are lying down. When the bladder is full, its muscular walls squeeze the urine further, but the urine is halted by a ring of muscle at the base of the bladder. Learning how to control this muscle gives us the choice about when to urinate.

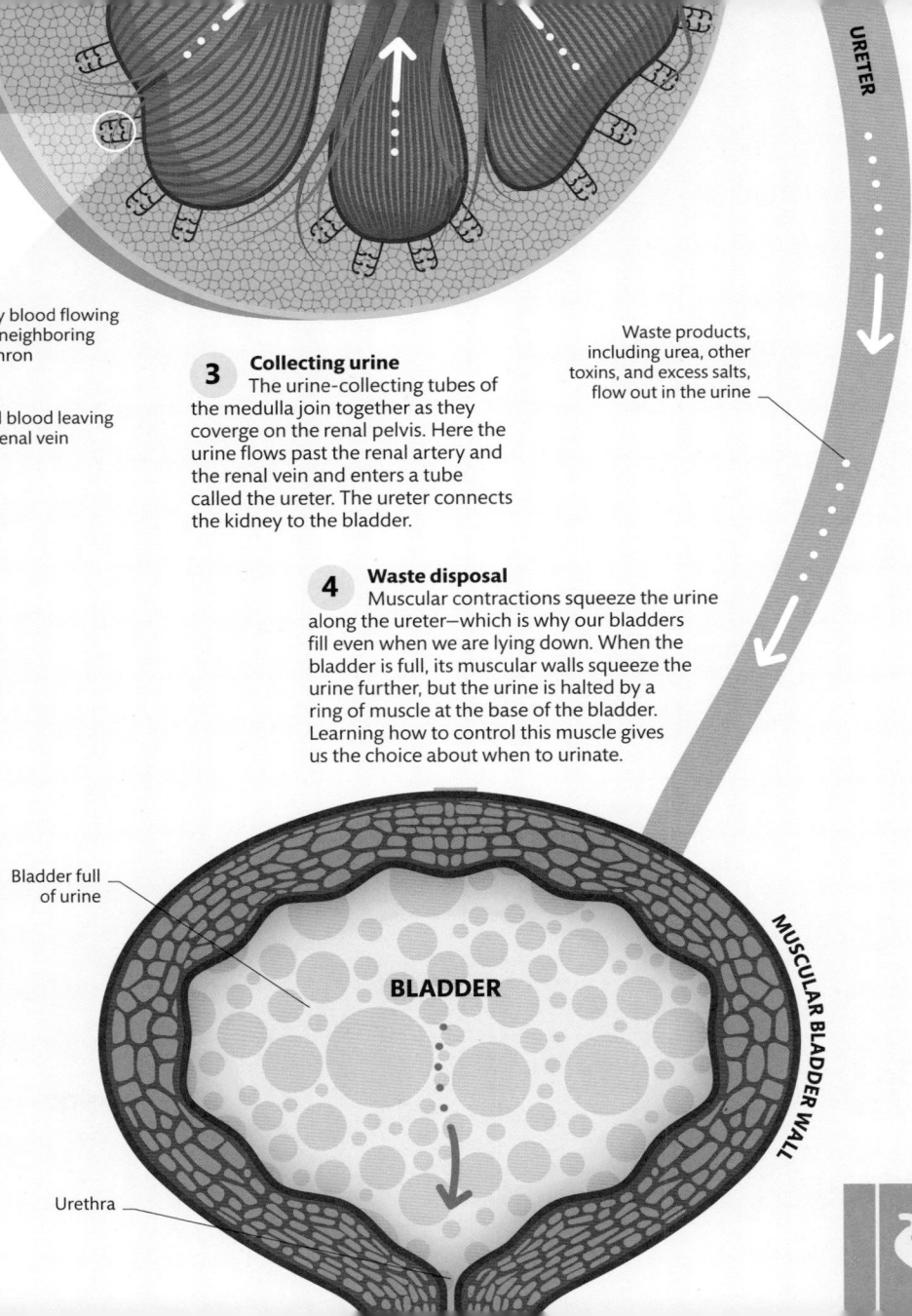

WHAT IF THE KIDNEYS FAIL?

If a person's kidneys are too weak to filter the blood, a dialysis machine can be used instead of the kidneys. The person's blood flows through a tube into the machine, gets cleaned and filtered, and then returns to their body.

Water balance

Water levels in the blood have to be kept within a certain range; otherwise, the body's cells become too shrunken (dehydrated) or too bloated (overhydrated) to work. The kidneys, the endocrine system, and the circulatory system work together to maintain a healthy balance in our bloodstream.

LOSING BALANCE

A number of commonly consumed substances uspet our water balance. Alcohol, for example, blocks the pituitary gland from releasing ADH. This means that the kidneys, which are working hard to get rid of the alcohol in the bloodstream, send more water out into the urine. Drinking just one glass of wine can cause the body to get rid of the equivalent of four wine glasses of water. Substances that make us produce a lot of urine are called "diuretics." Caffeine is another diuretic.

Too little water

We lose water constantly, but there are times when we lose a lot of water quickly—through sweating, vomiting, or diarrhea, for example. This results in both a decrease in blood volume and a rise in the level of salt relative to water in our blood. These act as triggers for balance to be restored.

Too much water

Far rarer than dehydration is overhydration, which can be caused by extreme water intake after exercise, drug abuse, or disease. This results in an increase in blood volume and a reduction in the level of salt relative to water in the blood.

1 Low water alert
The hypothalamus receives signals that blood pressure is low and salt levels are high. It responds by increasing the production of ADH (antidiurectic hormone), which is carried to the pituitary gland, where it is released into the blood.

1 High water alert
The hypothalamus receives signals that blood pressure is high and salt levels are low. It responds by producing less ADH. Since ADH instructs the kidneys to store water, a reduction in ADH means an increase in urination.

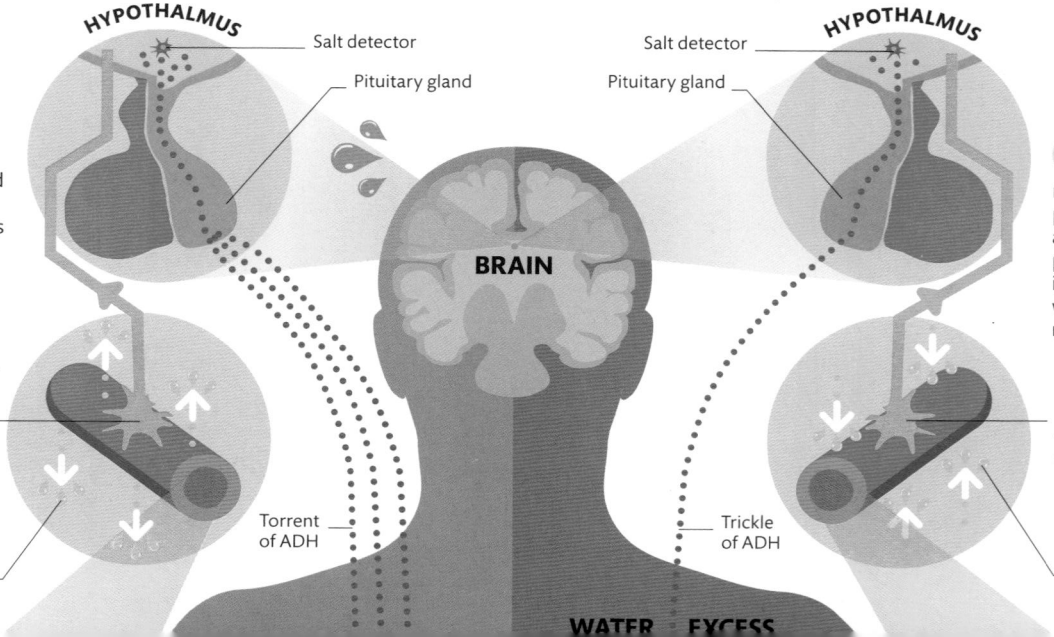

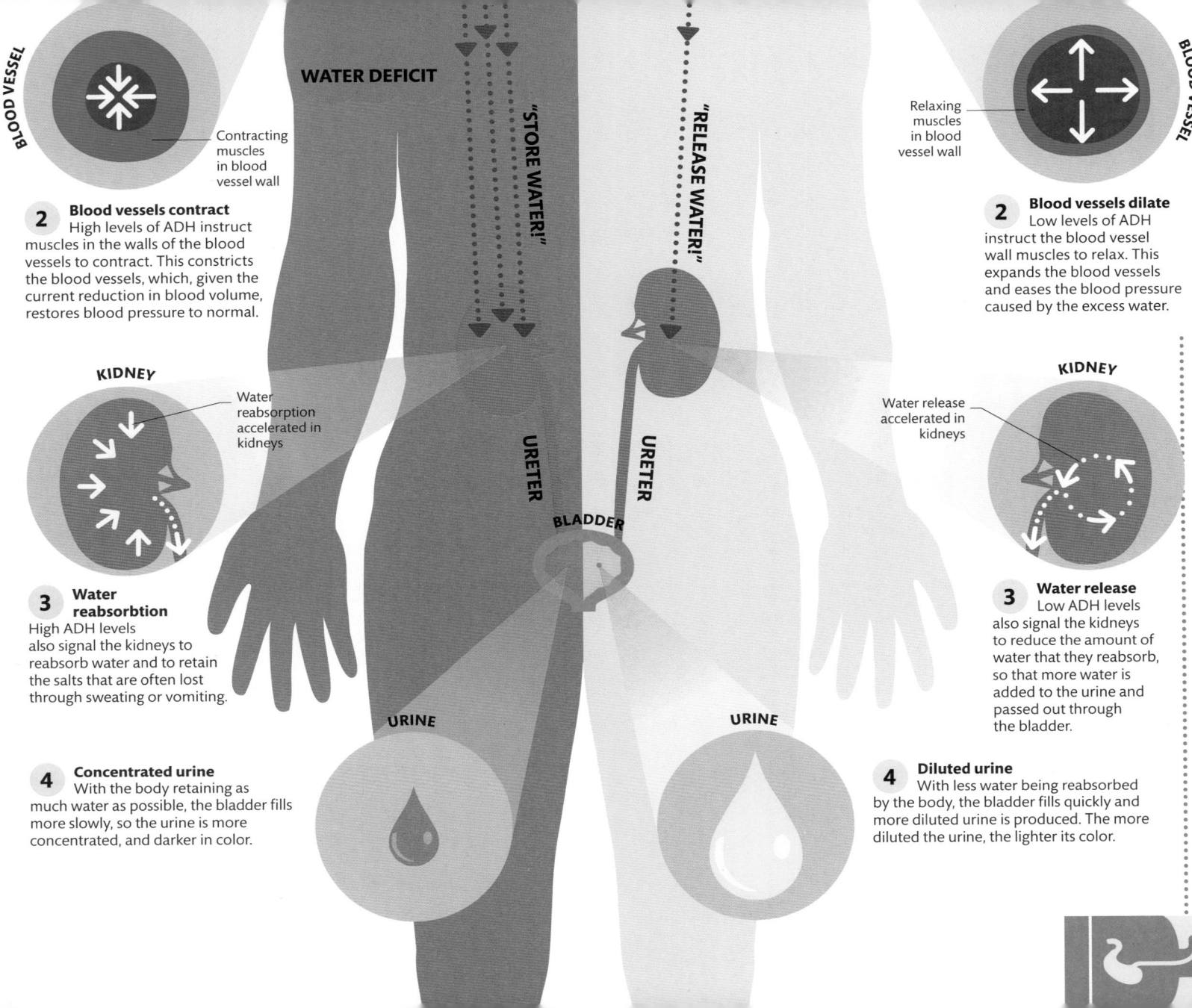

2 Blood vessels contract
High levels of ADH instruct muscles in the walls of the blood vessels to contract. This constricts the blood vessels, which, given the current reduction in blood volume, restores blood pressure to normal.

2 Blood vessels dilate
Low levels of ADH instruct the blood vessel wall muscles to relax. This expands the blood vessels and eases the blood pressure caused by the excess water.

3 Water reabsorbtion
High ADH levels also signal the kidneys to reabsorb water and to retain the salts that are often lost through sweating or vomiting.

3 Water release
Low ADH levels also signal the kidneys to reduce the amount of water that they reabsorb, so that more water is added to the urine and passed out through the bladder.

4 Concentrated urine
With the body retaining as much water as possible, the bladder fills more slowly, so the urine is more concentrated, and darker in color.

4 Diluted urine
With less water being reabsorbed by the body, the bladder fills quickly and more diluted urine is produced. The more diluted the urine, the lighter its color.

How the liver works

Once nutrients have entered the blood—via the mouth, stomach, and intestines—they are taken straight to the liver. Here, they are variously stored, dismantled, or turned into something new. At any one time, the liver holds some 10 percent of the body's blood supply.

Liver lobule

The liver is made up of thousands of tiny factories called lobules. Each of these contains thousands of chemical processors called hepatocytes. These do all the liver's work, albeit supported by Kupffer cells and stellate cells. Each lobule has a central outflowing vein and is six-sided, with each of its corners supporting two incoming blood supplies and an outflowing duct for bile.

Ins and outs of the liver
Blood arrives from two directions, then the liver outputs blood via the hepatic vein and bile through the bile duct.

- Blood from the intestines
- Blood from the heart
- Blood to the heart
- Bile to the gallbaldder

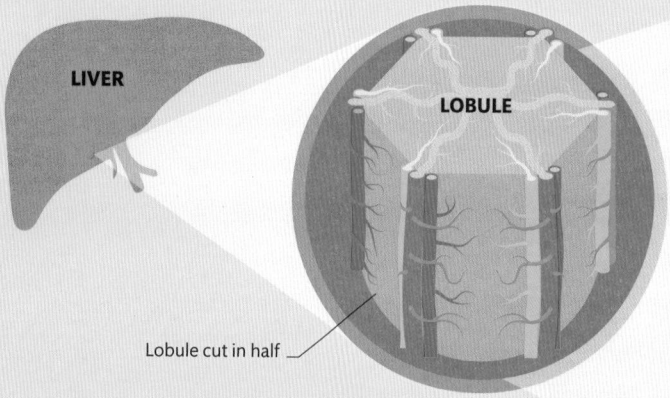

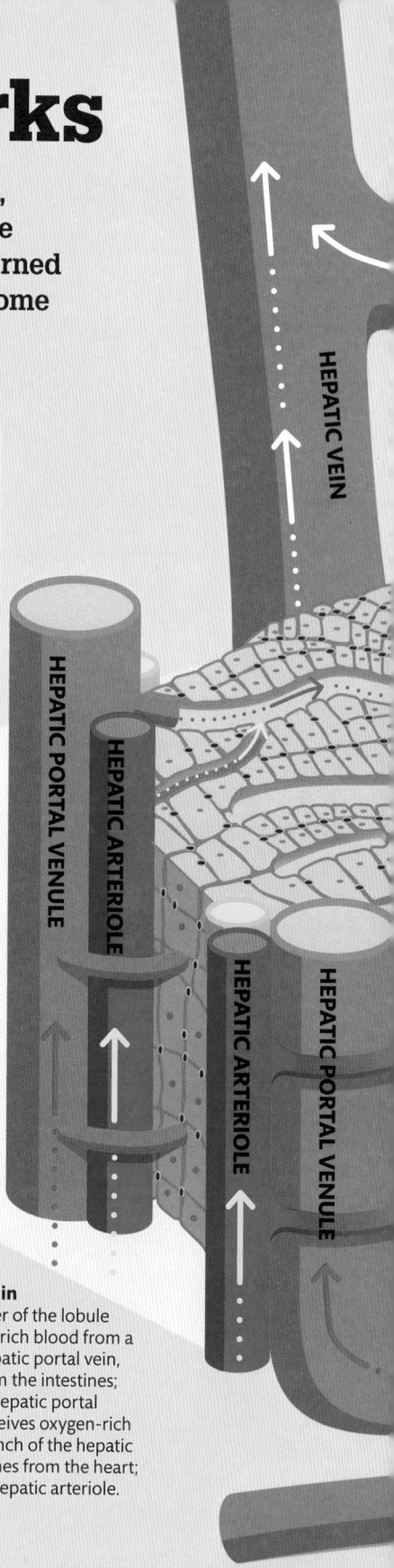

1 Nutrients in
Each corner of the lobule receives nutrient-rich blood from a branch of the hepatic portal vein, which comes from the intestines; this is called the hepatic portal venule. It also receives oxygen-rich blood from a branch of the hepatic artery, which comes from the heart; this is called the hepatic arteriole.

DOUBLE BLOOD SUPPLY

An unusual fact about the liver is that it has two blood supplies. Like all other organs, it receives oxygenated blood from the heart to give it energy, but it also receives blood from the intestines, which it cleans, stores, and processes.

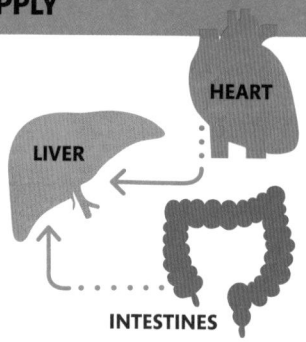

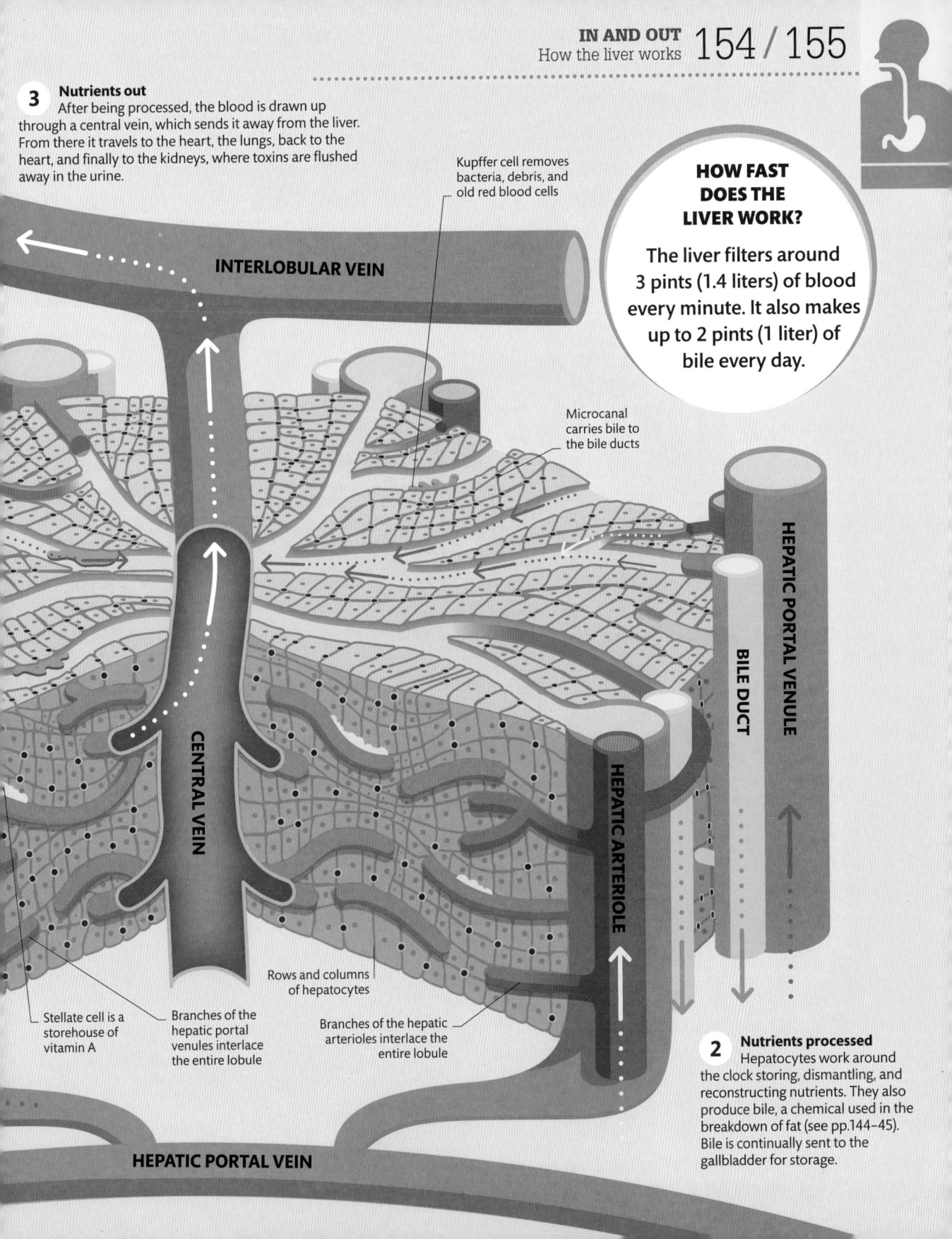

3 Nutrients out
After being processed, the blood is drawn up through a central vein, which sends it away from the liver. From there it travels to the heart, the lungs, back to the heart, and finally to the kidneys, where toxins are flushed away in the urine.

HOW FAST DOES THE LIVER WORK?

The liver filters around 3 pints (1.4 liters) of blood every minute. It also makes up to 2 pints (1 liter) of bile every day.

2 Nutrients processed
Hepatocytes work around the clock storing, dismantling, and reconstructing nutrients. They also produce bile, a chemical used in the breakdown of fat (see pp.144–45). Bile is continually sent to the gallbladder for storage.

What the liver does

The liver is perhaps best understood as a factory—a processing plant with three main departments; processing, manufacturing, and storage. Its raw materials are the nutrients absorbed by the blood during digestion—but which department they go to depends on the body's priorities.

WHAT ELSE DOES THE LIVER DO?

It produces blood clotting proteins, which ensure that we stop bleeding when injured. People with unhealthy livers tend to bleed easily.

Glucose from carbohydrates
In a process called glycogenolysis, the liver makes glucose out of carbohydrates when the body is low on energy.

Metabolizing fat
Excess carbohydrates and proteins are converted into fatty acids and released into the bloodstream for energy. This becomes vital when glucose is running out.

Detoxifying the blood
Pollutants, bacterial toxins, and defensive chemicals from plants are turned into less dangerous compounds, then sent to the kidneys to be flushed out of the body.

Processing

The liver spends most of its time processing nutrients. This involves making sure that the right nutrients are sent to the right parts of the body, and that back-ups are provided when needed. Crucially, this also means flushing out toxic subtances.

THE REGENERATING ORGAN

Unlike other organs, which create scar tissue at sites of injury, the liver creates brand new cells when it needs them. This is lucky, since it is constantly being bombarded by unhealthy, toxic chemicals. These chemicals—which include some prescribed medications—frequently damage the liver, but it holds its ground by regenerating itself. Incredibly, it can lose 75 percent of its mass and still regrow completely—all in a matter of weeks.

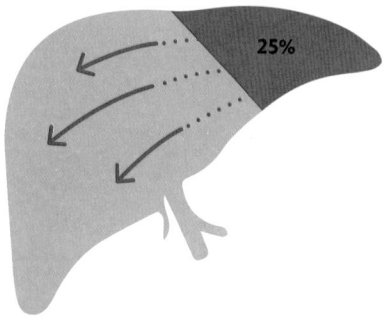

Bile production
Bile is constantly being produced by the liver and sent to the gallbladder for storage. It is made out of hemoglobin, which is released during the breakdown of old red blood cells.

Hormone production
The liver secretes at least three hormones, making it a key player in the endocrine system (see pp.190–91). The liver's hormones stimulate cell growth, encourage bone marrow production, and aid blood pressure control.

Manufacturing

The liver is a major manufacturing hub, turning simple nutrients into, among other things, chemical messengers (hormones), body tissue components (proteins), and a vital digestive fluid (bile). Since it is always busy, the liver also produces another precious commodity—an enormous amount of heat.

Protein synthesis
The liver produces many proteins that are then secreted into the blood. It does so particularly when certain amino acids (the building blocks of proteins) are missing from the diet.

Vitamins
The liver can store up to 2 years' worth of vitamin A, which is vital to the immune system. Vitamins B12, D, E, and K are also stored until needed.

Storage

A great deal of stockpiling goes on in the liver, mainly of vitamins, minerals, and glycogen (the stored form of glucose). This enables the body to survive without food for days and weeks on end, and ensures that any shortfall in dietary nutrients can quickly be corrected.

Minerals
Two vital minerals are stored in the liver: iron, which carries oxygen through our bodies; and copper, which keeps the immune system healthy. Copper is also used to make red blood cells.

Glycogen
Energy is stored as glycogen in the liver. When the body runs out of energy (see pp.158–59), the liver converts it to glucose and releases it into the bloodstream.

THE LIVER PERFORMS SOME **500 CHEMICAL FUNCTIONS** IN TOTAL

LIVER DAMAGE

The liver is the only organ that can regenerate itself. However, repeated exposure to damaging agents, such as alcohol, a drug, or a virus, can eventually injure the liver. This happens when it is inundated by toxins and never gets a chance to regenerate. In this strung-out state, the liver is finally scarred—a condition known as cirrhosis. A common cause of cirrhosis is drinking too much alcohol.

Energy balance

Most of the body's cells use glucose or fatty acids for energy. To maintain a regular supply of these, the body alternates between absorbing energy (by eating) and releasing it (after which we feel hungry again). In ideal conditions, this cycle repeats itself every few hours.

Filling the tanks

Glucose and fatty acids enter our bodies through the food we eat. As blood glucose levels rise, the pancreas releases the hormone insulin. This tells muscle and fat cells to absorb and store the glucose and fatty acids as energy for the future.

Food rich in sugar

Numerous sugar molecules indicate high blood sugar level after meal

Fatty acid molecule

Glucose molecule

Fatty acids being stored in a fat cell

3 Excess glucose stored
Most fatty acids are stored in fat cells, which serve as resevoirs of energy. These cells also absorb excess glucose and convert it into fatty acid molecules.

Excess glucose heading for storage in a fat cell

ABSORB!

2 Muscle burns glucose
Muscle cells, among others, convert glucose into energy for contracting. Muscle cells also absorb fatty acids. They burn the fatty acids when glucose levels are low.

Glucose being absorbed by a muscle cell

Fatty acid being absorbed by a muscle cell

ABSORB!

1 "Absorb!" signal sent
After a meal, the pancreas detects high levels of sugar in the blood. In response, it releases insulin, which circulates in the blood. This readies the muscle and fat cells to open and receive glucose, which all cells use for energy.

PANCREAS

DOES FAT MAKE YOU FAT?

Only when eaten with sugary foods or carbohydrates. These foods contain glucose, which signals the body to store nutrients, and so put on weight.

Burning the fuel

As the body's cells absorb nutrients, blood glucose levels start to fall. Unless more food is digested, these levels drop to a point where the body burns fat instead of glucose for energy. Once again, this process is organized by the pancreas.

Sparse sugar molecules indicate low blood sugar level

Fatty acids being burned in a muscle cell

Fatty acids released into the bloodstream

BURN!

BURN!

PANCREAS

3 Muscle cell burns fat
Here, a muscle cell receives fatty acids from a fat cell and breaks them down for energy.

2 Fat sent to muscle
Glucagon also tells fat cells to release their stored fatty acids into the bloodstream. These fatty acids can then be used as a source of energy by other cells.

1 "Burn!" signal sent
A few hours after eating, specialized cells in the pancreas detect a drop in blood glucose levels. The pancreas releases the hormone glucagon into the bloodstream. This signals the liver to release the glucose it has stored in the form of glycogen into the bloodstream (see pp.156–57).

ENERGY SUPPLY AND DEMAND

Food energy is measured in calories. A steak contains around 500 calories, as does a large bag of potato chips or 10 apples. A person at rest needs around 1,800 calories a day to maintain weight—more in or out tips the scales.

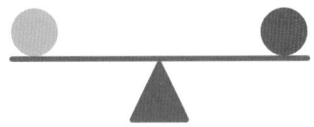

WEIGHT MAINTAINED

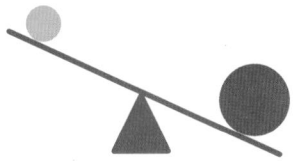

WEIGHT LOSS

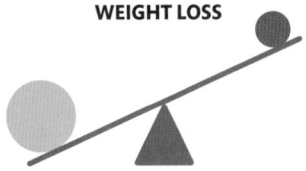

WEIGHT GAIN

CALORIES IN CALORIES OUT

The sugar trap

Calories are equal in terms of the amount of energy they contain, but where they come from—fat, protein, or carbohydrate—determines how they are used by the body. Some foods give us a steady source of energy; others can take us on a hormone roller-coaster ride.

ARE CALORIES BAD FOR YOU?

A calorie is the amount of energy your body will gain from eating the food that contains it, so no—we need energy to live! But if you eat too many calories, your body will store the excess as fat.

Lingering insulin

Foods that are quickly turned into sugars cause a spike in blood glucose levels (see p.158). Insulin spikes in response, causing glucose levels to plummet. The sugar crash leaves us tired and craving more sugar, while insulin lingers in our blood and prevents us from burning fat.

Rise and fall
The peak and crash of glucose and steady rise and fall of insulin levels in the blood is traced along mealtimes during a morning.

→ Glucose
→ Insulin

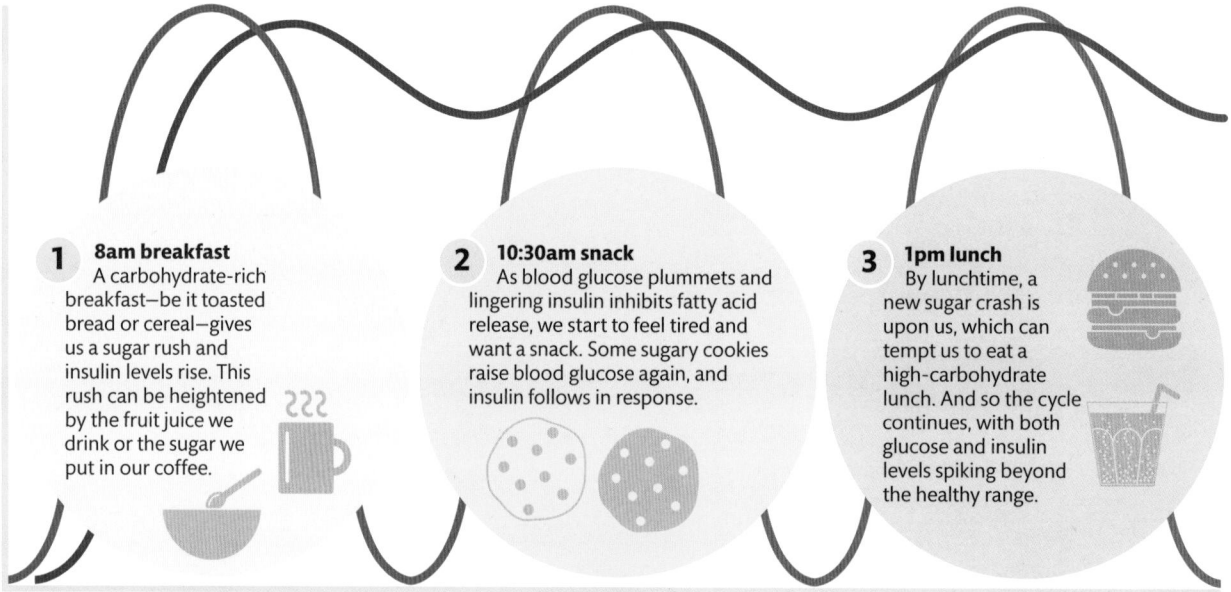

Putting on the pounds

The sugar trap quickly leads to weight gain, and being overweight can have serious health implications. These include insulin sensitivity, insulin resistance, type 2 diabetes (see pp.201), heart disease, some types of cancer, and stroke. To avoid obesity, it is vital to keep insulin levels low, and one way of doing that is through a low-carbohydrate diet.

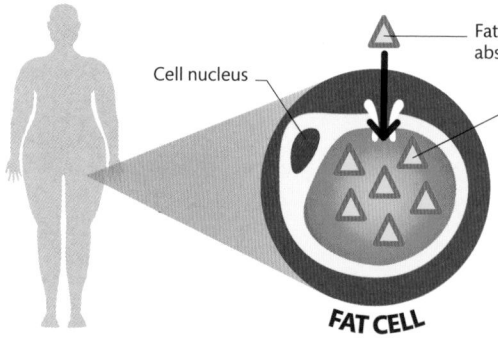

Storing fat
When we put on fat we don't increase the number of fat cells in our body. The same fat cells just get bigger as they accumulate more fat deposits.

HIGH-PROTEIN DIET

To cut out carbohydrates, some diet promoters recommend getting calories from protein and healthy fats instead. You can follow a diet in phases designed to train your body to start burning fat and rely less on carbohydrates.

Low-carbohydrate diets

A popular, if controversial, way out of the sugar trap is to limit our consumption of carbohydrates, which are otherwise broken down into sugars and stored as fat. By doing so, we avoid the glucose-insulin roller coaster that ends in sugar cravings and increased fat storage. Keeping sugar and insulin levels within a healthy range enables fat, rather than glucose, to be used as an energy source.

SUGAR IS NOW THOUGHT TO BE **MORE ADDICTIVE THAN COCAINE**

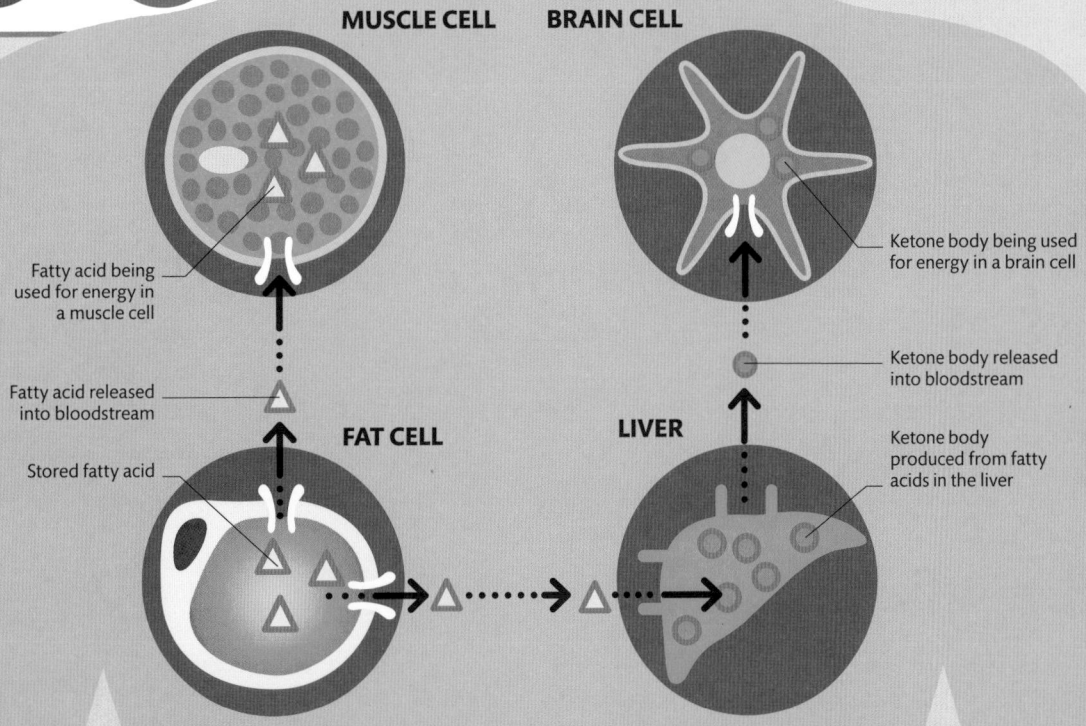

Releasing fatty acids
When blood glucose is maintained at a healthy level, insulin levels remain low. This allows the release of fatty acids from fat cells—a process that is otherwise inhibited by insulin.

Producing ketone bodies
Unlike other tissues, the brain can't use fatty acids as an energy source. So when blood glucose is low, the liver begins to convert fatty acids into ketone bodies—molecules that provide energy for brain cells.

Feast or fast?

Two of today's most popular diets don't involve calorie counting at all. Palaeolithic diets aim for an ancestral way of eating, removing the highly processed foods of today. Intermittent fasting, on the other hand, takes a more feast and fast approach, restricting when you eat rather than what you eat.

Back to basics

The theory behind paleolithic diets is that our bodies have not evolved to consume the highly processed, sugary, carbohydrate-rich foods that are abundant in supermarkets today. The diet promotes foods that are thought to have been available to our hunter-gatherer ancestors, who lived before the advent of farming, 10,000 years ago—although the lifestyle doesn't involve reverting back to cave life. Dieters used to getting their calcium from dairy foods need to find calcium-rich alternatives, or they put themselves in danger of calcium deficiency.

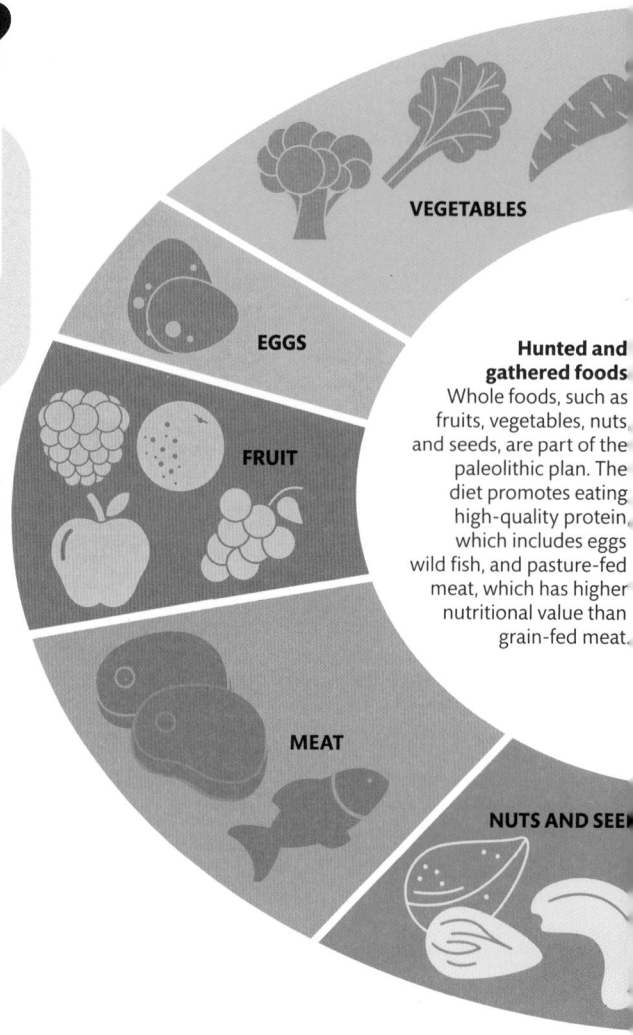

Hunted and gathered foods
Whole foods, such as fruits, vegetables, nuts and seeds, are part of the paleolithic plan. The diet promotes eating high-quality protein, which includes eggs wild fish, and pasture-fed meat, which has higher nutritional value than grain-fed meat.

Intermittent fasting

The idea behind intermittent fasting is to take regular breaks from eating, during which the body gets all its energy from stored fat, but not for so long that it starts to break down muscle protein for energy. There are two main intermittent fasting methods; the 16:8 and the 5:2.

The 16:8 method
Followers of this regime eat during an 8-hour period every day (say noon to 8pm). The other 16 hours you fast, but luckily a lot of this time is spent sleeping, which makes it more manageable.

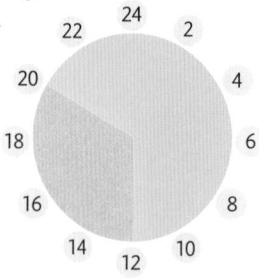

Key Eating Fasting

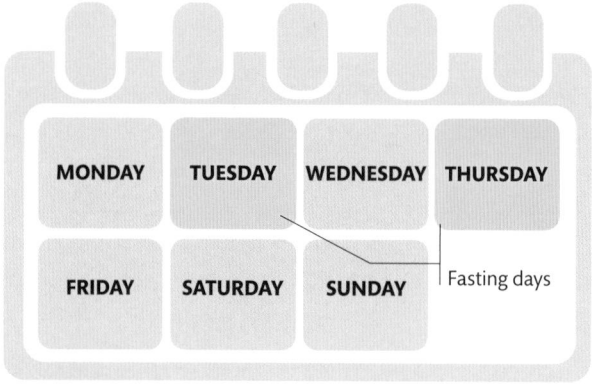

The 5:2 method
This regime restricts your daily energy intake to about 500 calories (about one meal) per day for two days of the week. You can eat as much as you like (within reason) for the other five days of the week.

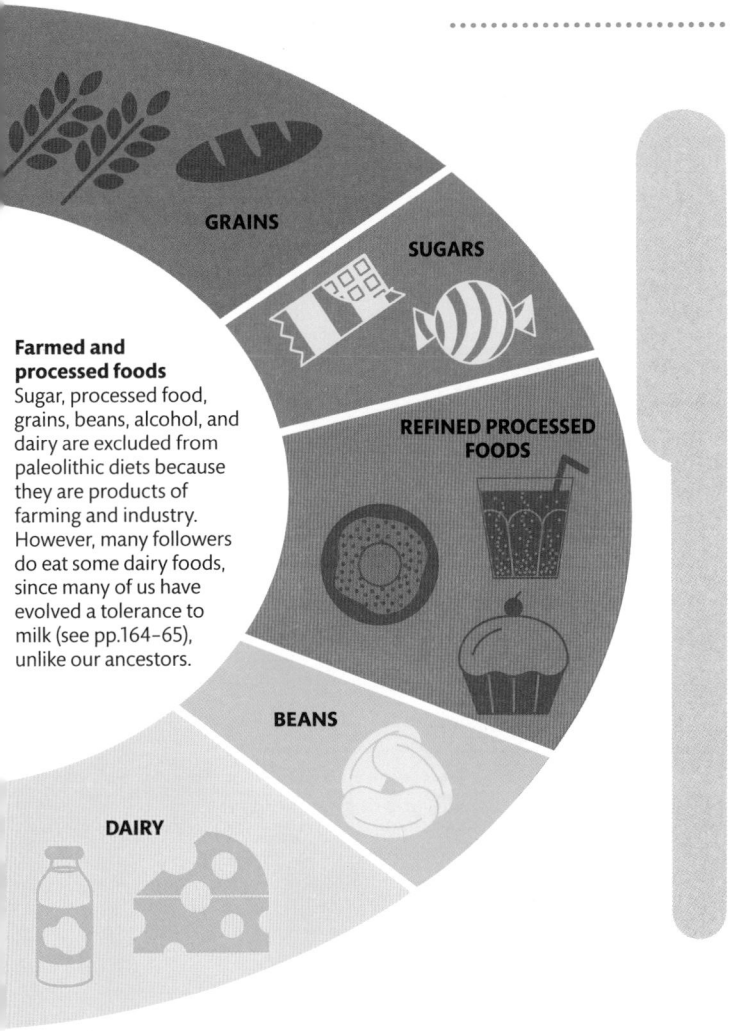

Farmed and processed foods
Sugar, processed food, grains, beans, alcohol, and dairy are excluded from paleolithic diets because they are products of farming and industry. However, many followers do eat some dairy foods, since many of us have evolved a tolerance to milk (see pp.164–65), unlike our ancestors.

ONE-THIRD OF THE WORLD'S ADULTS NOW PRODUCE, THE ENZYME THAT DIGESTS DAIRY SUGAR

The glycemic index

The glycemic index (GI) is a measure of how quickly carbohydrate-containing foods increase glucose levels in the blood. The lower a food's GI value, the less it affects blood sugar levels. An attraction of paleolithic diets is that they focus on low GI foods.

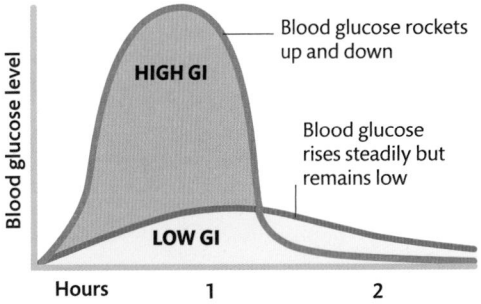

Blood glucose levels
High GI foods rapidly increase blood sugar levels, but this is followed by a rapid decrease, leaving us feeling hungry. Low GI foods gradually increase blood sugar levels, leaving us feeling full for longer.

Natural fat-burning

Exercising when your body is naturally burning fat may give your workout more punch. A run before breakfast, for example, takes advantage of the fact that your body is already burning fat after fasting all night. A run in the evening, however, is more likely to be fueled by blood glucose supplied by the day's food. For this reason, morning exercise is generally more effective for losing weight.

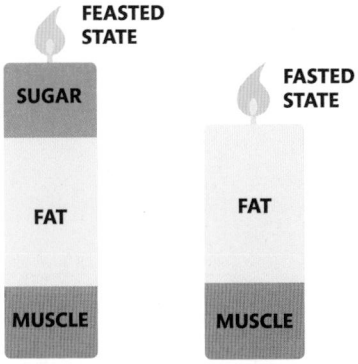

Evening
The body can run on the glucose from a meal for about 3–5 hours.

Morning
Once glucose is used up, the body starts to burn fat stores.

BRAIN HEALTH

There is evidence that fasting improves brain health. Intermittent fasting in particular puts the neurons under mild stress—much like our muscles are stressed by exercise. This stress causes the release of chemicals that help in the growth and maintenance of neurons.

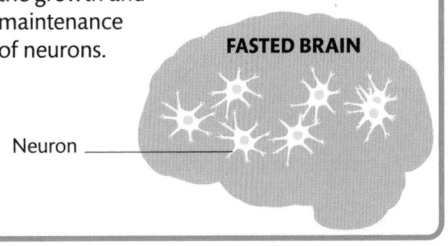

Digestive problems

Digestive problems can range from temporary discomfort after eating to life-long persistent disorders. In most cases, the treatment is simply to avoid the foods that cause the symptoms.

Lactose intolerance

Many adults lack the enzyme lactase, which is needed to break down lactose, the sugar found in milk. All healthy babies have it, but most of us stop producing this enzyme after weaning. Only about 35 percent of the world's population have acquired a mutation that allows them to produce lactase into adulthood.

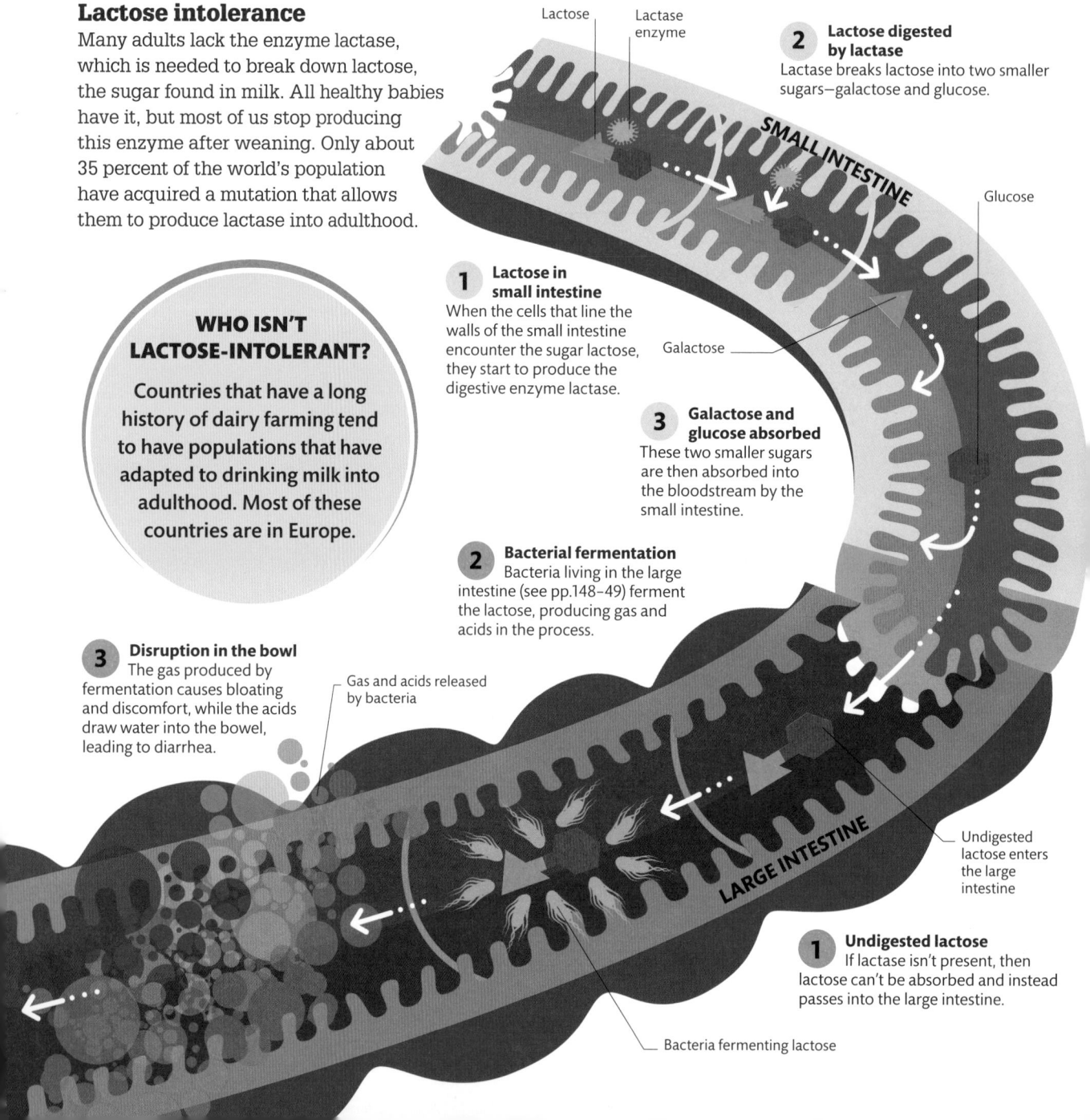

2 Lactose digested by lactase
Lactase breaks lactose into two smaller sugars—galactose and glucose.

1 Lactose in small intestine
When the cells that line the walls of the small intestine encounter the sugar lactose, they start to produce the digestive enzyme lactase.

3 Galactose and glucose absorbed
These two smaller sugars are then absorbed into the bloodstream by the small intestine.

2 Bacterial fermentation
Bacteria living in the large intestine (see pp.148–49) ferment the lactose, producing gas and acids in the process.

3 Disruption in the bowl
The gas produced by fermentation causes bloating and discomfort, while the acids draw water into the bowel, leading to diarrhea.

1 Undigested lactose
If lactase isn't present, then lactose can't be absorbed and instead passes into the large intestine.

WHO ISN'T LACTOSE-INTOLERANT?

Countries that have a long history of dairy farming tend to have populations that have adapted to drinking milk into adulthood. Most of these countries are in Europe.

BRINGING IT UP

One way the body avoids digestive problems is by vomiting. When we eat something rotten or poisonous, the stomach, the diaphragm, and the abdominal muscles all contract, forcing the food back up through the esophagus and out through the mouth.

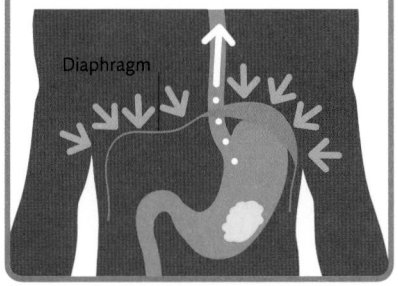

Irritable bowel syndrome

IBS is a long-term condition that can cause stomach cramps, bloating, diarrhea, and constipation. It is poorly understood, but seems to be triggered by stress, lifestyle, and certain types of food.

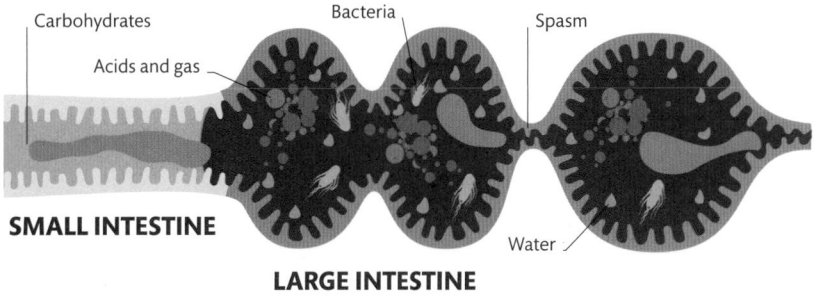

1 Bacterial fermentation
Carbohydrates that are poorly absorbed may increase the amount of water in the intestinal tract. Once in the large intestine, these carbohydrates are fermented by bacteria, producing acids and gas.

2 Bowel spasms
IBS causes bowel spasms, which can block the waste and gas from passing through. Alternatively, it can cause the waste to move too quickly, preventing water reabsorption and causing diarrhea.

Gluten intolerance

Many people experience abdominal pain, fatigue, headaches, and even numbness of the limbs when they eat gluten—a protein found in grains such as wheat, barley, and rye. These symptoms are indicators of various gluten-related disorders, ranging from gluten sensitivity to celiac disease.

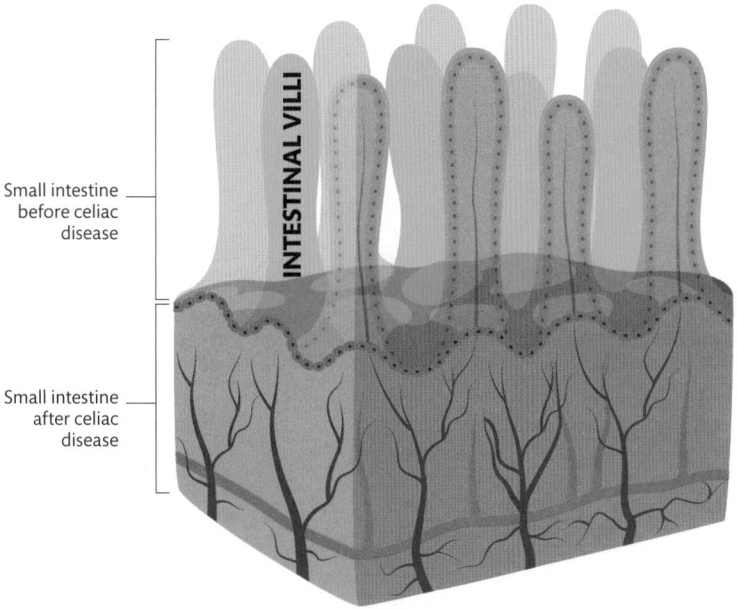

Gluten sensitivity
Lethargy, mental fatigue, cramps, and diarrhea are all symptoms of gluten sensitivity, which is only cured by avoiding all gluten products—including rye bread, beer, and pasta. Gluten sensitivity does not damage the intestines like celiac disease does.

Celiac disease
Celiac disease is a serious genetic disorder that causes the body's immune system to attack itself when it encounters gluten. This immune response causes damage to the lining of the small intestine, inhibiting the absorption of nutrients. Left unchecked, it can totally destroy the small intestine's little fingerlike projections, or villi.

Take dummies with you everywhere you go!

Whether you are excited about e-books, want more from the web, must have your mobile apps, or are swept up in social media, dummies makes everything easier.

Find us online!

dummies.com

dummies
A Wiley Brand

Leverage the powe

Dummies is the global leader in the reference category and one of the most trusted and highly regarded brands in the world. No longer just focused on books, customers now have access to the dummies content they need in the format they want. Together we'll craft a solution that engages your customers, stands out from the competition, and helps you meet your goals.

Advertising & Sponsorships

Connect with an engaged audience on a powerful multimedia site, and position your message alongside expert how-to content. Dummies.com is a one-stop shop for free, online information and know-how curated by a team of experts.

- Targeted ads
- Video
- Email Marketing
- Microsites
- Sweepstakes sponsorship

20 MILLION PAGE VIEWS EVERY SINGLE MONTH

15 MILLION **UNIQUE** VISITORS PER MONTH

43% OF ALL VISITORS ACCESS THE SITE VIA THEIR MOBILE DEVICES

700,000 NEWSLETTER SUBSCRIPTIONS TO THE INBOXES OF *300,000* UNIQUE INDIVIDUALS EVERY WEEK

# of dummies

Custom Publishing

Reach a global audience in any language by creating a solution that will differentiate you from competitors, amplify your message, and encourage customers to make a buying decision.

- Apps
- Books
- eBooks
- Video
- Audio
- Webinars

Brand Licensing & Content

Leverage the strength of the world's most popular reference brand to reach new audiences and channels of distribution.

For more information, visit dummies.com/biz

dummies®
A Wiley Brand

PERSONAL ENRICHMENT

9781119187790
USA $26.00
CAN $31.99
UK £19.99

9781119179030
USA $21.99
CAN $25.99
UK £16.99

9781119293354
USA $24.99
CAN $29.99
UK £17.99

9781119293347
USA $22.99
CAN $27.99
UK £16.99

9781119310068
USA $22.99
CAN $27.99
UK £16.99

9781119235606
USA $24.99
CAN $29.99
UK £17.99

9781119251163
USA $24.99
CAN $29.99
UK £17.99

9781119235491
USA $26.99
CAN $31.99
UK £19.99

9781119279952
USA $24.99
CAN $29.99
UK £17.99

9781119283133
USA $24.99
CAN $29.99
UK £17.99

9781119287117
USA $24.99
CAN $29.99
UK £16.99

9781119130246
USA $22.99
CAN $27.99
UK £16.99

PROFESSIONAL DEVELOPMENT

9781119311041
USA $24.99
CAN $29.99
UK £17.99

9781119255796
USA $39.99
CAN $47.99
UK £27.99

9781119293439
USA $26.99
CAN $31.99
UK £19.99

9781119281467
USA $26.99
CAN $31.99
UK £19.99

9781119280651
USA $29.99
CAN $35.99
UK £21.99

9781119251132
USA $24.99
CAN $29.99
UK £17.99

9781119310563
USA $34.00
CAN $41.99
UK £24.99

9781119181705
USA $29.99
CAN $35.99
UK £21.99

9781119263593
USA $26.99
CAN $31.99
UK £19.99

9781119257769
USA $29.99
CAN $35.99
UK £21.99

9781119293477
USA $26.99
CAN $31.99
UK £19.99

9781119265313
USA $24.99
CAN $29.99
UK £17.99

9781119239314
USA $29.99
CAN $35.99
UK £21.99

9781119293323
USA $29.99
CAN $35.99
UK £21.99

Learning Made Easy

ACADEMIC

9781119293576
USA $19.99
CAN $23.99
UK £15.99

9781119293637
USA $19.99
CAN $23.99
UK £15.99

9781119293491
USA $19.99
CAN $23.99
UK £15.99

9781119293460
USA $19.99
CAN $23.99
UK £15.99

9781119293590
USA $19.99
CAN $23.99
UK £15.99

9781119215844
USA $26.99
CAN $31.99
UK £19.99

9781119293378
USA $22.99
CAN $27.99
UK £16.99

9781119293521
USA $19.99
CAN $23.99
UK £15.99

9781119239178
USA $18.99
CAN $22.99
UK £14.99

9781119263883
USA $26.99
CAN $31.99
UK £19.99

Available Everywhere Books Are Sold

dummies.com

dummies
A Wiley Brand

Small books for big imaginations

9781119177173
USA $9.99
CAN $9.99
UK £8.99

9781119177272
USA $9.99
CAN $9.99
UK £8.99

9781119177241
USA $9.99
CAN $9.99
UK £8.99

9781119177210
USA $9.99
CAN $9.99
UK £8.99

9781119262657
USA $9.99
CAN $9.99
UK £6.99

9781119291336
USA $9.99
CAN $9.99
UK £6.99

9781119233527
USA $9.99
CAN $9.99
UK £6.99

9781119291220
USA $9.99
CAN $9.99
UK £6.99

9781119177302
USA $9.99
CAN $9.99
UK £8.99

Unleash Their Creativity

dummies.com

dummies
A Wiley Brand

FIT AND HEALTHY

Body battleground

Humans are attacked on a daily basis by a host of marauding invaders, for whom the body is an ideal place to feed and reproduce. Ranged against them are the body's defense forces. Any harmful microbe, or pathogen, that breaks through the outer barriers is met with a quick, local response at the site of the infection. If this doesn't work, a second team is called into action.

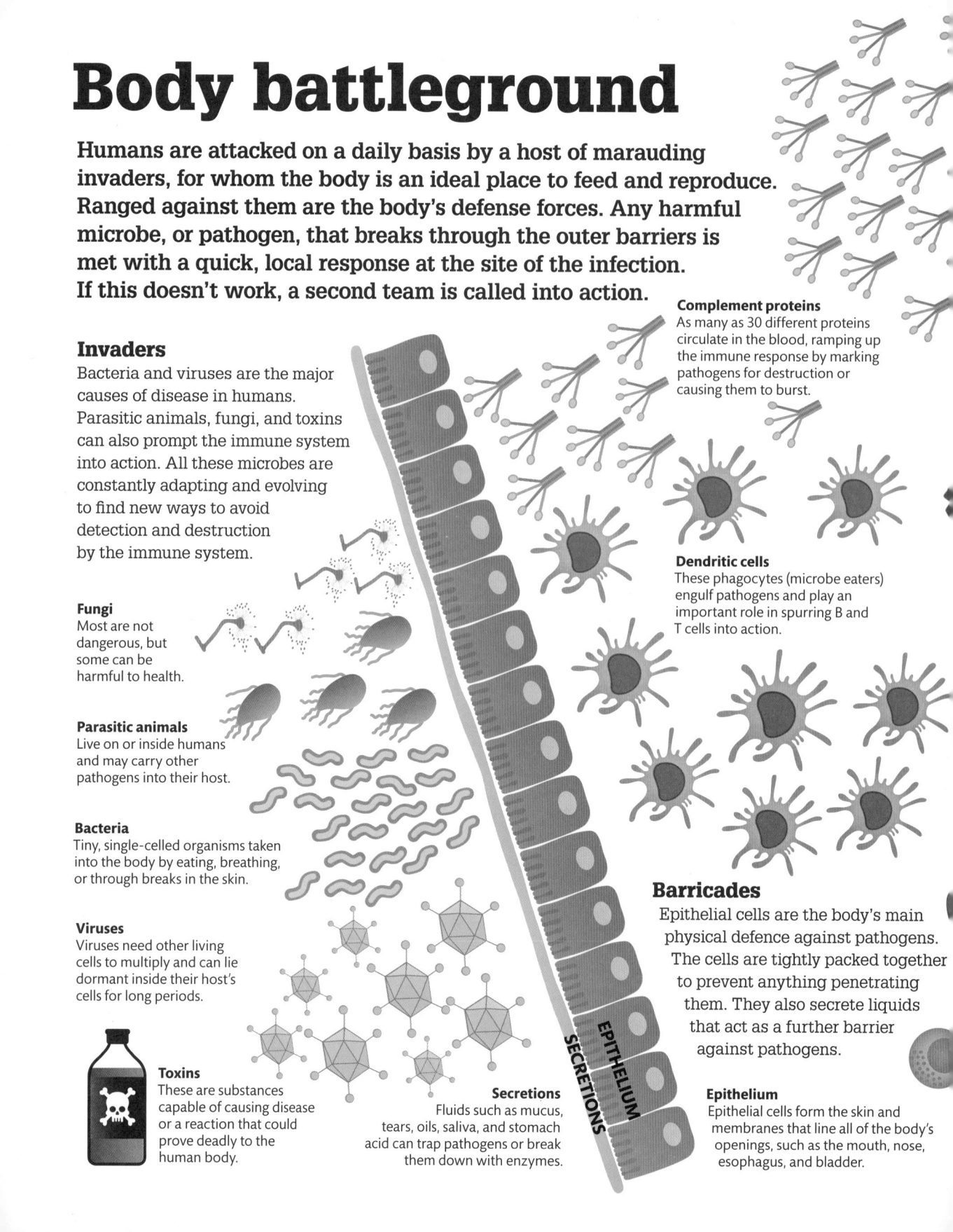

Invaders

Bacteria and viruses are the major causes of disease in humans. Parasitic animals, fungi, and toxins can also prompt the immune system into action. All these microbes are constantly adapting and evolving to find new ways to avoid detection and destruction by the immune system.

Fungi
Most are not dangerous, but some can be harmful to health.

Parasitic animals
Live on or inside humans and may carry other pathogens into their host.

Bacteria
Tiny, single-celled organisms taken into the body by eating, breathing, or through breaks in the skin.

Viruses
Viruses need other living cells to multiply and can lie dormant inside their host's cells for long periods.

Toxins
These are substances capable of causing disease or a reaction that could prove deadly to the human body.

Complement proteins
As many as 30 different proteins circulate in the blood, ramping up the immune response by marking pathogens for destruction or causing them to burst.

Dendritic cells
These phagocytes (microbe eaters) engulf pathogens and play an important role in spurring B and T cells into action.

Barricades

Epithelial cells are the body's main physical defence against pathogens. The cells are tightly packed together to prevent anything penetrating them. They also secrete liquids that act as a further barrier against pathogens.

Secretions
Fluids such as mucus, tears, oils, saliva, and stomach acid can trap pathogens or break them down with enzymes.

Epithelium
Epithelial cells form the skin and membranes that line all of the body's openings, such as the mouth, nose, esophagus, and bladder.

Frontline troops

Pathogens that break through the barriers are met with an immediate response known as the innate immune system. This is a group of cells and proteins that respond to alarm signals from damaged or infection-stressed cells. Some target and mark invading organisms for destruction, while others (phagocytes) eat up the pathogens.

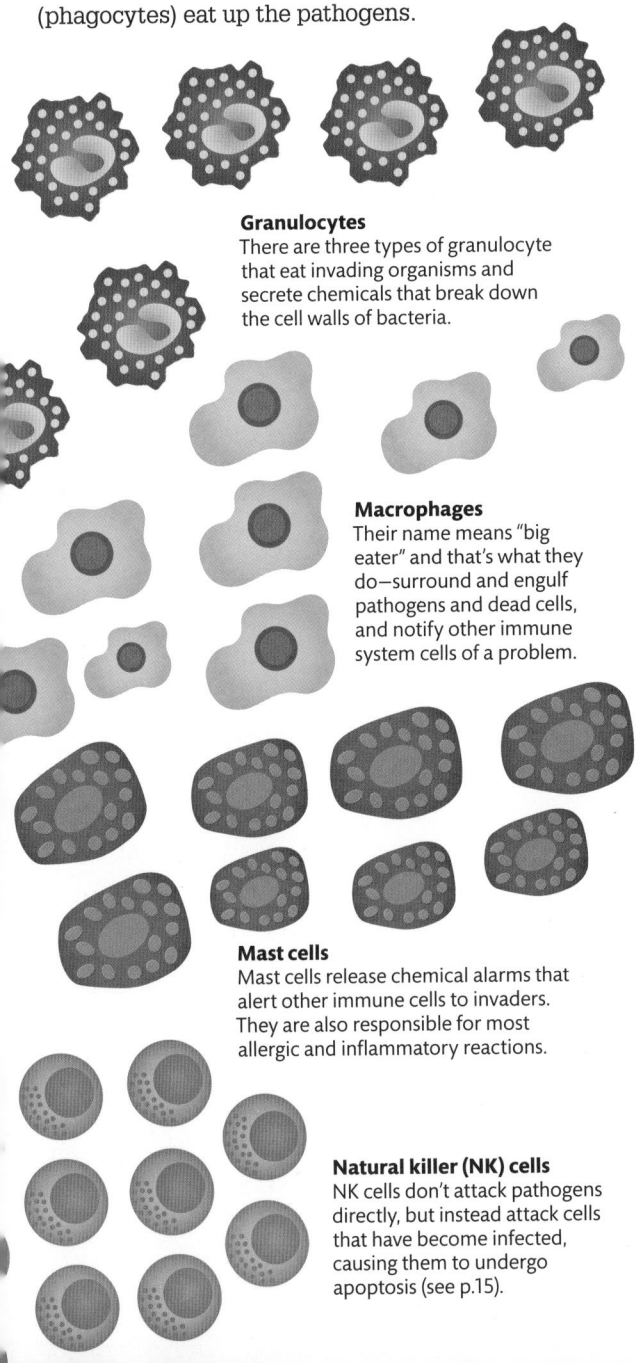

Granulocytes
There are three types of granulocyte that eat invading organisms and secrete chemicals that break down the cell walls of bacteria.

Macrophages
Their name means "big eater" and that's what they do—surround and engulf pathogens and dead cells, and notify other immune system cells of a problem.

Mast cells
Mast cells release chemical alarms that alert other immune cells to invaders. They are also responsible for most allergic and inflammatory reactions.

Natural killer (NK) cells
NK cells don't attack pathogens directly, but instead attack cells that have become infected, causing them to undergo apoptosis (see p.15).

HOW MANY INFECTIOUS DISEASES CAN THE IMMUNE SYSTEM RESPOND TO?

It is thought that B cells alone can produce enough different antibodies to deal with 1 billion different types of pathogen.

Killer cavalry

If the front-line response hasn't contained the infection within 12 hours, the adaptive immune system swings into action. This system remembers previous exposures to the pathogen to launch a specific, targeted response.

B cells
B cells are a special type of cell that can be trained to produce antibodies in response to the presence of a particular pathogen. They can multiply rapidly to increase the response.

Antibodies
Antibodies are Y-shaped proteins produced by B cells. They stick to the surface of invaders and mark them out for destruction by phagocytes.

T cells
T cells are another type of trainable cell that directly attack infected or cancerous cells and prompt phagocytes to eat pathogens. Some T cells also stimulate B cells to produce antibodies.

Friend or foe?

The immune system has to distinguish the harmful pathogens that invade our body from the body's own cells and friendly microbes—in other words, recognize friends and foes. The body puts its most potent immune cells—B and T cells—through safety checks to prevent them from attacking us.

Self and nonself

Every cell in the body is coated in groups of molecules that are unique to each individual. The main function of these molecules is to display fragments of protein made by the body and friendly microbes so that the immune system learns to tolerate them and recognize them as "self."

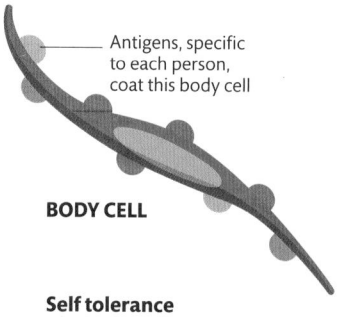

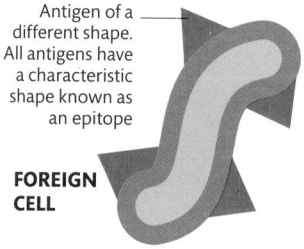

Self tolerance
All body cells carry "self" surface marker proteins, or antigens, allowing them to live in harmony with other cells. If the immune system loses its ability to recognize self markers, it can lead to autoimmune diseases.

Nonself markers
Foreign cells carry their own surface marker proteins, which trigger an immune response. Even the proteins you eat may be identified as foreign unless they are broken down first by the digestive system.

TRANSPLANTS

Compatibility is examined before an organ transplant is given, because if it is not a close enough match the recipient's immune system may attack the donated tissue and start to destroy it. Transplant recipients may have to take immunosuppressant drugs to try to minimize this complication.

Starting point
Both B cells (which produce antibodies to kill invaders, see pp.178–79) and T cells (which kill invaders directly, see pp.180–81) start life as stem cells in the bone marrow.

1 Bone marrow
B cells mature and are tested in the bone marrow. Any that bond with self proteins in the marrow are deactivated and killed by apoptosis (see p.15).

BONE

B cell receptor

B CELL

2 B cell
If a B cell passes the self test, it is released from the bone marrow into the lymphatic system. This is a network of vessels that runs parallel to blood vessels and carries immune cells around the body.

ONLY 2 PERCENT OF T CELLS PASS THEIR TRAINING—THE REST ARE REJECTED BECAUSE THEY MIGHT ATTACK US!

DO IDENTICAL TWINS HAVE THE SAME IMMUNE SYSTEM?

No. Immunity is shaped by what each person is exposed to in life, so it is very individual.

Tested to destruction

When the T cells and B cells of the immune system are forming, they generate random receptors and put them on their surface. Because this process is random, it is possible that these receptors might bind strongly with self," or friendly, antigens. Therefore, these cells go through vigorous testing before being released into the body. Those that bind to the body's own proteins are destroyed.

1 Thymus
T cells move to the thymus (a specialized lymph gland found in front of the heart), where they mature. Their receptors are tested to make sure they don't form strong bonds with self proteins.

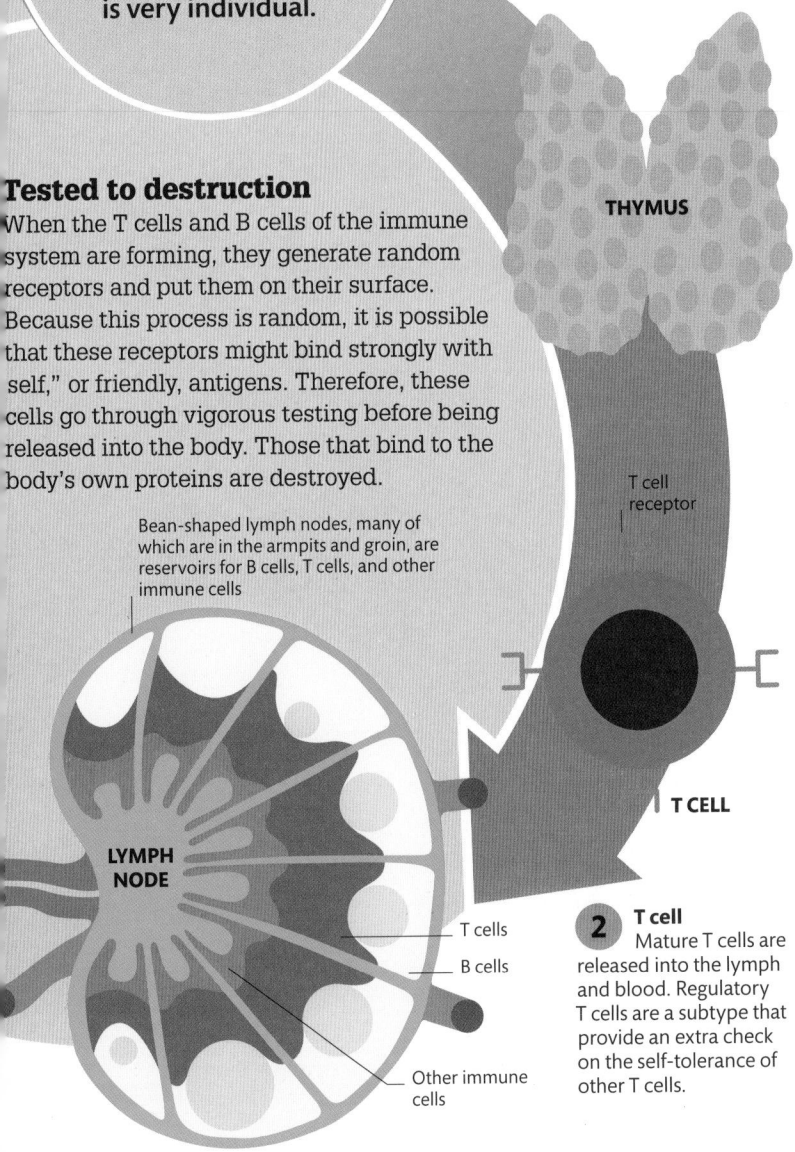

2 T cell
Mature T cells are released into the lymph and blood. Regulatory T cells are a subtype that provide an extra check on the self-tolerance of other T cells.

Destination
If invaders are present in the body's circulation, eventually, they have to pass the lymph nodes, where B cells and T cells lie in wait. The cells activate when they encounter an alien antigen that matches their receptors.

Compatibility

Compatibility tests look at the likelihood of a recipient's immune system attacking donated tissue. Red blood cells carry extra self markers called blood groups. Two of them, the ABO and Rhesus groups, prompt an immune reaction to donated blood from a different group. People with blood group O, for example, will launch a response to blood from any other group because they carry both anti-A and anti-B antibodies.

Blood group A
The red blood cells display A antigens on their surface and antibodies to B antigens are found in the blood plasma.

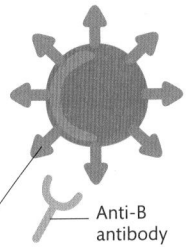

Blood group B
The red blood cells display B antigens on their surface and the plasma has antibodies to A antigens.

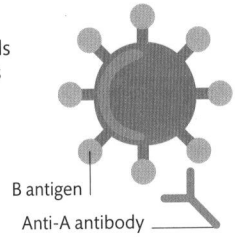

Blood group AB
The red blood cells display both A and B antigens on their surface, but there are no antibodies in the blood plasma.

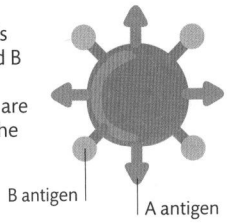

Blood group O
The red blood cells display neither A nor B antigens on their surface, but the blood plasma carries both types of antibody.

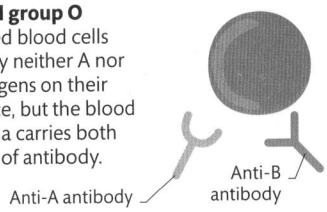

Germs are us

The microbes that live peacefully in and on our body are a big part of staying healthy. These microbes—mostly bacteria and fungi—have benefits that range from keeping our skin healthy by eating dead cells to helping us digest food.

Your local neighborhood

Just as towns may be built around a particular resource, microbes collect around specific areas of the body. On the skin, for example, they are most abundant around sweat glands and hair follicles, where they are more likely to find the nutrients they need to survive. The conditions in each area of the body—moist, dry, acidic—also determines which species can live there. Skin has the greatest diversity of microbes. Those on the oily back are different to those on the drier front.

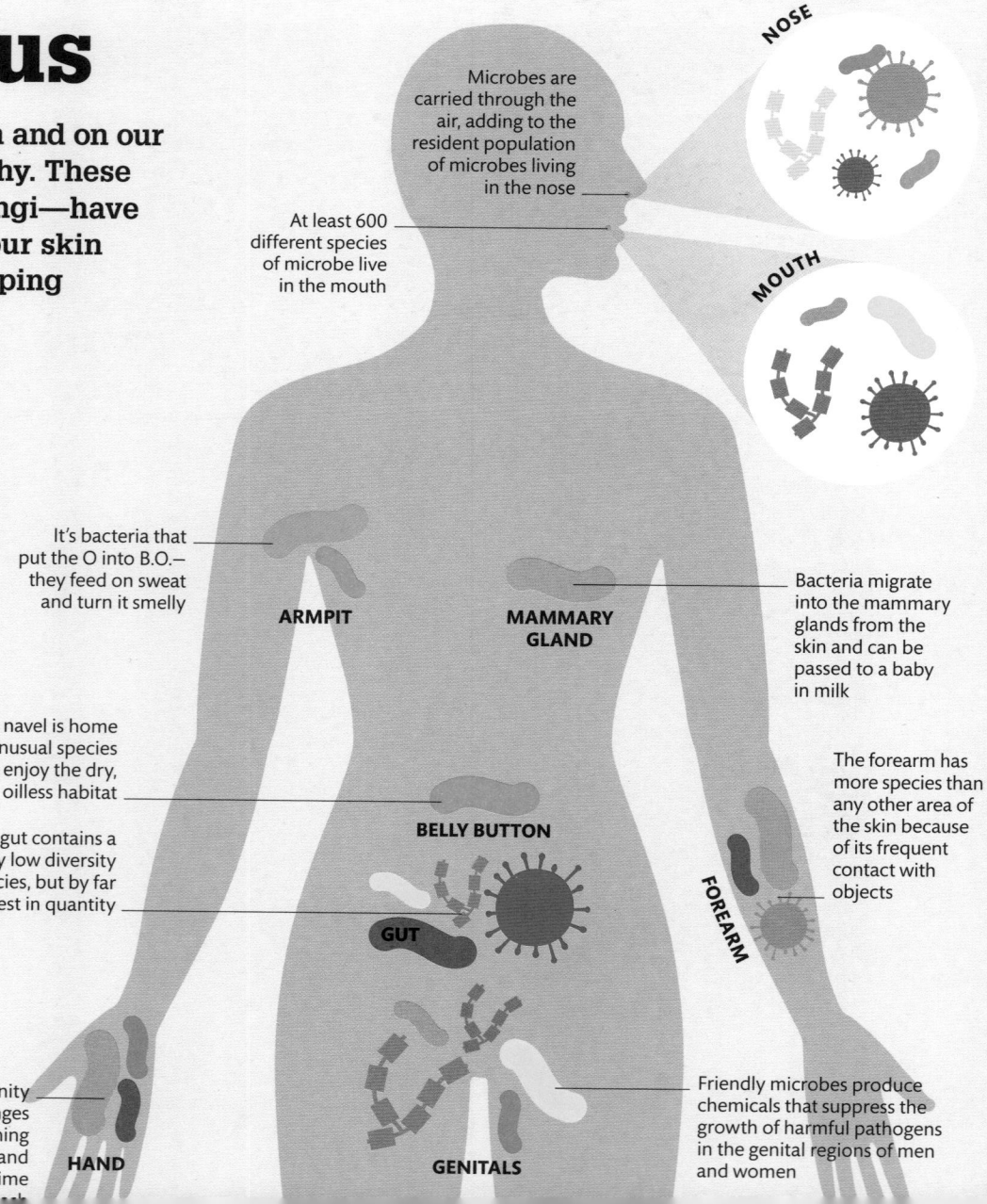

AM I A HABITAT FOR RARE WILDLIFE?

Quite possibly. In a study of 90 belly buttons, researchers found 1,400 species of bacteria that had never been found on human bodies before, some of them new to science.

What's living where
The graphic shows the main types of organism found in or on regions of the body. Large icons indicate species that comprise more than 50 percent of the population.

Bacteria
- Bacteriodetes
- Proteobacteria
- Staphylococcaceae
- Firmicutes
- Corynebacteria
- Actinobacteria

Fungi
- *Malassezia*
- *Candida*
- *Aspergillus*
- Other fungi

Viruses
- Living on bacteria
- Living in our cells

The skin hosts large quantities of microbes, but most are harmless

SKIN

Naturally moist hotspots are dominated by species that thrive in warm, wet conditions

BACK OF KNEE

SOLES OF FEET

Feet are dominated by fungi—around 100 species thrive in their cool and damp environment

Beneficial microbes

Science is still revealing the different species that live within the human microbiome, let alone their many benefits. Some benefits are direct, such as eating dead skin and changing the chemical environment to prevent harmful microbes from growing. Others are less obvious, such as the calming effect some gut bacteria have on the immune system by reducing inflammation. Medicines, such as antibiotics, can also have devastating effects, wiping out the good microbes as well as the bad.

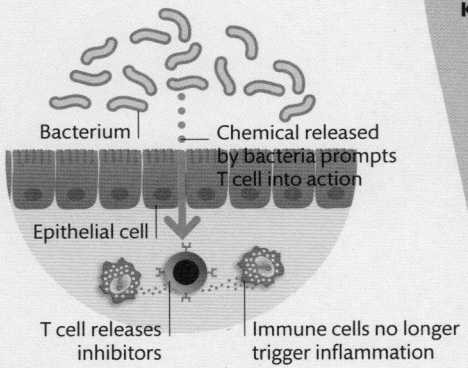

Happy bacteria = healthy gut
Eating the right foods helps good bacteria to thrive. They produce chemicals that damp down inflammation in the gut, which would allow bad bacteria to penetrate the epithelial wall.

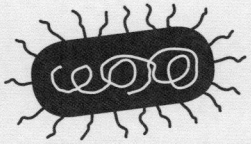

MICROBIAL CELLS OUTNUMBER HUMAN CELLS BY 10 TO 1

ARE WE TOO CLEAN?

It's possible that our obsession with antibacterial cleansers is taking its toll on friendly microbes. Some studies have shown that excessive handwashing can lead to the growth of more harmful microbes—but this is debatable, since other studies have shown the opposite.

Birthday presents

Babies start to build their own microbiome at birth by picking up some of their mother's microbes as they pass through the birth canal. These bacteria start to produce chemicals that encourage other beneficial microbes to colonize. Many factors can influence the development of the microbiome; different species will colonize depending on how the baby is delivered (cesarian babies have different bacteria), whether a baby is breast fed, and who it has contact with.

Damage limitation

When a physical barrier such as skin is damaged, the immune system works quickly to repair it and defend the body against infection. The local immune cells swing into action against the first invaders, calling for more specialist reinforcements if there are more than they can cope with.

THERE ARE **375,000** IMMUNE CELLS IN **EVERY DROP OF BLOOD**

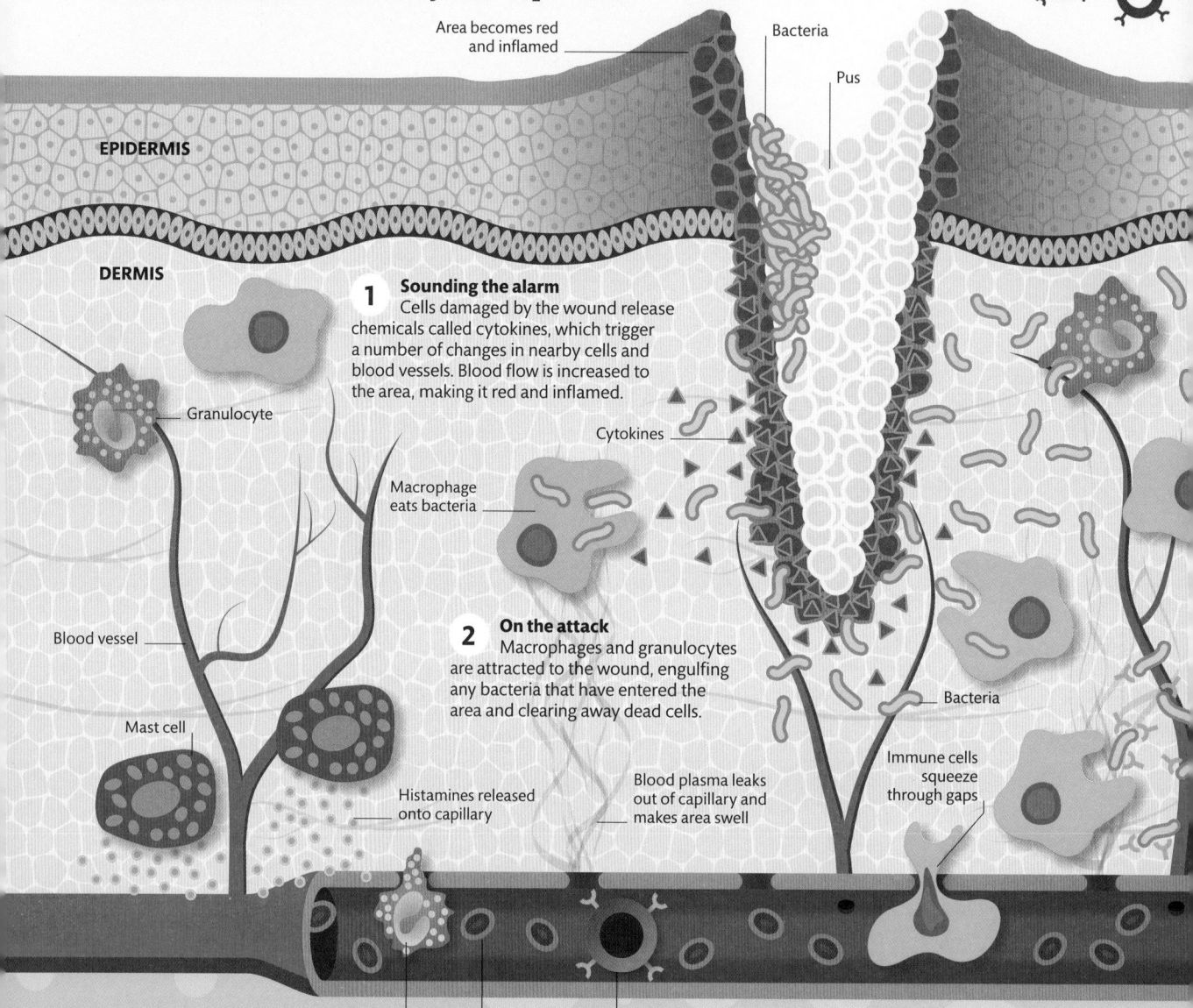

1 Sounding the alarm
Cells damaged by the wound release chemicals called cytokines, which trigger a number of changes in nearby cells and blood vessels. Blood flow is increased to the area, making it red and inflamed.

2 On the attack
Macrophages and granulocytes are attracted to the wound, engulfing any bacteria that have entered the area and clearing away dead cells.

3 Opening the walls
The chemicals released by the damaged cells and local immune cells make the capillary walls more permeable, which allows immune cells in the blood to pass through more easily.

Call to arms

A number of immune cells, such as macrophages, mast cells, and granulocytes, live in the dermis. If the skin is cut, mast cells detect the injured cells and release histamines that cause nearby blood vessels to swell. This increases blood flow to the area, making the wound feel hot, but it also brings other immune cells to the site quickly. The formation of pus is an indication that bacteria have gotten into the wound—pus is the accumulated remains of dead immune cells.

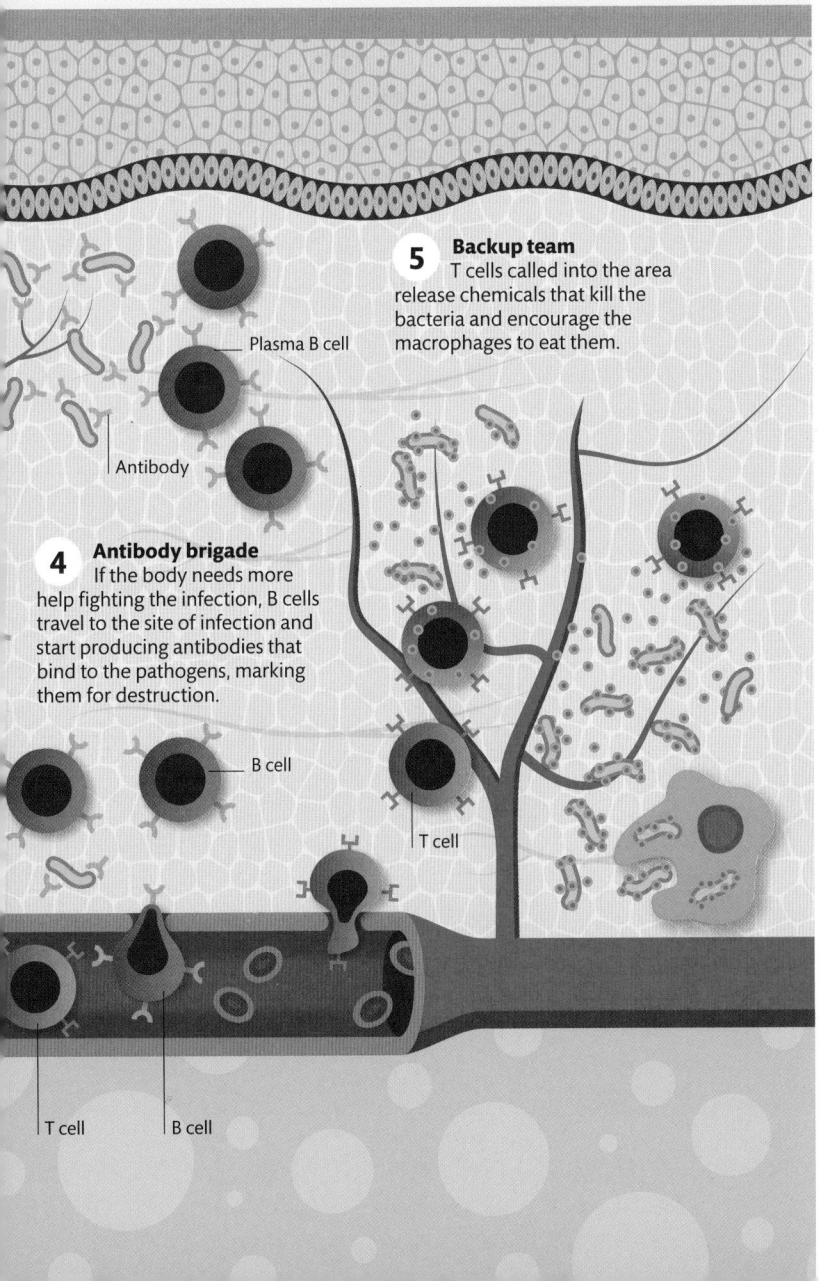

WHY DO CUTS TAKE LONGER TO HEAL WHEN WE'RE OLDER?

Blood vessels can become more fragile as you get older, which makes it more difficult to deliver immune cells to the wound.

MAGGOT THERAPY

If a wound in the skin isn't healing properly or responding to conventional treatment, maggots may be the answer. These little fly larvae are particularly precise in digesting dead cells while leaving the healthy cells alone. As they eat, the maggots secrete antimicrobial chemicals that protect the maggot but which are also effective at killing bacteria, even those resistant to antibiotics. These secretions also help inhibit inflammation of the wound, contributing to the healing process.

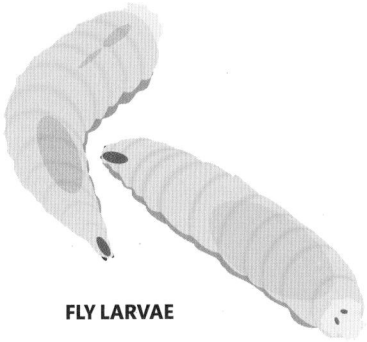

FLY LARVAE

Bacteria
Bacteria are microscopic organisms that are usually harmless, but can sometimes cause disease. Bacteria are responsible for some globally important diseases, such as tuberculosis and pneumonia.

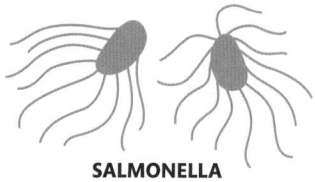

SALMONELLA
(food poisoning)

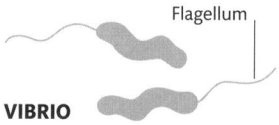

VIBRIO
(cholera)

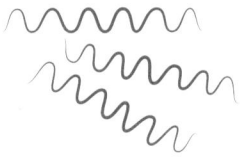

TREPONEMA
(syphilis, yaws)

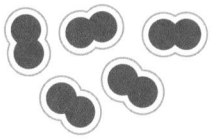

STREPTOCOCCUS
(pneumonia, bronchitis)

Viruses
Viruses are the smallest and simplest organisms of all, made up of only their genetic material (DNA or RNA) in a protein coat. Unlike other pathogens, viruses need the host's cells to live and replicate.

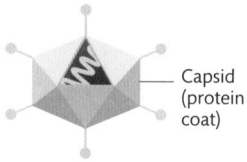

ADENOVIRUS
(tonsilitis, conjunctivitis)

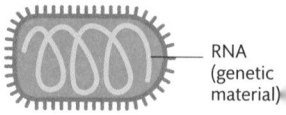

LYSSAVIRUS
(rabies)

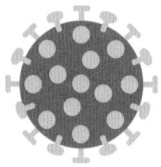

LENTIVIRUS
(HIV/AIDS)

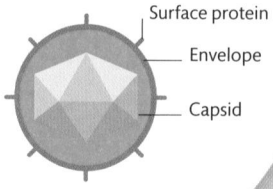

HERPESVIRUS
(hepatitis B, cold sores)

Antibiotics
Commonly used for bacterial infections, antibiotics break down the walls of bacteria or interrupt their growth. However, they can't distinguish the good bacteria from the bad.

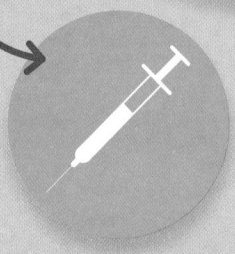

Vaccination
The best way of preventing the spread of viral infections is through vaccination. A vaccine primes the immune system to recognize the virus and launch an immediate attack (see pp.184–85).

Infectious diseases

Bacteria, viruses, parasites, and fungi live in and on us all the time. Most are harmless, but certain species are pathogens—they can cause an illness if a change in conditions allows them to thrive. Other diseases are passed to us from people or animals. A fever is almost always a sign that an infection is taking hold.

Unwanted visitors

Organisms that live off the body's cells or tissues are called parasites. There are five main types: bacteria, viruses, fungi, and animals and protozoans. When they find favorable conditions they multiply rapidly but may produce harmful products or effects that make us feel sick, prompting our immune system to swing into action.

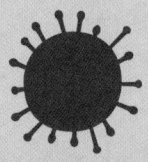

A SINGLE SNEEZE CONTAINS 100,000 GERMS

Animals and protozoans
We also face attacks from tiny animals and single-celled organisms called protozoans that live on or inside the body. Some are large enough to see with the naked eye, such as worms, or they may be microscopic, such as Giardia, the protozoan that causes diarrhea.

NEMATODE
(Guinea worm, threadworm)

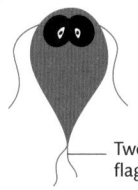

GIARDIA
(diarrhea)

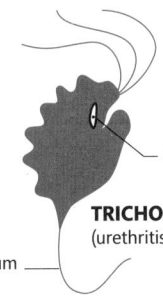

TRICHOMONAS
(urethritis, vaginitis)

Fungi
Fungi are always present in and on the body, but sometimes pathogenic species take hold and cause diseases such as athlete's foot or thrush.

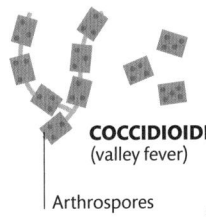

COCCIDIOIDES
(valley fever)

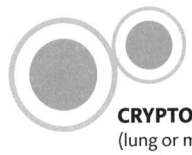

CRYPTOCOCCUS
(lung or meningial cryptococcosis)

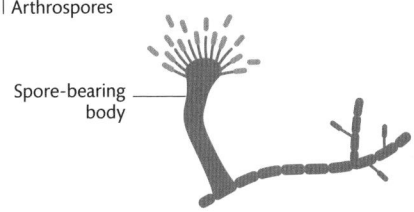

ASPERGILLUS
(lung infections)

Prevention
The best strategy against this type of infection is to avoid activities and areas where there are known health hazards, be wary of unsafe food and water sources, and take recommended precautionary drugs.

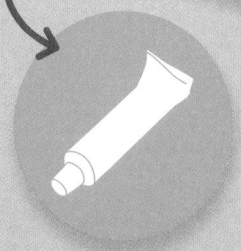

Antifungal medications
Fungal infections are treated according to whether they are internal or external. The active ingredients either attack the fungus directly by breaking down its cell walls, or prevent it from growing.

How diseases spread

There are many infectious diseases but some affect relatively few individuals and are local to a small area—only diseases that spread easily by person-to-person contact are said to be contagious. Many pathogens travel between people by less direct means—through the air or in water, on objects someone has touched, or in contaminated food. Zoonotic diseases are animal infections that can spread to humans, usually through bites.

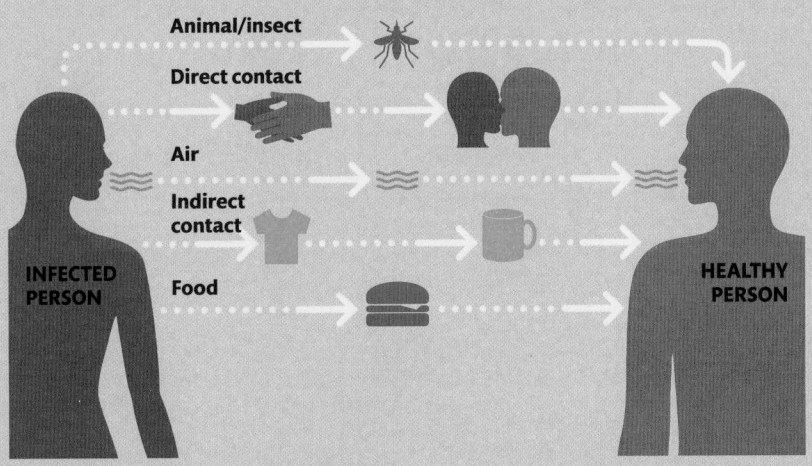

Looking for trouble

If an infection becomes too great for the initial immune system to deal with, a second, more targeted force springs into action. B cells learn to recognize harmful microbes that have attacked the body in the past. They can then produce antibodies that will surround the pathogen and tag it for destruction by other immune cells.

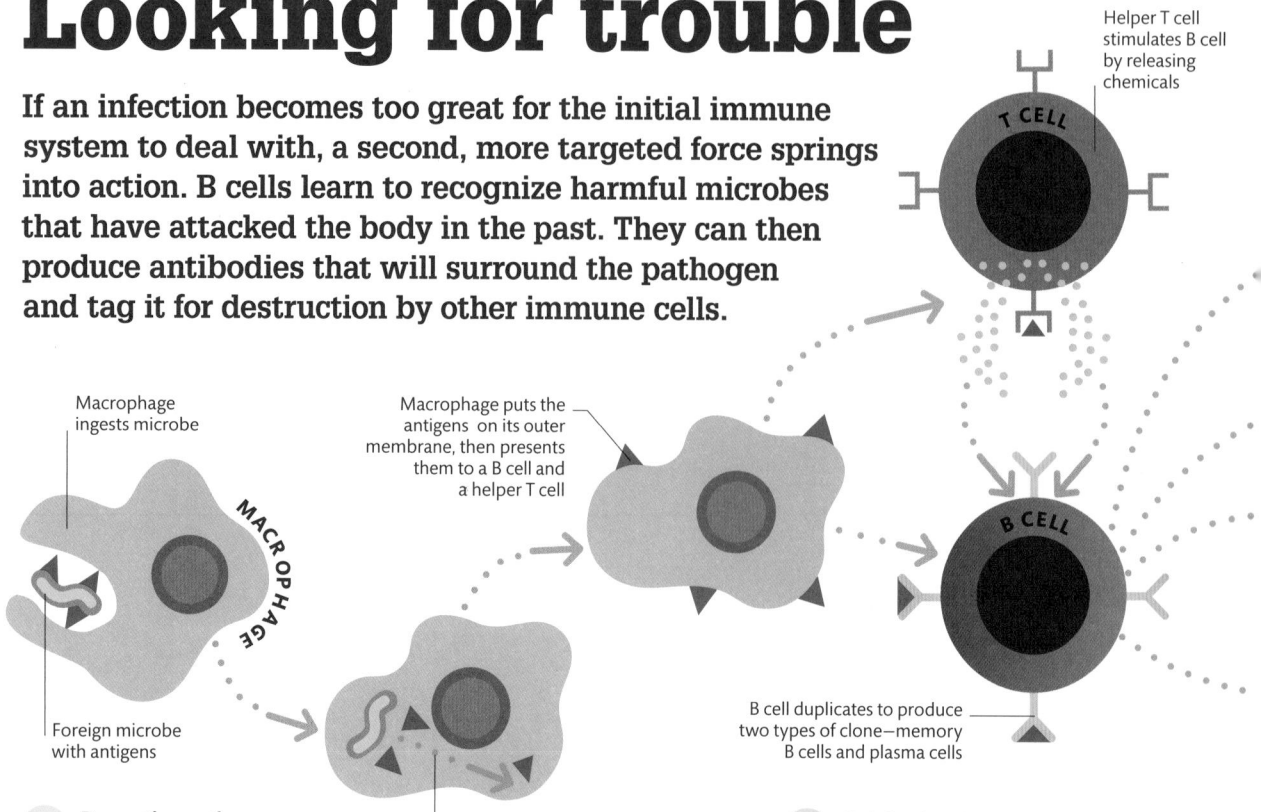

1 Presenting antigens
When a macrophage ingests a pathogenic microbe, it breaks it up and puts the microbe's antigens (surface proteins) onto its cell wall. This is known as an antigen-presenting cell.

2 Helping hand
The B cell starts to get ready when it binds to an antigen, but it isn't fully activated until a helper T cell recognizes and binds to that same antigen. The helper cell then releases chemicals that prompt the B cell to produce antibodies.

Activating antibodies

B cells are a type of white blood cell that constantly patrol the blood vessels or lie waiting in the lymph nodes (see pp.170–171). When a B cell encounters an antigen it recognizes, it becomes primed and ready to clone itself. This can happen only when another cell of the immune system, the helper T cell, recognizes and binds to that same antigen, triggering the B cell to clone itself and release antibodies.

A SINGLE **B CELL** MAY HAVE **UP TO 100,000 ANTIBODIES** ON ITS OUTER SURFACE

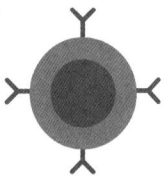

TESTING FOR ANTIBODIES

Blood tests show the levels of immunoglobulins (another name for antibodies) present during infections. IgM is a large antibody that the body produces at the first sign of infection, but it quickly disappears. IgG is a more specific, lifelong antibody that is produced during a later infection. A high IgM value shows you have a current infection, whereas IgG simply means you have been infected by a pathogen in the past.

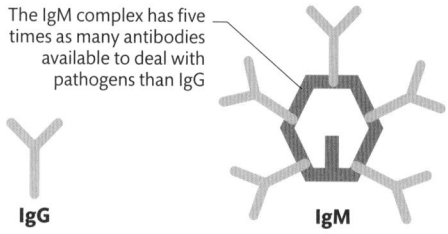

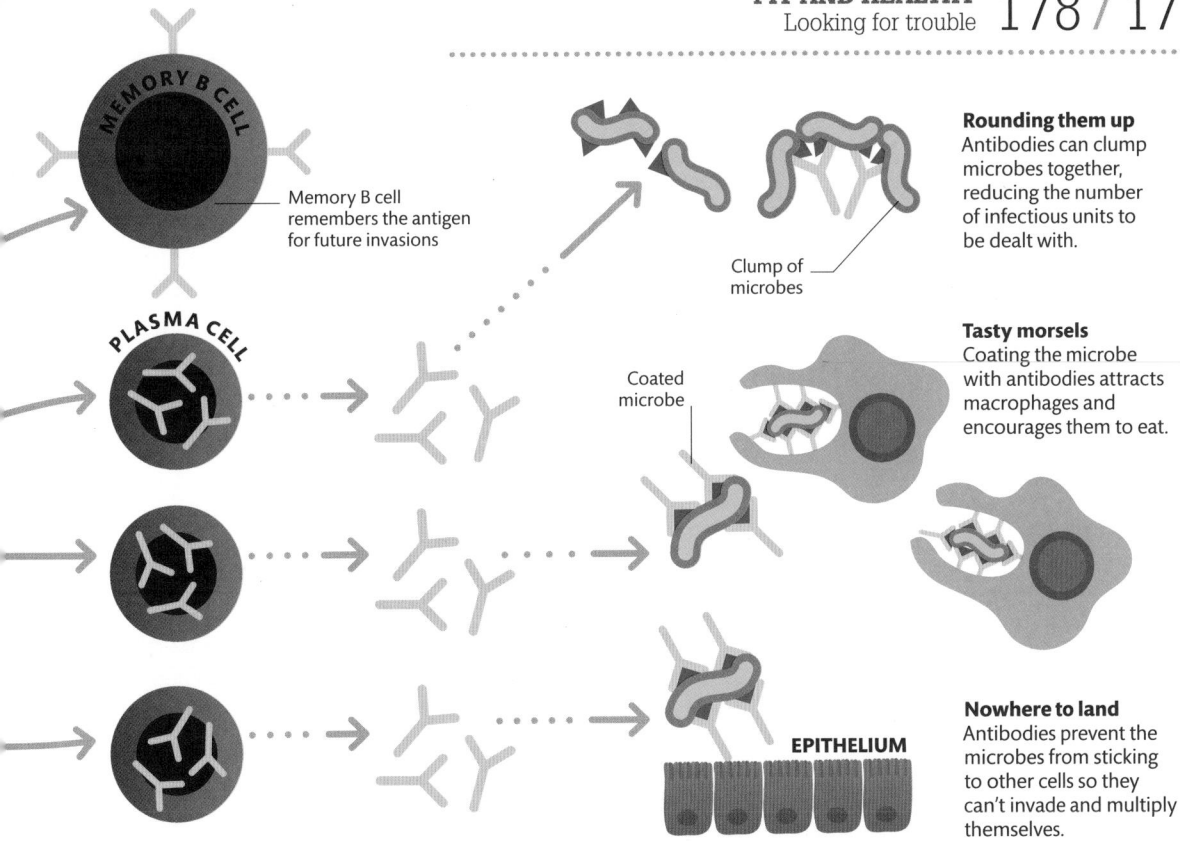

Rounding them up
Antibodies can clump microbes together, reducing the number of infectious units to be dealt with.

Tasty morsels
Coating the microbe with antibodies attracts macrophages and encourages them to eat.

Nowhere to land
Antibodies prevent the microbes from sticking to other cells so they can't invade and multiply themselves.

3 Antibody release
The B cell clones itself. Some of these clones become memory cells, but most become plasma cells, which produce antibodies that are specific to the invader's antigens. These antibodies are then released into the blood.

4 Neutralizing pathogens
The antibodies bind to the invading microbes, neutralizing them and marking them for destruction by other immune cells.

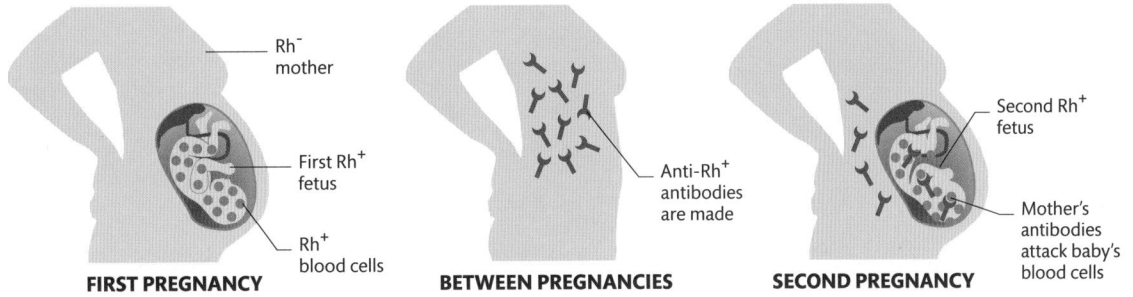

Rhesus babies

The Rhesus factor (Rh) is a protein on the surface of red blood cells—people who have it are called Rh+. When an Rh− mother is exposed to the blood of her Rh+ fetus (from the father's Rh+ gene) during birth, she makes antibodies against it. These antibodies may attack future Rh+ embryos, but an injection of anti-Rh+ antibodies early in the pregnancy usually reduces this danger.

Not-so-safe haven
Antibodies produced in response to the baby's blood mingling with the mother's during birth will prompt her immune system to attack the next Rh+ child she conceives. This is because her antibodies can actually cross the placenta into the baby's blood.

Assassination squad

The immune system can prime some cells to go out into the body and attack the invasion one-on-one. These are known as T cells. They hunt down infected and abnormal cells, then destroy them.

Keeping control

T cells are a type of white blood cell that play a key role in dealing with infections. Circulating in the blood and lymph, the T cells look for foreign antigens on the surface of body cells. These characteristic proteins show that the cells have been invaded by a microbe or that they have developed a dangerous abnormality. T cells also marshal the actions of other immune cells and prime B cells to produce antibodies.

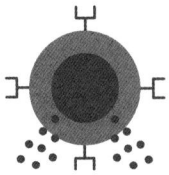

REGULATOR T CELLS ARE VITAL IN PREVENTING AUTOIMMUNE DISEASES

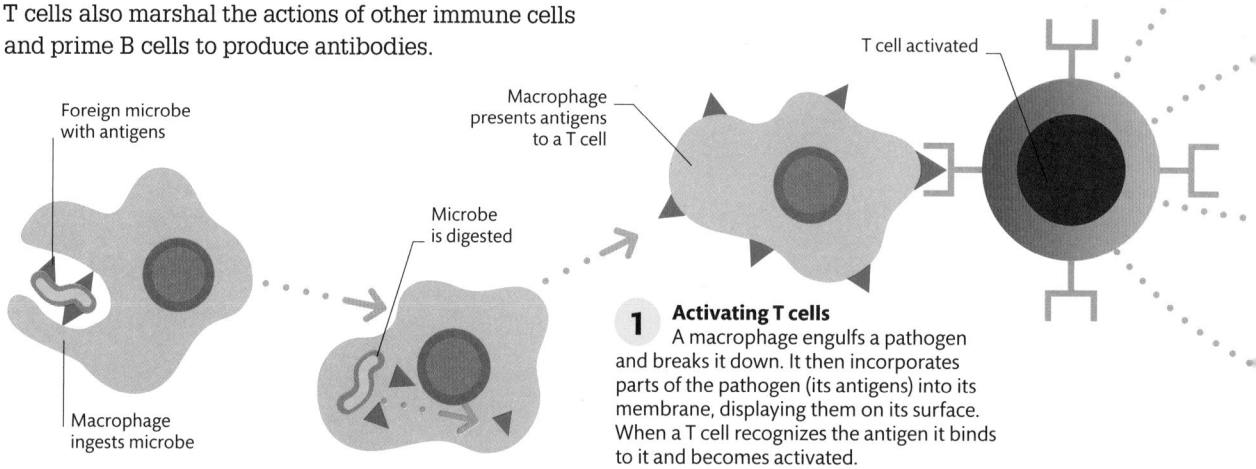

1 Activating T cells
A macrophage engulfs a pathogen and breaks it down. It then incorporates parts of the pathogen (its antigens) into its membrane, displaying them on its surface. When a T cell recognizes the antigen it binds to it and becomes activated.

Cornering cancer

Immunotherapy is a treatment designed to help the immune system fight cancer. There are many different ways of doing this. All of them either make the cancer cells more easily identified by the immune system or boost the immune system by multiplying cells or cytokines in the lab before injecting them back into the patient.

Cancer vaccines
Vaccines form one of the methods of immunotherapy being developed. They prompt the immune system to target only cancerous cells.

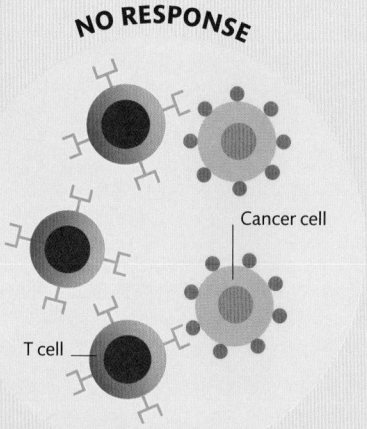

1 No threat
Cancer is the uncontrolled division of abnormal cells. The immune system may not recognize these cells as abnormal because they are the body's own cells.

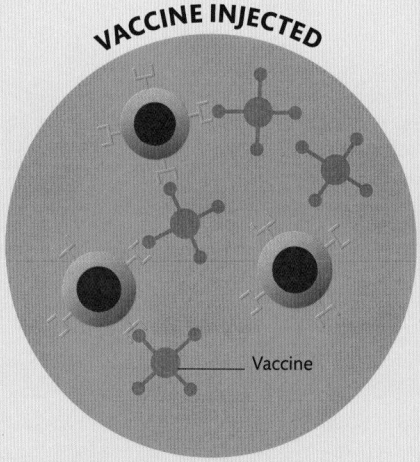

2 Identifying the adversary
Cancerous cells have "self" antigens on their surface but also produce their own antigens. A vaccine is designed to match the shape of the cancer antigen.

2 T cells in action
Once the T cell has been activated it begins to clone itself. These clones then become one of four different types of cells in the T cell family.

Killers
These are the muscle of the family, actively seeking and killing infected cells.

Helpers
Spur the B cells and other T cells into action. Also encourage other immune cells to eat invaders.

Regulators
Calm the immune system and are critical in recognizing friendly microbes.

Memory
Remember microbes that have attacked the body in the past.

Killer T cell releases cytokines

T cell moves away to seek another target

Pores open and cell starts to swell

INFECTED CELL

CELL BURSTS

3 Killer T cells
The killer T cell recognizes and binds to an antigen displayed by an infected cell. It releases chemicals that open up pores in the membrane of the infected cell and cause it to break down and be eaten by macrophages.

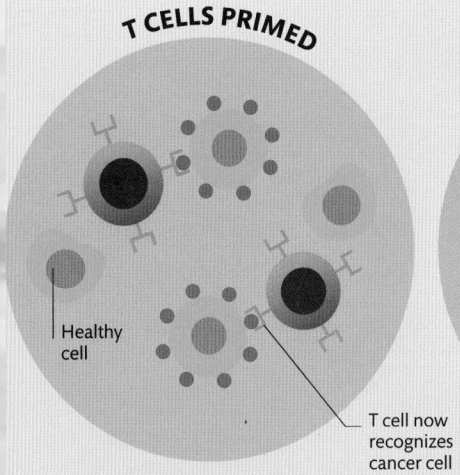

3 Trained to kill
The vaccine trains the T cells to recognize and bind to the antigens being displayed by the cancerous cells in the body.

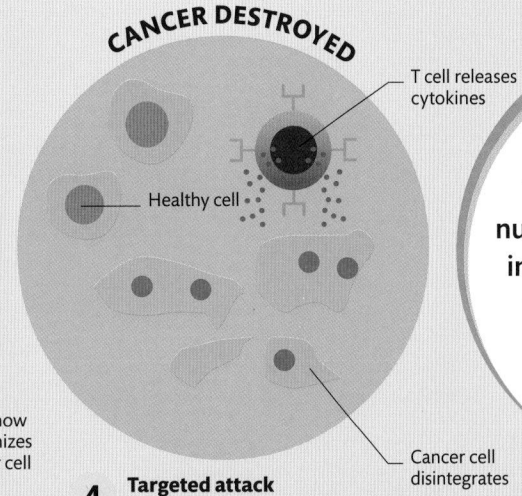

4 Targeted attack
The T cells can now target and attack the cancerous cells, distinguishing them from healthy cells of the same type.

WHAT IS A T-CELL COUNT?

This is a measure of the number of T cells circulating in your blood. Both higher and lower than normal T-cell counts can be an indicator of disease.

Colds and flu

The reason why you are assailed by colds again and again is because the virus mutates each time, and your immune system fails to recognize it when you catch your next cold. Usually, the symptoms you experience are your immune system reacting to the virus, not directly caused by the virus itself.

Cold or flu?

Many of the symptoms of colds and flu are similar, and that makes them hard to differentiate. There are many viruses that cause the common cold, and the influenza virus is caused by three virus subtypes. Generally, the symptoms of a cold are much milder than those of the flu.

Common cold
Frequent sneezing, a mild to moderate fever, low energy, and fatigue are all symptoms of the common cold. There are more than 100 viruses responsible for the common cold, which can be caught at any time of year.

Shared symptoms
Both the common cold and flu are classed as upper respiratory tract infections. Either illness may cause a runny nose, sore throat, cough, headache, an aching body, shaking, and chills.

Flu
Influenza is caused by virus types A, B, and C. Having the flu may induce a moderate to high fever and constant fatigue. It is generally caught in the winter months and can develop into more serious conditions such as pneumonia.

How a virus invades a cell

Viruses need to invade healthy cells to replicate. A virus tricks the cell into making copies of it. A cell's nucleus is where instructions to make body proteins are stored. Viruses are surrounded by a coat of protein, and the virus can hijack cells to make these viral proteins instead of normal body proteins. Once they have replicated, the virus will then enter other cells in your body and the cycle continues. This process is the same for both the common cold and the flu.

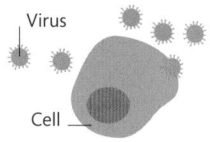

1 The virus attaches itself to your cell and the cell engulfs the virus.

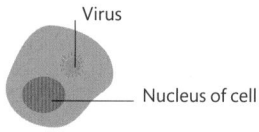

2 Substances in the cell begin to strip the virus's outer coat of protein.

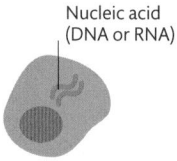

3 Nucleic acid from the virus is released, ready to be replicated.

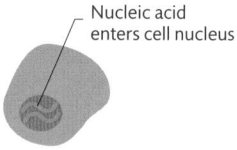

4 Your cell replicates the viral nucleic acid under the false pretense that it is your own DNA.

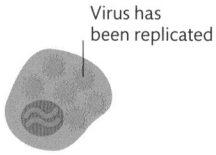

5 The cell ignores its own chemical needs and switches to making new viral nucleic acids, which become copies of the virus.

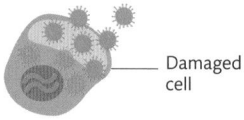

6 The virus is released from the host cell. This can destroy the cell, and the viruses go on to invade other cells.

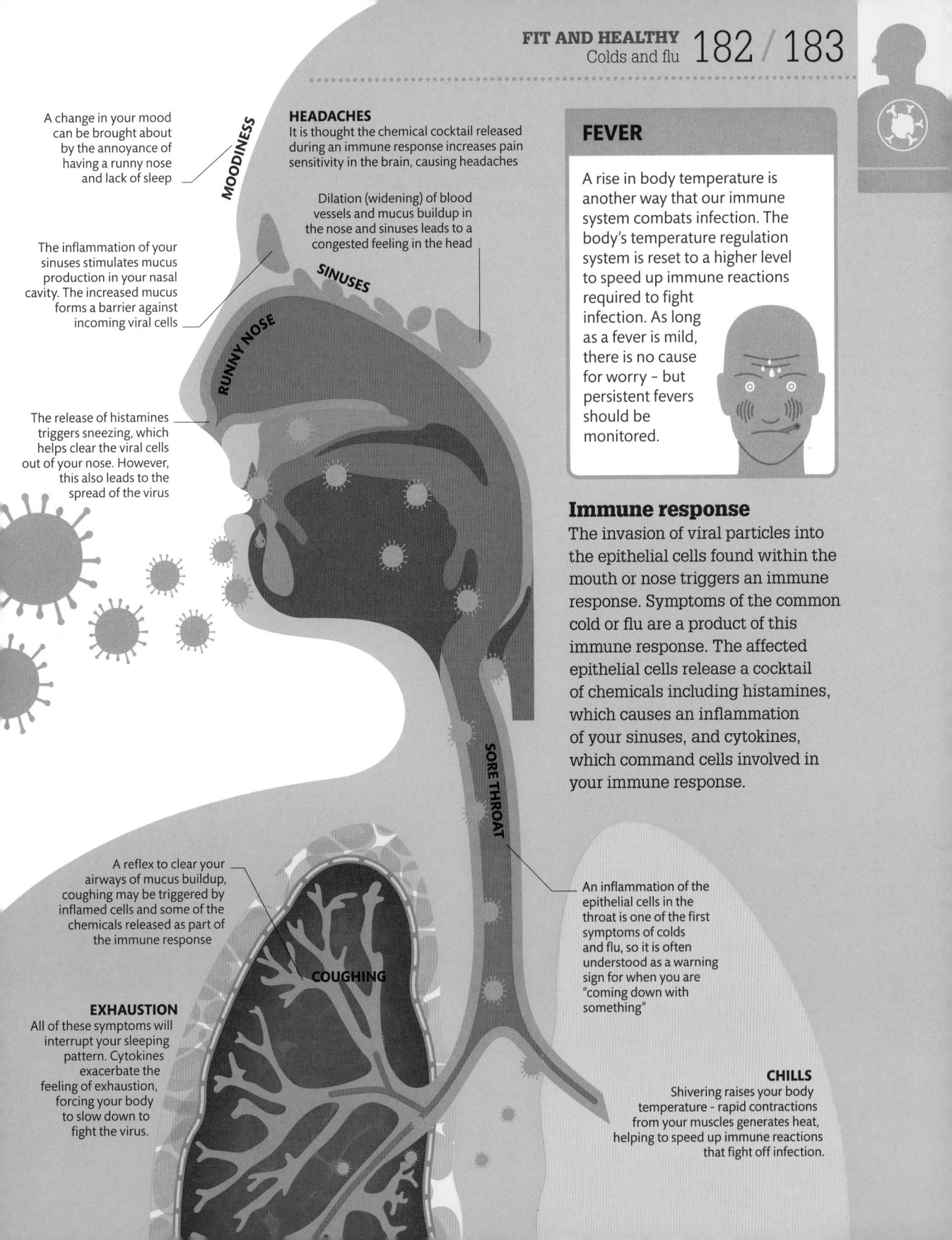

FEVER

A rise in body temperature is another way that our immune system combats infection. The body's temperature regulation system is reset to a higher level to speed up immune reactions required to fight infection. As long as a fever is mild, there is no cause for worry – but persistent fevers should be monitored.

Immune response

The invasion of viral particles into the epithelial cells found within the mouth or nose triggers an immune response. Symptoms of the common cold or flu are a product of this immune response. The affected epithelial cells release a cocktail of chemicals including histamines, which causes an inflammation of your sinuses, and cytokines, which command cells involved in your immune response.

Vaccine action

One of the most effective ways of preventing the spread of infectious disease is to prime the immune system through vaccination. A vaccine trains the immune system to launch a fast and furious attack on a pathogen.

Herd immunity

Vaccinating a significant portion (around 80 percent) of a population can help provide immunity even to those who have not been vaccinated. When the disease is passed to vaccinated individuals, their primed immune system destroys it, preventing it from spreading further. This can help protect people who can't be vaccinated due to age or illness. Widespread vaccination can eliminate diseases entirely, such as smallpox.

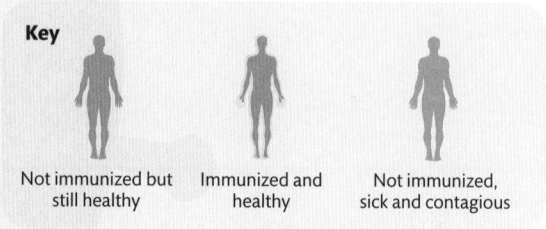

Safety first

Contagious diseases can be contained if a sufficient number of people are vaccinated. Vaccination also helps people who have an existing medical condition that may be worsened by the effects of the disease.

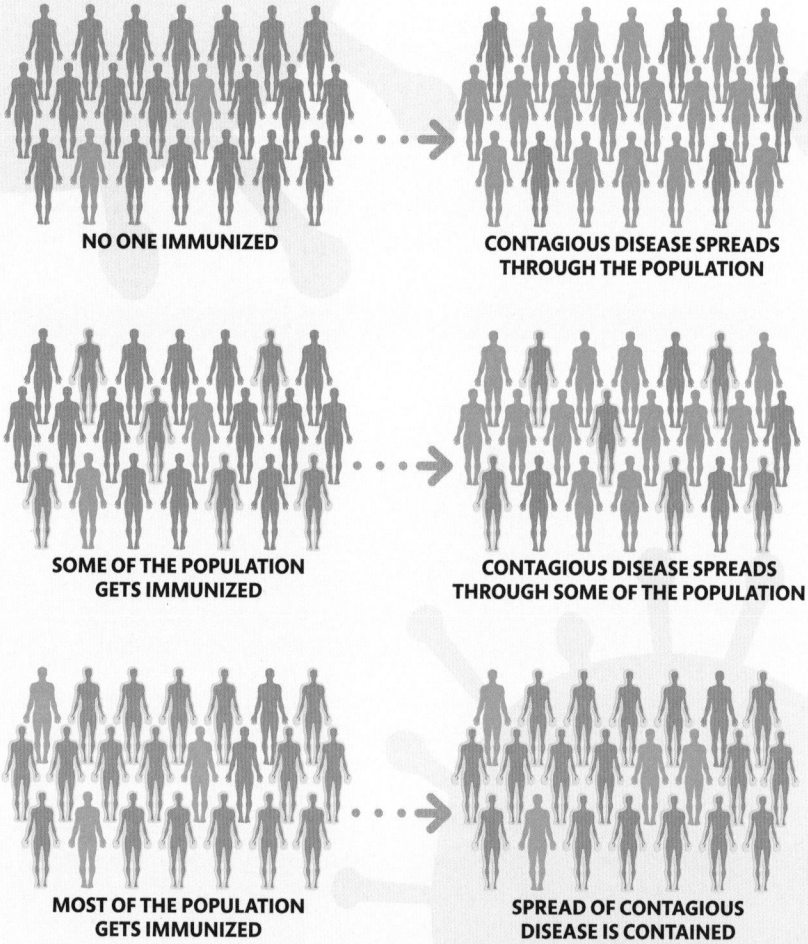

TO VACCINATE OR NOT?

Controversy exists over the use of vaccines. Fears over possible side effects have led some parents to refuse to have their children vaccinated, which has resulted in outbreaks of preventable diseases, such as measles and pertussis. If only a small portion of the population is vaccinated, herd immunity breaks down.

Types of vaccines

Each vaccine is developed for a specific pathogen and is designed to kickstart the immune system. This is done by injecting a harmless version of the pathogen that the immune system will remember if attacked by the real pathogen. This can be difficult—killing the pathogen may make it safe, but the vaccine may not produce an immune response. There are also some diseases that progress too quickly for the immune's memory system to respond in time, so booster immunizations are given to keep reminding the immune system.

WHY DO VACCINES MAKE YOU FEEL ILL?

Vaccinations stimulate an immune response, which can produce symptoms in some people—but it means the vaccine is doing what it's supposed to.

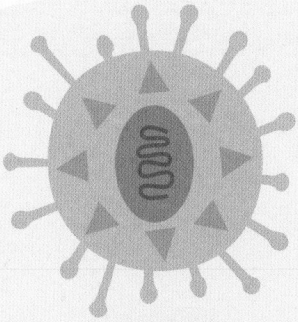

Inactivated
The pathogen is killed using heat, radiation, or chemicals. Used for influenza, cholera, and bubonic plague vaccines.

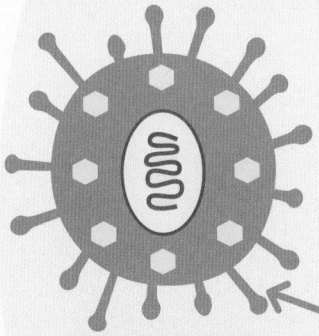

Related microbe
A pathogen that causes disease in another species, but few or no symptoms in humans, is sometimes used. For example, tuberculosis vaccine is made from a bacterium that infects cattle.

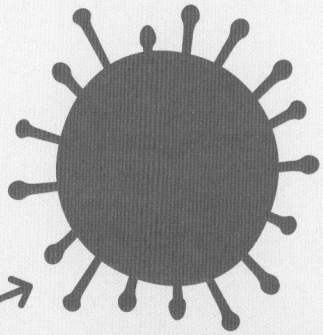

Alive, but not dangerous
The pathogen is kept alive but the parts that make it harmful are removed or disabled. Used for measles, rubella, and mumps vaccines.

ORIGINAL DISEASE-CAUSING PATHOGEN

DNA
DNA from the pathogen is injected into the body, whose own cells take up this DNA and start to produce proteins from the pathogen, which triggers an immune response. Used for Japanese encephalitis vaccine.

Tame toxins
Toxic compounds released by the pathogen, which are responsible for the illness, are deactivated using heat, radiation, or chemicals. Used for tetanus and diphtheria vaccines.

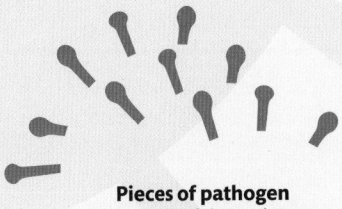

Pieces of pathogen
Fragments of the pathogen, such as proteins on the surface of the cell, are used instead of the whole pathogen. Used for vaccines against hepatitis B and human papilloma virus (HPV).

Immune problems

Sometimes the immune system is too reactive—launching attacks on things that aren't harmful and even attacking the body's own cells. Allergies, hay fever, asthma, and eczema are all caused by an oversensitive immune system. Alternatively, the immune system may not be reactive enough, leaving the body vulnerable to infection.

ANAPHYLACTIC SHOCK

Sometimes the immune system launches an extreme panic attack when it encounters an allergen such as a sting or a nut. Symptoms include itchy eyes or face, followed quickly by extreme swelling in the face, hives, and difficulty swallowing and breathing. This is a medical emergency that needs to be treated with an injection of epinephrine, which constricts blood vessels to reduce swelling and relax the muscles around the airways.

ARE FOOD ALLERGIES AN IMMUNE RESPONSE?

Yes. Similar to hay fever, allergies to certain foods cause an inflammatory response from the mouth to the gut. Severe allergies may result in anaphylaxis.

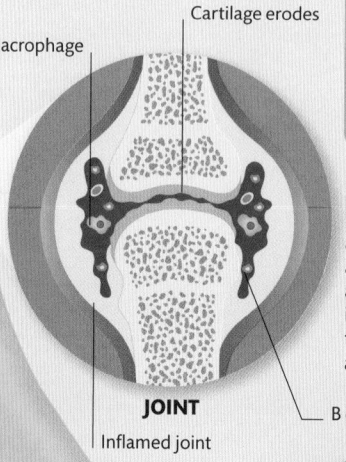

Rheumatoid arthritis
If the immune system attacks cells around a joint, causing an inflammatory response, an autoimmune disease called rheumatoid arthritis can result. The joint swells, gets inflamed, and is very painful. Eventually, there is permanent damage to the joints and surrounding tissues.

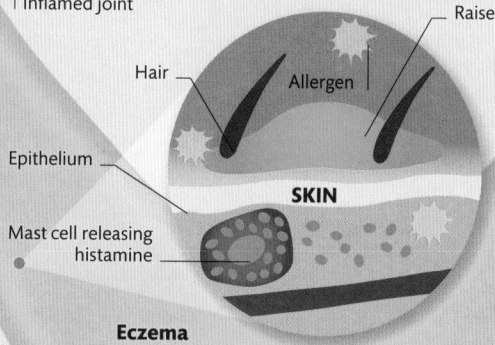

Eczema
The causes of eczema are unclear, but it is thought to be a miscommunication between the immune system and the skin. It is probably triggered by an irritant (allergen) on the skin that stimulates the immune system beneath to launch an inflammatory response, causing swelling and redness.

Immunity overload

Most immune problems are a combination of genetic and environmental factors. While immune conditions are usually triggered by exposure to environmental factors, such as pollen, foods, or irritants on the skin or in the air, some people are genetically more susceptible to developing them. Even autoimmune diseases (when the immune system attacks healthy body tissue by mistake), such as rheumatoid arthritis, can be made worse by irritants that cause inflammation elsewhere in the body. People with a hypersensitive immune system may experience several conditions; for example, many people with asthma also suffer from allergies.

Allergies and our modern lifestyle
More people in developed countries suffer from allergies, and incidences have been rising since World War II. The specific reasons are open to debate, but there is agreement that it is likely to do with the immune system being exposed to fewer microbes during childhood.

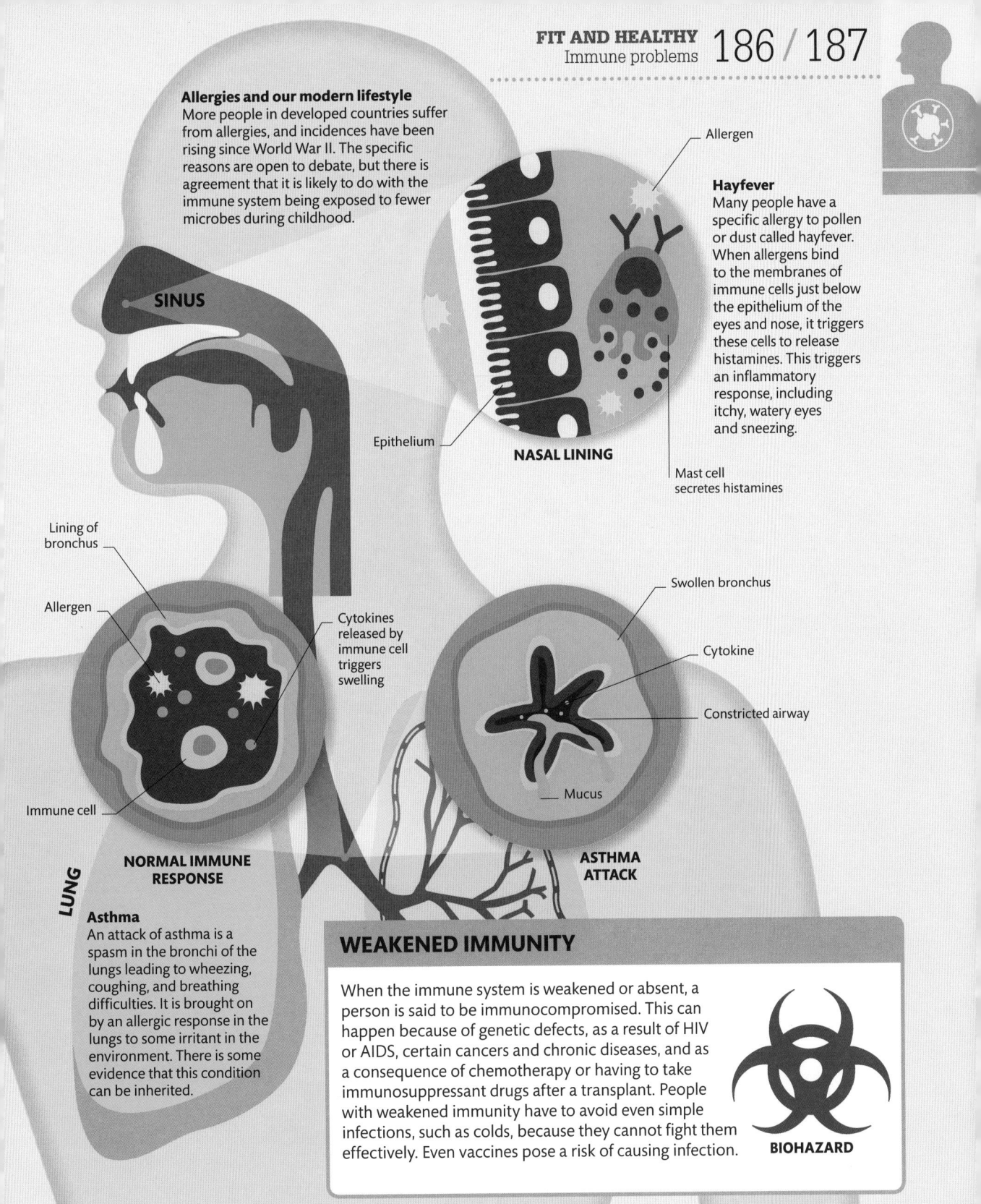

Hayfever
Many people have a specific allergy to pollen or dust called hayfever. When allergens bind to the membranes of immune cells just below the epithelium of the eyes and nose, it triggers these cells to release histamines. This triggers an inflammatory response, including itchy, watery eyes and sneezing.

Asthma
An attack of asthma is a spasm in the bronchi of the lungs leading to wheezing, coughing, and breathing difficulties. It is brought on by an allergic response in the lungs to some irritant in the environment. There is some evidence that this condition can be inherited.

WEAKENED IMMUNITY

When the immune system is weakened or absent, a person is said to be immunocompromised. This can happen because of genetic defects, as a result of HIV or AIDS, certain cancers and chronic diseases, and as a consequence of chemotherapy or having to take immunosuppressant drugs after a transplant. People with weakened immunity have to avoid even simple infections, such as colds, because they cannot fight them effectively. Even vaccines pose a risk of causing infection.

BIOHAZARD

CHEMICAL BALANCE

Chemical regulators

Some of the organs of the endocrine system are dedicated specifically to hormone production, while others, such as the stomach and the heart, have other more familiar functions too. Each receives information from the body and responds by secreting either more or less of a certain hormone. The hormones act as messengers, telling cells to either "keep the balance" or giving instructions to bring about short-term or long-term changes, such as puberty.

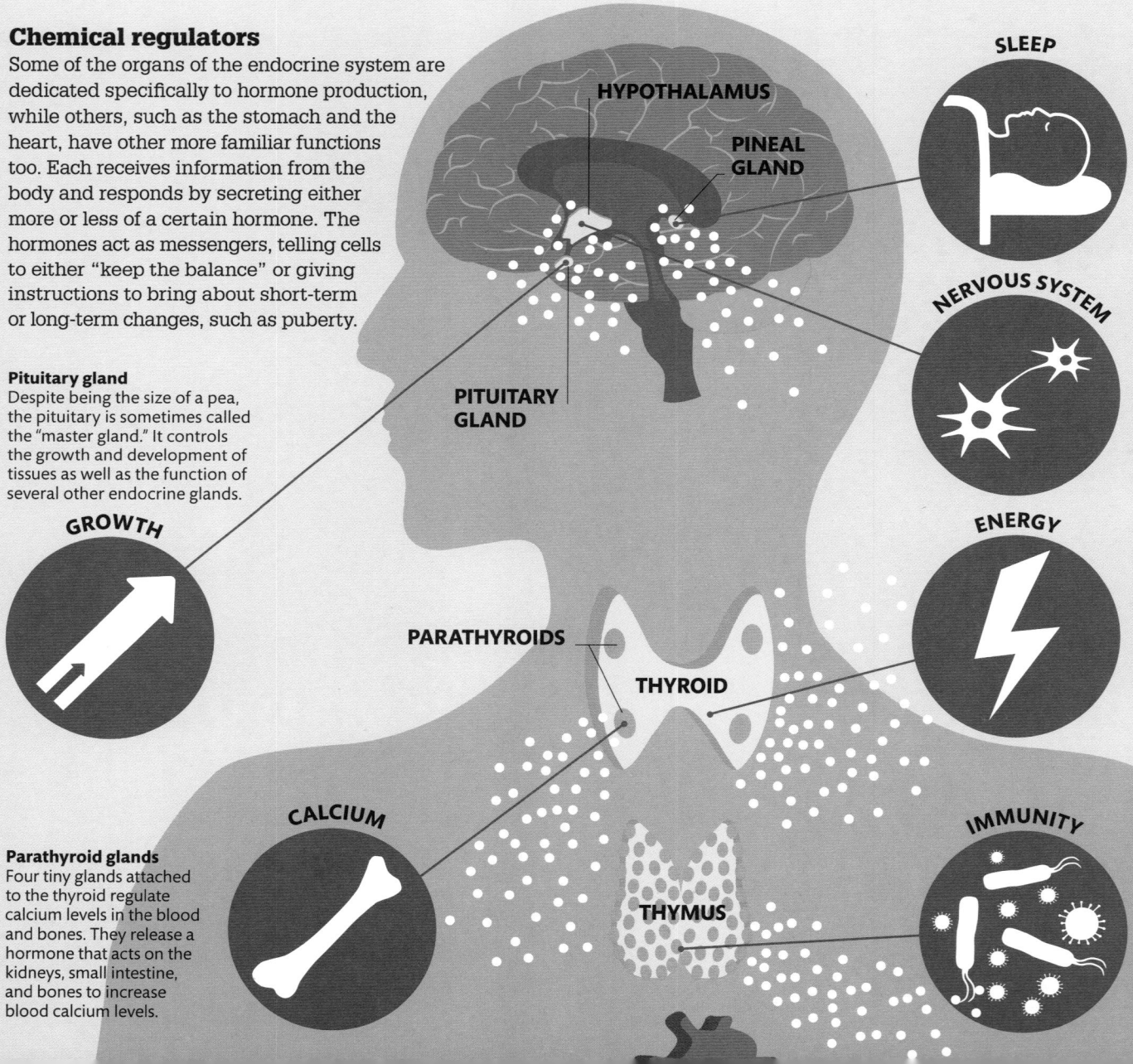

Pineal gland
When light levels decrease, the pineal gland releases melatonin, which makes you sleepy. It works in close partnership with the hypothalamus.

Hypothalamus
The hypothalamus is a part of the brain that links the nervous system to the endocrine system. It sits above the pituitary gland and works with it closely. Among other things, it controls thirst, fatigue, and body temperature.

Pituitary gland
Despite being the size of a pea, the pituitary is sometimes called the "master gland." It controls the growth and development of tissues as well as the function of several other endocrine glands.

Thyroid gland
The thyroid secretes hormones that control growth and metabolic rate. It also secretes calcitonin, which encourages calcium storage in the bones.

Parathyroid glands
Four tiny glands attached to the thyroid regulate calcium levels in the blood and bones. They release a hormone that acts on the kidneys, small intestine, and bones to increase blood calcium levels.

Thymus
The thymus secretes the hormone that stimulates the production of the pathogen-fighting T cells. The gland is most active in babies and adolescents, and shrinks with the onset of adulthood.

Hormone factories

Molecules known as hormones travel throughout the body, triggering changes in tissues that regulate everything from sleep and reproduction to digestion, growth, and pregnancy. They are secreted into the bloodstream by organs that are collectively known as the endocrine system.

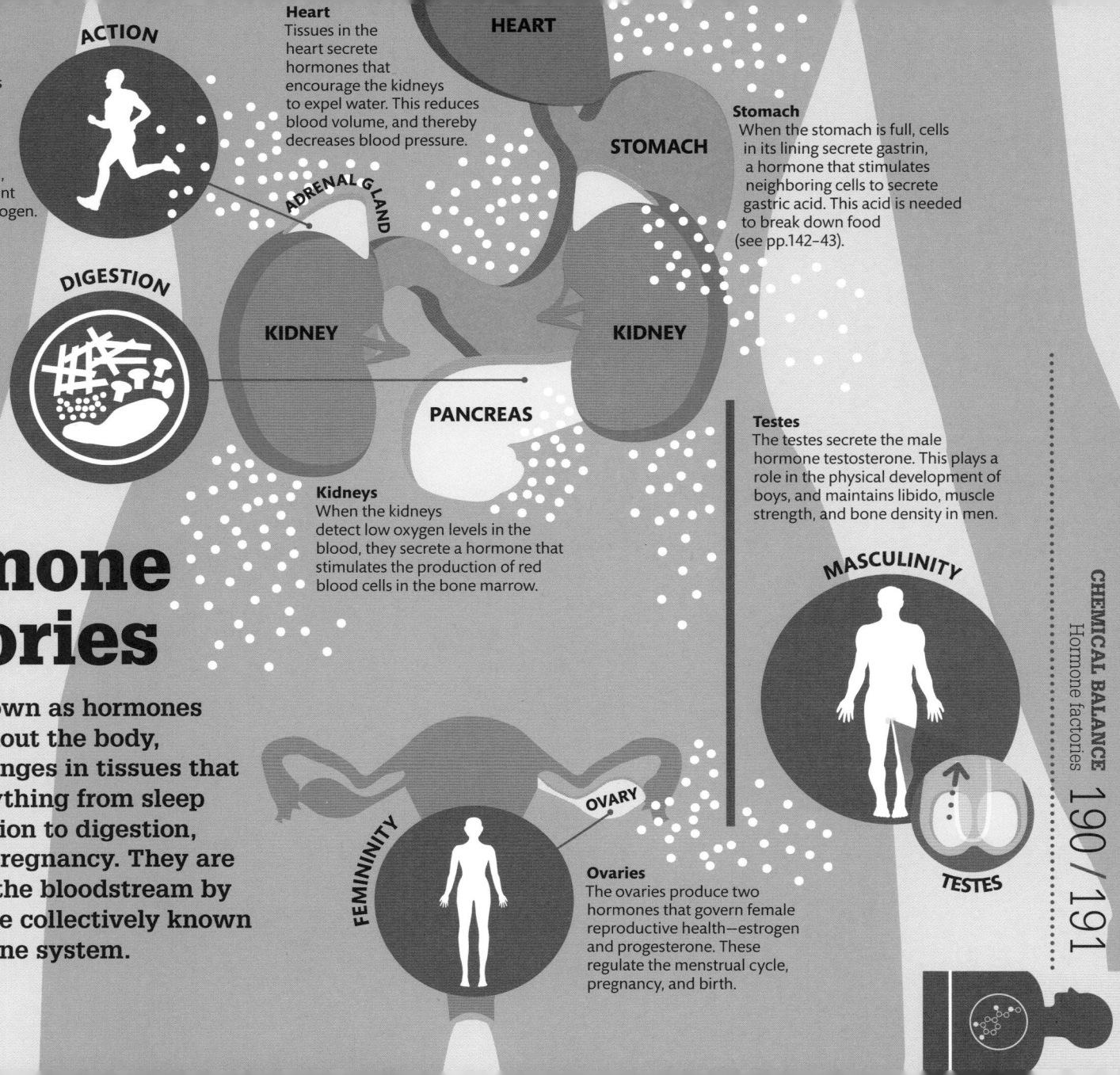

Adrenal glands
These produce hormones that govern the "fight or flight" response, such as epinephrine. They also help regulate blood pressure and metabolism, and secrete a small amount of testosterone and oestrogen.

Heart
Tissues in the heart secrete hormones that encourage the kidneys to expel water. This reduces blood volume, and thereby decreases blood pressure.

Stomach
When the stomach is full, cells in its lining secrete gastrin, a hormone that stimulates neighboring cells to secrete gastric acid. This acid is needed to break down food (see pp.142–43).

Pancreas
As well as producing digestive enzymes, the pancreas makes insulin and glucagon – hormones that control blood glucose levels (see pp.158–59).

Kidneys
When the kidneys detect low oxygen levels in the blood, they secrete a hormone that stimulates the production of red blood cells in the bone marrow.

Testes
The testes secrete the male hormone testosterone. This plays a role in the physical development of boys, and maintains libido, muscle strength, and bone density in men.

Ovaries
The ovaries produce two hormones that govern female reproductive health—estrogen and progesterone. These regulate the menstrual cycle, pregnancy, and birth.

How hormones work

Hormones are molecules that act as messengers between the body's organs and tissues. They are released indiscriminately into the bloodstream, so they travel throughout the body, but they can only affect cells that have receptors to pick them up—and each hormone has its own particular receptor. Some receptors float in the cytoplasm of target cells, others line the cell membrane.

Protein, produced by the estrogen trigger, in turn makes oxytocin, which prepares the body for childbirth

PANCREAS

NUCLEUS

ESTROGEN TARGET CELL

Hormone receptor

Estrogen passes through the cell membrane

The receptor-hormone pair floating in the cell's nucleus, where it triggers a gene to make a specific protein

Estrogen binds to receptor

CELL MEMBRANE

CYTOPLASM

ESTROGEN MOLECULES

OVARY

Straight to the nucleus

Some hormones can pass straight through the outer membrane of a target cell. The receptors for these hormones lie in wait in the cytoplasm of the cell. Once the hormone passes through the membrane, it binds to the receptor, and together they cross into the cell nucleus. Here the receptor-hormone pair binds to the DNA and activates a specific gene.

Estrogen
Estrogen is a hormone produced by the ovaries. It targets most body cells, binding to estrogen receptors, which then trigger genes that help and maintain female reproductive organs.

Hormone triggers

Endocrine glands secrete hormones in response to some sort of trigger. These triggers can be of three kinds; changes in the blood, nerve signals, or instructions from other hormones. However, these triggers themselves are often responses to messages from the outside world. When it gets dark, for example, the hormone melatonin is released to help us go to sleep (see pp.198–99).

Triggered by blood
Some hormones are released when sensory cells detect changes in the blood or other body fluids. The parathyroids, for example, release the hormone PTH in response to low calcium levels in the blood (see pp.194–95).

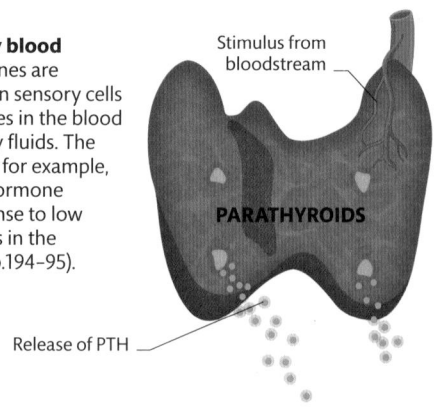

TARGET CELLS CAN HAVE BETWEEN 5,000 AND 100,000 HORMONE RECEPTORS

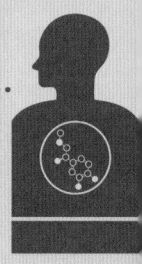

WHAT IS HORMONE THERAPY?

Hormones can be used to trigger changes throughout the body. Sex hormones, for example, can be manipulated to change individuals to the gender they identify with.

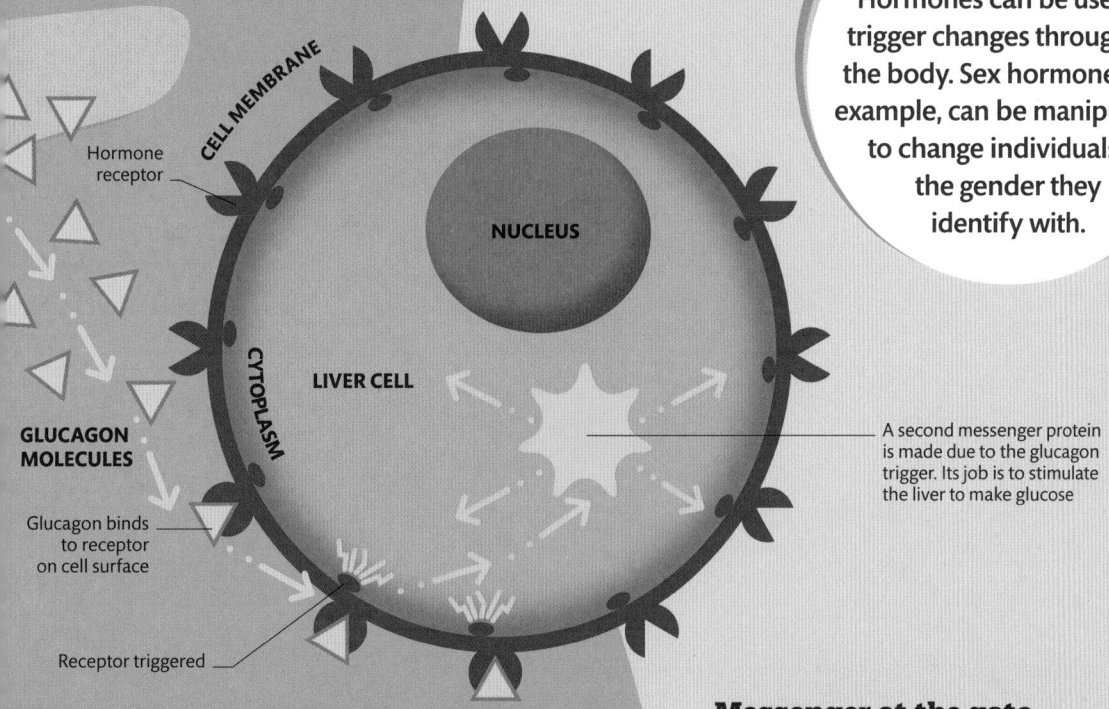

Glucagon
Glucagon, released by the pancreas, targets liver cells, where it binds to receptors on the cell surface. This prompts the cell's molecular machinery to start converting glycogen into glucose (see pp.156–57).

Messenger at the gate

Another class of hormones can't pass through the outer membrane of a cell. These hormones bind to receptors on the surface of the cell instead. This triggers the cell to produce a "second messenger" protein, which causes further changes within the cell.

Triggered by nerves
Many endocrine glands are stimulated by nerve impulses. When we experience physical stress, for example, an impulse is sent along nerves to the adrenal gland, causing it to secrete the fight-or-flight hormone epinephrine (see pp.240–41).

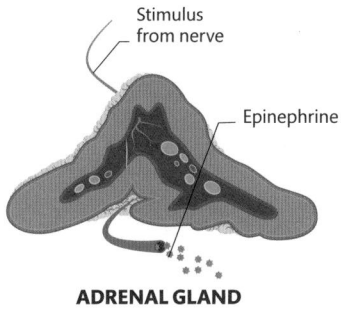

Triggered by hormones
Hormones can also be released in response to other hormones. The hypothalamus, for example, produces a hormone that travels to the pituitary gland and prompts it to release a second hormone—growth hormone—which in turn stimulates growth and metabolism.

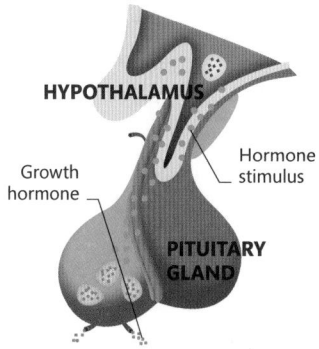

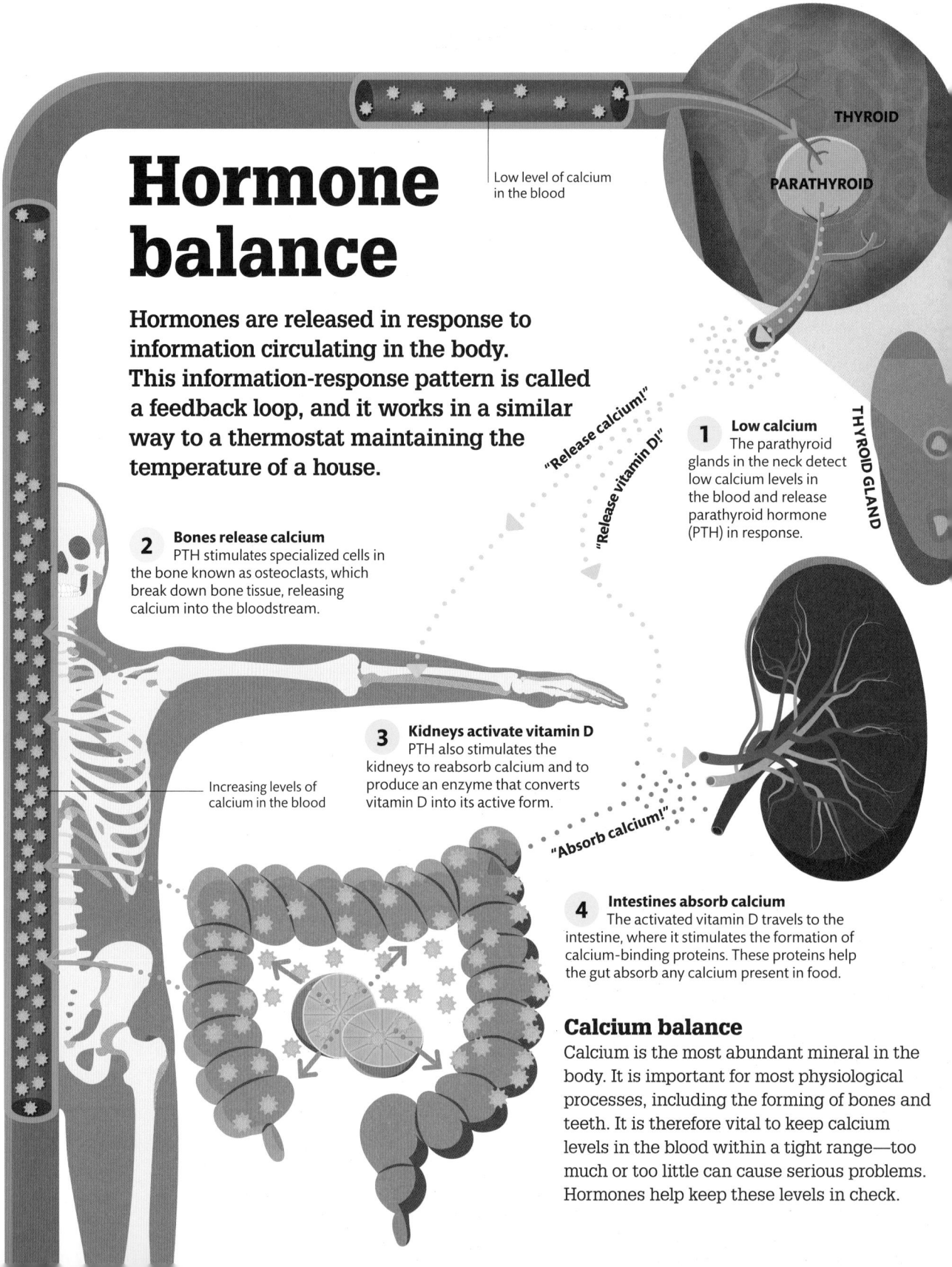

Hormone balance

Hormones are released in response to information circulating in the body. This information-response pattern is called a feedback loop, and it works in a similar way to a thermostat maintaining the temperature of a house.

1 Low calcium
The parathyroid glands in the neck detect low calcium levels in the blood and release parathyroid hormone (PTH) in response.

2 Bones release calcium
PTH stimulates specialized cells in the bone known as osteoclasts, which break down bone tissue, releasing calcium into the bloodstream.

3 Kidneys activate vitamin D
PTH also stimulates the kidneys to reabsorb calcium and to produce an enzyme that converts vitamin D into its active form.

4 Intestines absorb calcium
The activated vitamin D travels to the intestine, where it stimulates the formation of calcium-binding proteins. These proteins help the gut absorb any calcium present in food.

Calcium balance

Calcium is the most abundant mineral in the body. It is important for most physiological processes, including the forming of bones and teeth. It is therefore vital to keep calcium levels in the blood within a tight range—too much or too little can cause serious problems. Hormones help keep these levels in check.

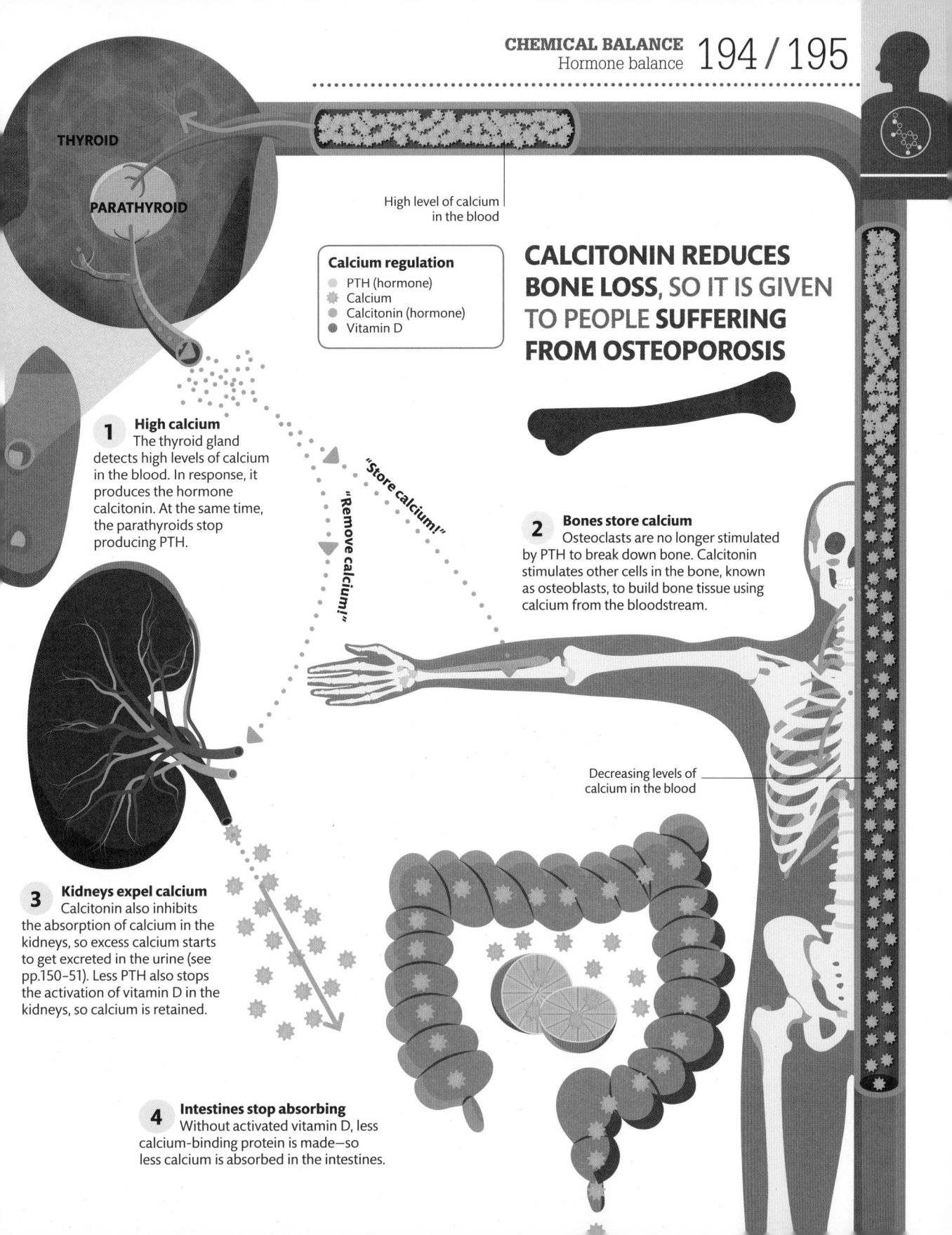
THYROID
PARATHYROID
High level of calcium in the blood
Calcium regulation
PTH (hormone)
Calcium
Calcitonin (hormone)
Vitamin D
CALCITONIN REDUCES BONE LOSS, SO IT IS GIVEN TO PEOPLE SUFFERING FROM OSTEOPOROSIS
1 High calcium
The thyroid gland detects high levels of calcium in the blood. In response, it produces the hormone calcitonin. At the same time, the parathyroids stop producing PTH.
"Store calcium!"
"Remove calcium!"
2 Bones store calcium
Osteoclasts are no longer stimulated by PTH to break down bone. Calcitonin stimulates other cells in the bone, known as osteoblasts, to build bone tissue using calcium from the bloodstream.
Decreasing levels of calcium in the blood
3 Kidneys expel calcium
Calcitonin also inhibits the absorption of calcium in the kidneys, so excess calcium starts to get excreted in the urine (see pp.150–51). Less PTH also stops the activation of vitamin D in the kidneys, so calcium is retained.
4 Intestines stop absorbing
Without activated vitamin D, less calcium-binding protein is made—so less calcium is absorbed in the intestines.

Hormonal changes

Hormones often get blamed for our behavior when the body is undergoing significant change—the moods of a teenager, for example. However, our daily behavior can also affect our hormones, and that in turn can have serious health implications.

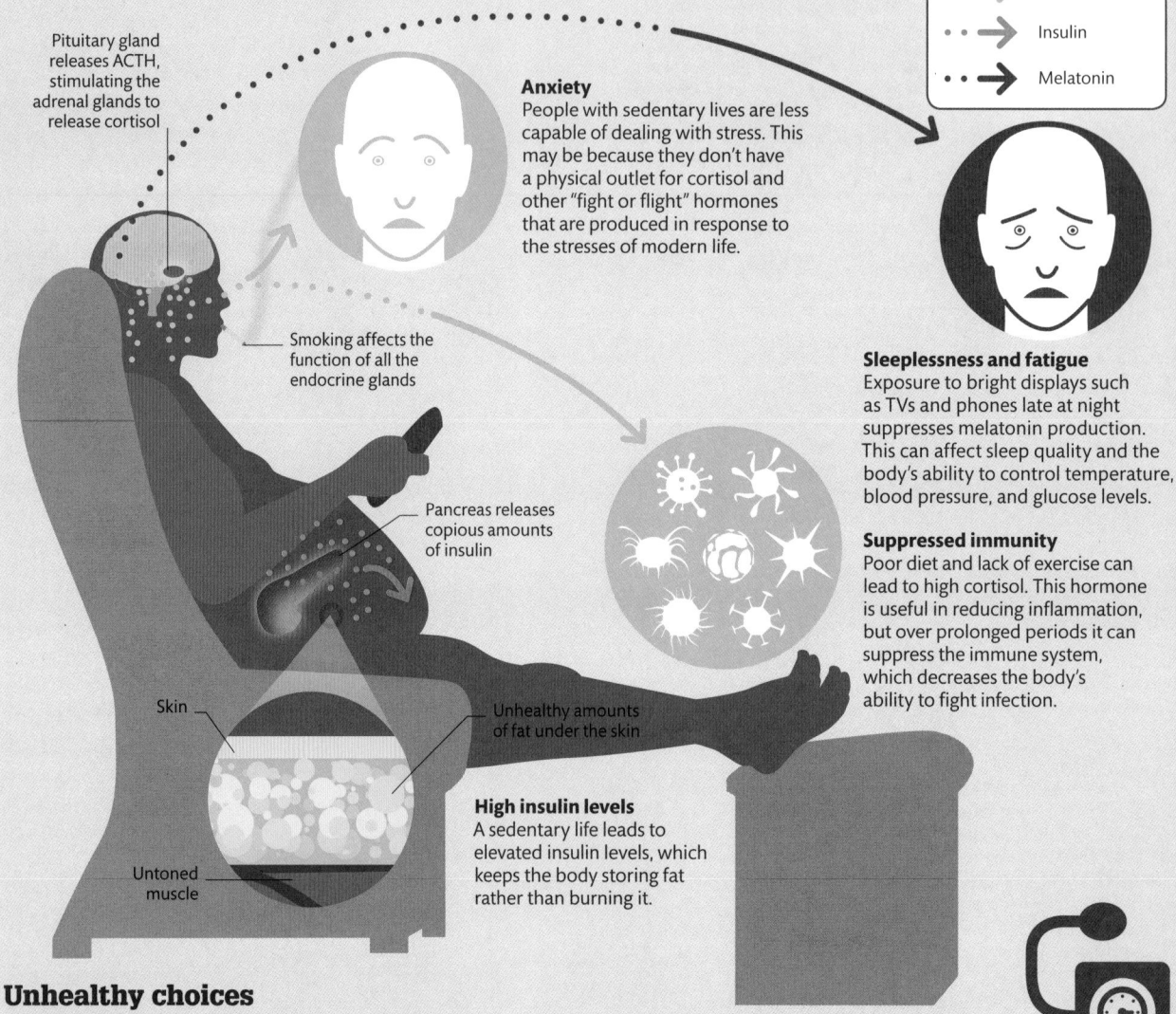

Unhealthy choices

Poor food choices and a sedentary life cause hormone changes that perpetuate that same unhealthy lifestyle. Lower activity levels lead to fewer "feel good" hormones. This can lead to poor food choices, which affect hormones that regulate blood sugar, leading to weight gain and less exercise.

HUGGING RELEASES THE HORMONE **OXYTOCIN**. THIS **REDUCES BLOOD PRESSURE** SO THE RISK OF **HEART DISEASE FALLS**

Healthy lifestyle

Regular exercise is one of the most effective ways to trigger changes in hormones that lead to a healthier mind and body. Some of the hormones that help equip us for physical activity by regulating temperature, maintaining water balance, and adapting to increased oxygen demands are also so-called "feel good" hormones, which greatly improve mood.

EXERCISE BUZZ

Exercise increases the release of neurotransmitters, which are the chemical messengers of the nervous system. They transmit signals at junctions between nerve cells, called synapses. The increase promotes the repair and maintenance of the brain. Some neurotransmitters, such as dopamine, also provide a feeling of happiness.

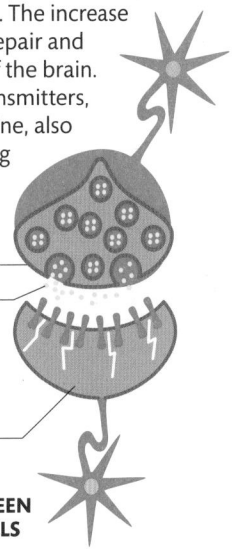

SYNAPSE BETWEEN TWO NERVE CELLS

Pituitary gland releases growth hormone during and after exercise

Strong bones

Bone density
Testosterone and growth hormone are released during exercise in men and women. As well as promoting sex drive, testosterone improves bone density. Growth hormone also promotes the laying down of bone and continues its work in the night after exercise, encouraging the body to recover and promoting general maintenance.

Testosterone release during exercise

Skin

Minimal fat

Lean muscle

Healthy insulin levels
Insulin is inhibited during exercise, forcing our cells to burn fat as an energy source instead of glucose. Insulin levels remain suppressed for a long time after exercise, meaning we burn fat even as we rest.

Good musculature, thanks to growth hormone and testosterone

Muscle mass
Testosterone stimulates the building of lean muscle mass, and increases our overall metabolism. Growth hormone promotes the growth of muscle tissue and helps the body burn fat.

Hormones and health
Three hormones play a role in improving our health and our state of our mind.

- Growth hormone
- Insulin
- Testosterone

Daily rhythms

The body has a built-in time-keeping system that drives our daily rhythms—particularly those of eating and sleeping. At the core of this is the daily chemical conversion of the wakeful hormone serotonin into the sleep hormone melatonin—a process that takes about 24 hours.

The daily cycle

Many hormones go through rhythmic fluctuations every day. These oscillations happen independently of any external prompting. Even in a black room with no windows, the body gets a serotonin surge in the morning, which wakes it up. However, these rhythms are not hard-wired—they are constantly readjusted and can be changed radically when we travel to a different time zone.

The circadian clock

Our bodies run on a (roughly) 24-hour hormone cycle, known as a circadian rhythm. The biological processes that govern it are called the circadian clock, which is what governs all the body's rhythms. One of the main cogs in this clock is a very small region of the brain known as the suprachiasmatic nucleus (SCN). Located very near the optic nerves, the SCN uses the amount of light entering the eye to calibrate the circadian clock.

3 Hunger hormones
Hunger hormones rise and fall throughout the day. Levels of ghrelin, the appetite increaser, rise during fasting, increasing hunger in the morning. Leptin, the appetite suppressor, signals when you are "full."

2 Stress-managing cortisol
As you start the day, the body produces the steroid hormone cortisol, which helps the body deal with stress by increasing blood sugar levels and kick-starting metabolism.

1 Wakeful serotonin
Light stimulates the suprachiasmatic nucleus to convert melatonin into serotonin—a hormone that helps get the brain and body going (especially the intestines).

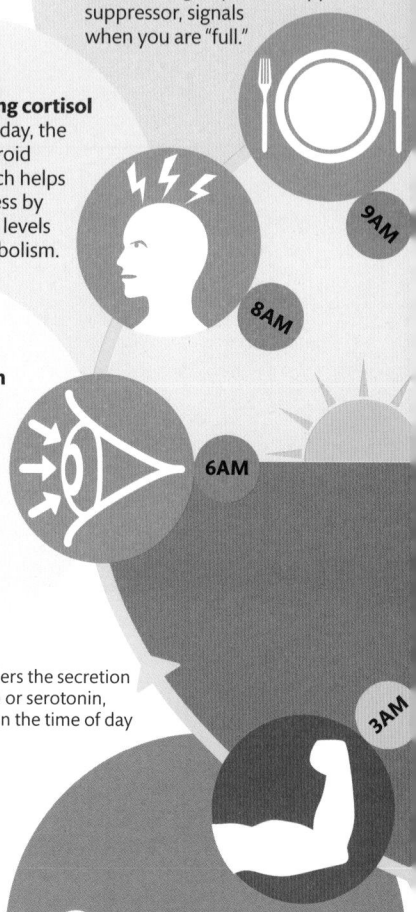

10 Testosterone surge
Men experience a rise in testosterone levels at night, regardless of whether or not they are asleep—a fact that might explain late-night fights at bars.

Internal timepiece
The SCN drives a two-way chemical conversion between the hormone serotonin, which wakes us up, and melatonin, which puts us to sleep.

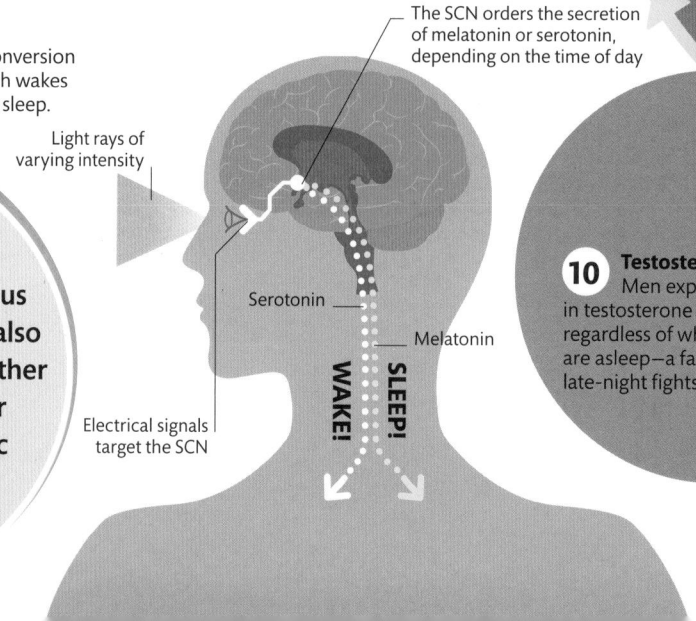

CAN STRESS MAKE YOU ILL?

Stress hormones prepare us for fight or flight, but they also take a toll on some of our other systems, particularly our immune system. Chronic stress can therefore lead to disease.

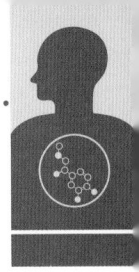

4 Cortisol peaks
After the morning surge of cortisol, the body gets another dose around noon. From then on, cortisol plays a smaller role in the system. Melatonin is at its lowest level at this time.

Cortisol
Melatonin
12AM

5 Aldosterone surge
Midafternoon sees a peak in the hormone aldosterone. This helps keep the blood pressure steady by increasing water reabsorption in the kidneys.

3PM

JET LAG

Air travel transports us into new time zones faster than the body can adjust. It takes time for the new rhythm of daylight to reset the body clock. Some hormone cycles are more flexible than others—cortisol can take 5–10 days to adapt. While our rhythms adjust, the body feels hungry and sleepy at all the wrong times—a phenomenon called jet lag. Shift workers experience this regularly, and the long-term health consequences are not yet fully understood.

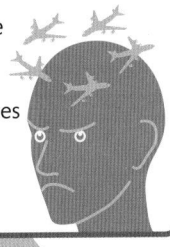

6 Sleepy melatonin
Decreasing light levels prompt the conversion of serotonin into melatonin. This slowly prepares the body for sleep and finally causes sleepiness itself.

6PM

Thyroid gland
8PM

7 Stimulating thyroid
In the evening, levels of thyroid-stimulating hormone abruptly increase. This stimulates growth and repair, but also inhibits neuronal activity, possibly preparing the body for sleep.

9PM

8 Growth hormone
The first two hours of sleep see a burst in growth hormone, which helps children grow and adults regenerate. It's also released in the day, but more is produced at night, when the body can focus on repair.

12PM
Melatonin
Cortisol

9 Melatonin peaks
Melatonin levels in the blood are highest around midnight. This is also when cortisol levels are at their lowest. This ensures that the body rests completely overnight.

A BRISK WALK AT LUNCHTIME HELPS BOOST SEROTONIN LEVELS

Diabetes

Insulin is the key that opens muscle and fat cells to receive glucose, which the body needs for energy. Without insulin, glucose remains in the blood, and the cells don't get the energy they need, which has serious health consequences. If insulin fails to work, the result is diabetes, a disease that has two forms—type 1 and type 2—and currently affects 382 million people globally.

MANAGING DIABETES

Sugary foods and certain carbohydrates cause fat to be laid down in the body's cells, and fat interferes with insulin. The more fat present, therefore, the greater the risk of type 2 diabetes. A healthy, balanced diet not only reduces this risk, it is also a vital part of managing the disease once it develops. Generally, diabetic diets aim at keeping blood glucose levels as normal as possible, avoiding foods that cause sharp rises and falls in glucose. This also helps in calculating insulin dosages, which may be a part of treatment.

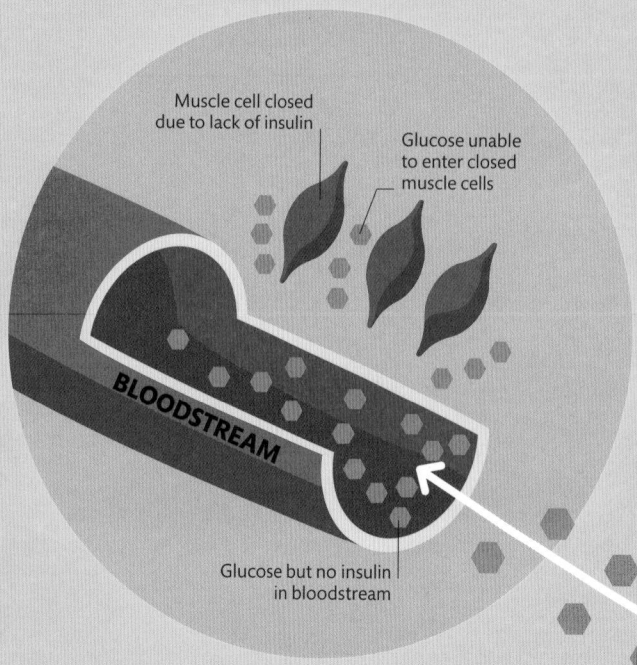

1 Glucose on the rise
During digestion, glucose is released into the bloodstream. The rise in glucose levels triggers mechanisms that will lower them—including the release of insulin from the pancreas (see pp.158–59).

3 No entry for glucose
Without insulin, glucose can't enter the body's cells. Instead, it builds up in the blood, and the body reacts by trying to get rid of it by other means, such as urination.

Glucose molecule

PANCREAS

2 No insulin available
However, in type 1 diabetes, the insulin-producing cells of the pancreas have been destroyed by the body's own immune cells. As a consequence, no insulin is released to counter the rising glucose levels.

Type 1 diabetes

In type 1 diabetes, the body's immune system attacks the insulin-producing cells of the pancreas, leaving the pancreas unable to produce any insulin. The symptoms emerge over a matter of weeks but can be reversed once treated with insulin. Although people can develop type 1 diabetes at any age, most are diagnosed before the age of 40, particularly in childhood. Type 1 accounts for 10 percent of all diabetes cases.

The symptoms of diabetes

The symptoms of type 1 and type 2 diabetes are similar. The glucose that the kidneys can't get rid of starts to build up in the body, so the body tries to flush it out—so thirst, water intake, and urination increase accordingly. Meanwhile, the body's cells are being starved of glucose, which causes fatigue throughout the body. Weight loss also occurs, due to the body burning fat instead of glucose.

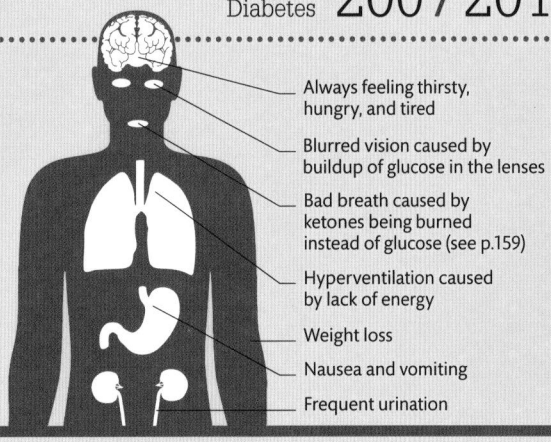

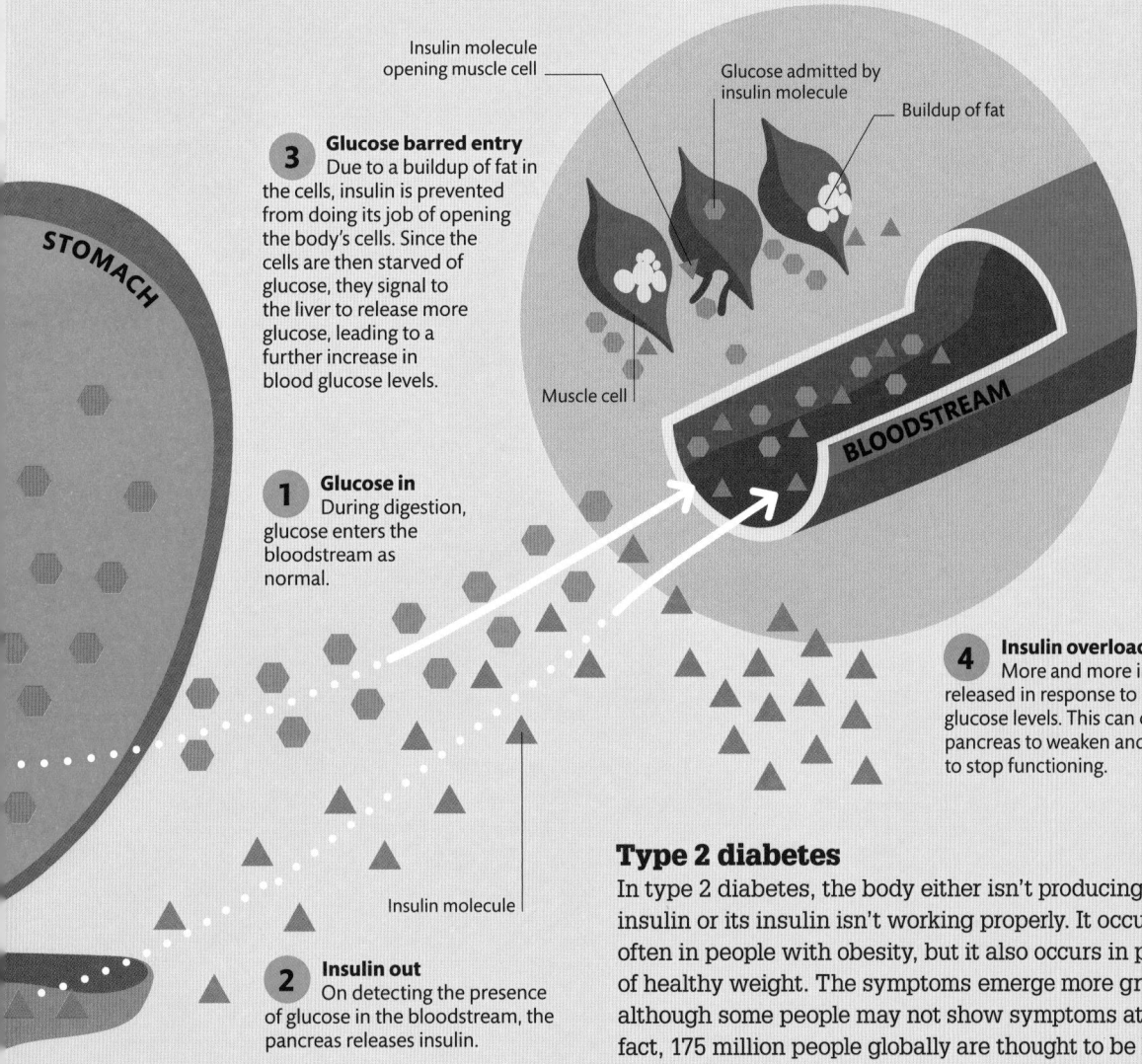

3 Glucose barred entry
Due to a buildup of fat in the cells, insulin is prevented from doing its job of opening the body's cells. Since the cells are then starved of glucose, they signal to the liver to release more glucose, leading to a further increase in blood glucose levels.

1 Glucose in
During digestion, glucose enters the bloodstream as normal.

4 Insulin overload
More and more insulin is released in response to rising blood glucose levels. This can cause the pancreas to weaken and eventually to stop functioning.

2 Insulin out
On detecting the presence of glucose in the bloodstream, the pancreas releases insulin.

Type 2 diabetes

In type 2 diabetes, the body either isn't producing enough insulin or its insulin isn't working properly. It occurs more often in people with obesity, but it also occurs in people of healthy weight. The symptoms emerge more gradually, although some people may not show symptoms at all. In fact, 175 million people globally are thought to be living with undiagnosed type 2 diabetes. Type 2 accounts for 90 percent of all diabetes cases.

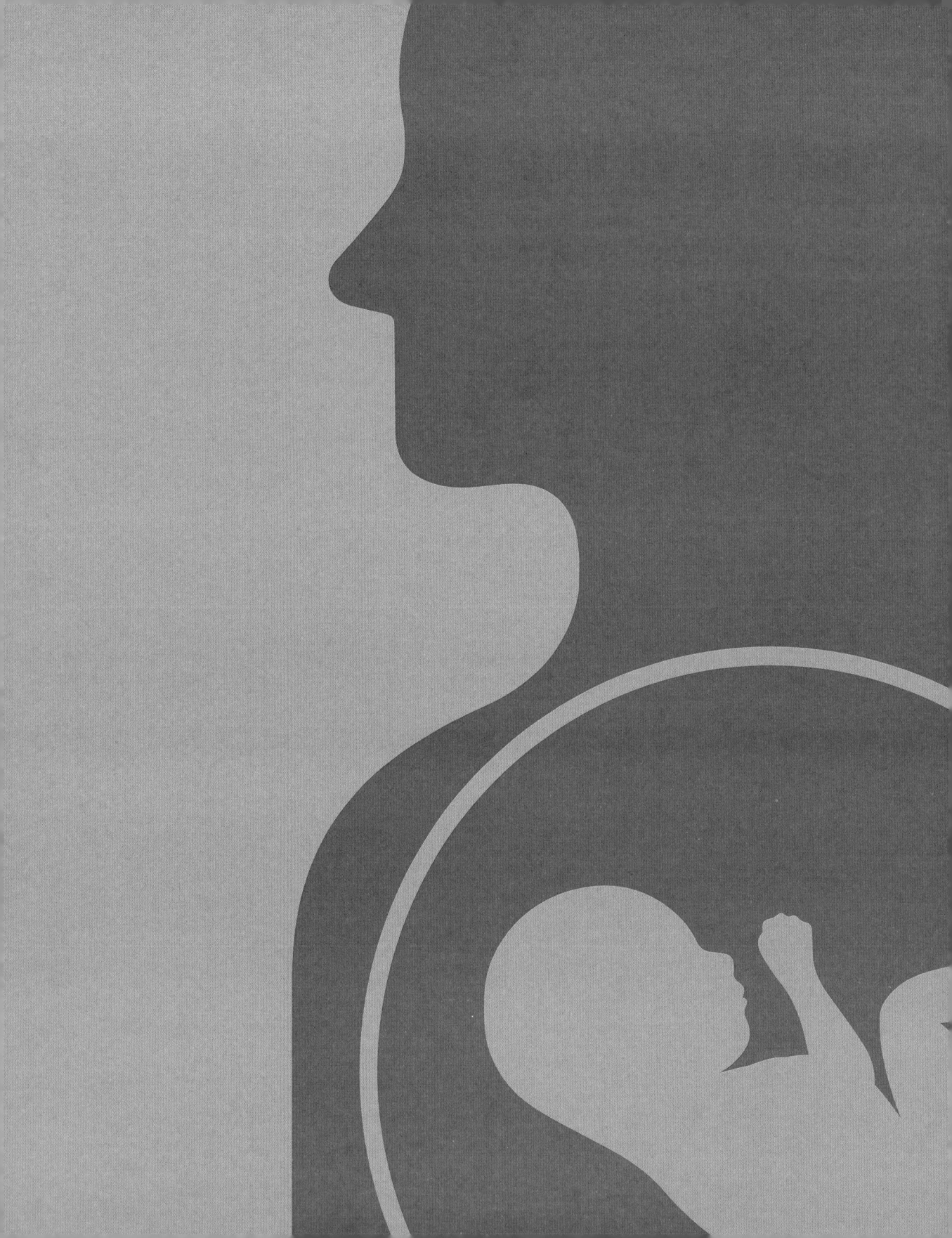

THE CIRCLE OF LIFE

Sexual reproduction

You are driven by your genes to reproduce, so that your genes continue to multiply in generations to come. Evolutionarily speaking, this is why we have sex. Millions of sperm compete against one another to find one egg and begin the process of creating a new individual.

Bringing sperm and egg together

The main aim of sex is to bring genes from the male and the female together. The male inserts millions of packets of genes in the form of sperm into the female in an attempt to fertilize one of her eggs. If successful, the male's and female's genes mix, generating a new, unique combination of genes in the offspring. To achieve this, both male and female individuals become sexually aroused by one another, which causes some physical changes. Genital organs in both genders enlarge due to increased blood flow, the penis becomes erect, and the vagina secretes a lubricating fluid to aid the penis's entry.

SEMEN NORMALLY CONTAINS 1–8 BILLION SPERM PER FLUID OUNCE (140–300 MILLION SPERM PER MILLILITER)

WHY DO WOMEN HAVE ORGASMS?

Sensitive nerve endings in the clitoris send pleasurable signals to the brain, causing the vagina to contract tightly around the penis, thus ensuring that the male ejaculates as much sperm as possible.

Seminal vesicle adds fluid to sperm

Prostate gland adds further fluid to sperm to produce semen

Bulbourethral gland neutralizes acidity of urine in urethra to prevent harm to sperm

Sperm travels through penis in the urethra

Sperm matures in epididymis

HOW DO ERECTIONS WORK?

The penis contains two cylinders of spongelike tissue, called the corpora cavernosa. When small arteries at the base of the penis dilate, or widen, blood flows into the penis and the corpora cavernosa expand to form rigid cylinders. This compresses small drainage veins so that blood cannot flow away and the penis hardens. After ejaculation, the pressure reduces and drainage veins reopen, allowing blood to flow out and the penis softens.

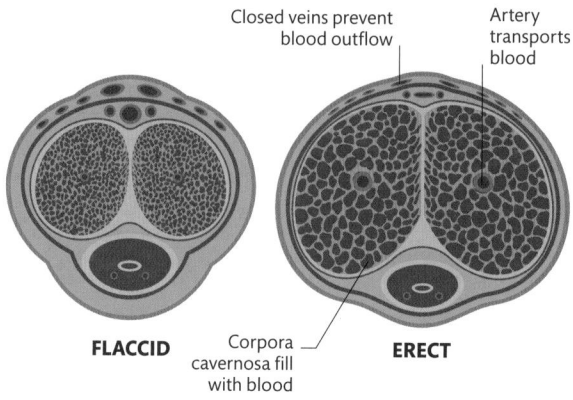

The perilous journey of sperm

During sex, the erect penis is inserted into the vagina. The penis releases semen during orgasm and sperm start their journey to find an egg. Millions of sperm, aided by whiplike movements of their tails, swim up from the vagina, through the cervix, and into the uterus. Sperm is carried in fluid currents caused by the movement of hairlike cells lining the fallopian tubes. Only 150 or so sperm find their way to the upper fallopian tubes, where fertilization usually occurs. The remaining sperm are naturally flushed out of the vagina.

MALE BLADDER

Vas deferens carries sperm from testis to penis

If sperm meets an egg, it is usually here in upper fallopian tube

FALLOPIAN TUBE

FEMALE BLADDER

OVARY

CLITORIS

UTERUS

PENIS

Sperm travels into the uterus

URETHRA

ERECTILE TISSUE

Cervix

STIS

SCROTUM

VAGINA

Penis ejaculates sperm into the vagina

Scrotum contains both testes outside body, because sperm production requires a cooler temperature

LARGEST CELL IN THE BODY

An egg (called an ovum) is the largest cell in the human body and just visible to the naked eye. It is protected by a thick, transparent shell. Sperm cells are one of the smallest types of cells in the body, averaging about 1/500 in (0.05 mm) long, but most of this is its tail.

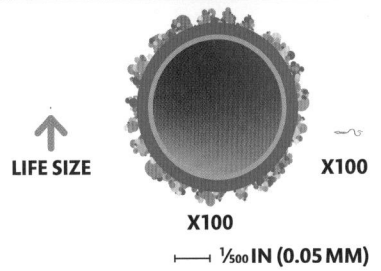

Monthly cycle

Every month, a woman's body prepares for the possibility of pregnancy. Stored in the ovaries, half a million dormant eggs await their turn for ovulation. When hormone levels reach their peak, an egg bursts from an ovary, ready for fertilization. Thick tissue in the uterus lining awaits the egg, if it is fertilized.

Menstrual cycle

The menstrual cycle is controlled by the pituitary gland in the brain. Beginning at puberty, follicle-stimulating hormone (FSH) is produced by the pituitary gland. FSH prompts the production of estrogen and progesterone hormones in the ovaries. The pituitary gland releases a monthly pulse of FSH and also luteinising hormone (LH), triggering a monthly cycle. A single matured egg is released from the ovary and the lining of the uterus—the endometrium—will thicken and then shed. If the egg is fertilized and then implants into the endometrium, the cycle ceases. Later in life, when the number of dormant eggs in the ovaries reaches a point where they cannot produce enough hormones to regulate the menstrual cycle, menopause is triggered, and the cycle stops.

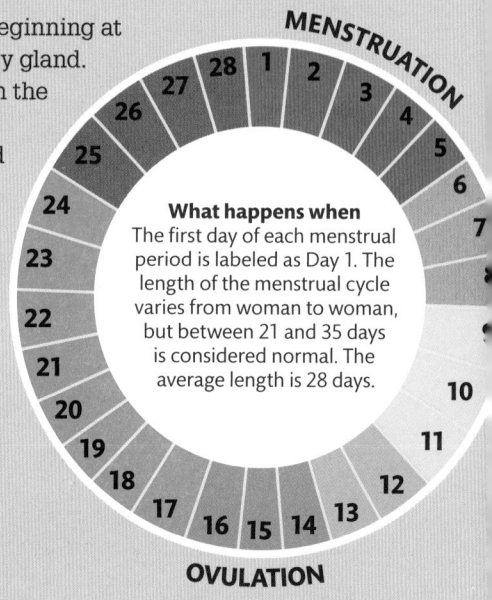

What happens when
The first day of each menstrual period is labeled as Day 1. The length of the menstrual cycle varies from woman to woman, but between 21 and 35 days is considered normal. The average length is 28 days.

MENSTRUAL CRAMPS

The muscles in the lining of the uterus naturally contract during a menstrual period, constricting tiny arteries to limit bleeding. If the contractions are intense or prolonged, they press against nearby nerves, causing pain.

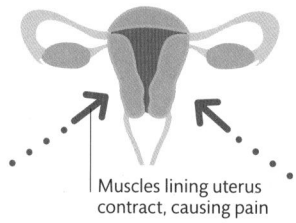

1 Menstrual bleeding
If a fertilized egg does not implant in the endometrium, falling progesterone levels cut down its blood supply, causing the outer layer to be shed as menstrual bleeding. This can serve as an indicator that pregnancy has not occured.

2 Endometrium grows
During the first 2 weeks of the menstrual cycle, steadily rising estrogen levels cause the endometrium to grow.

3 Hormone surge
Estrogen is produced by cells in the follicle that surrounds a maturing egg in the ovary. When estrogen levels peak, this causes a surge of FSH and LH to be released from the pituitary gland, which triggers ovulation.

ESTROGEN

Bleeding from vagina as endometrium sheds

FSH AND LH

Slight rise in FSH levels stimulates production of estrogen and progesterone

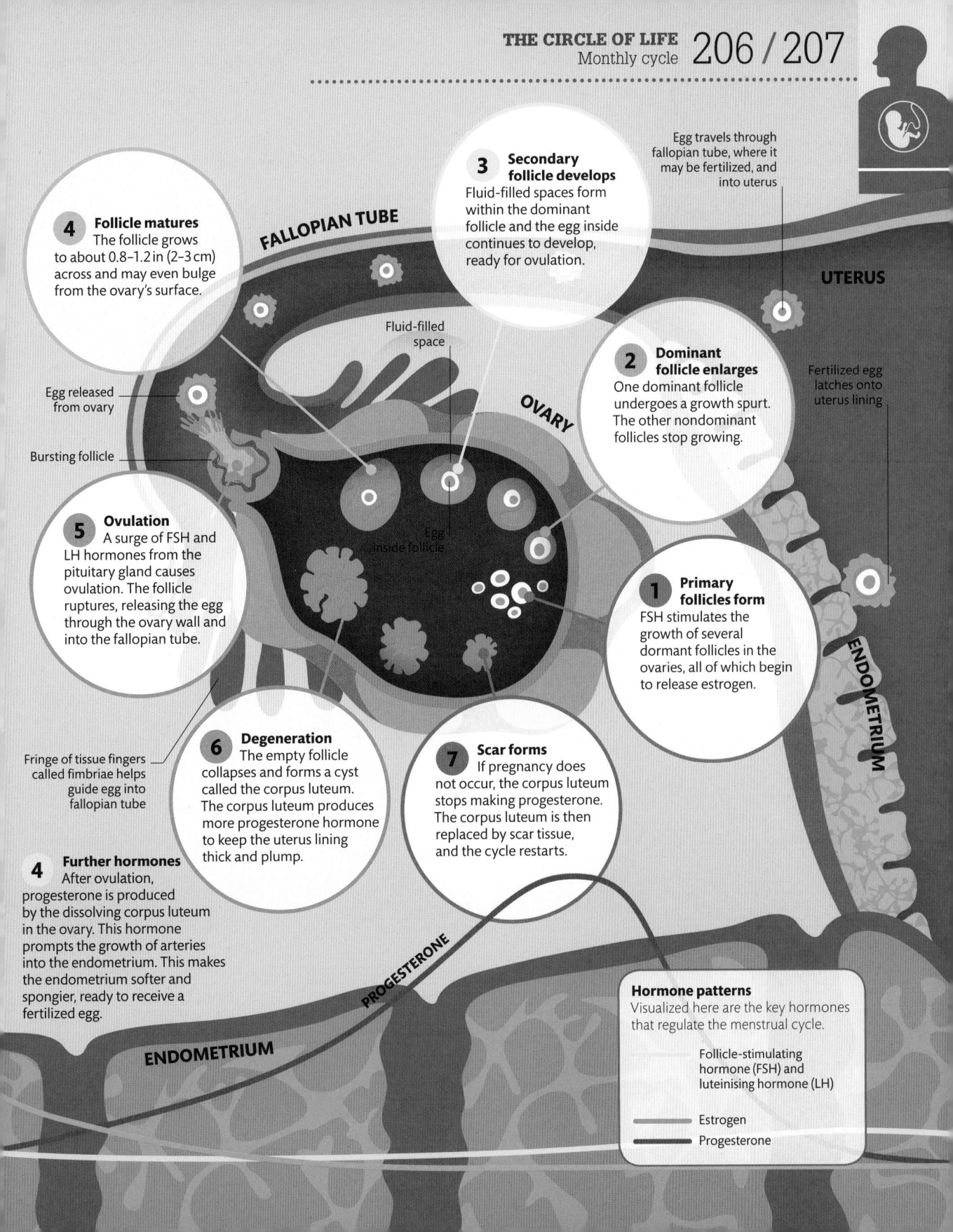
FALLOPIAN TUBE
UTERUS
OVARY
ENDOMETRIUM
ENDOMETRIUM
PROGESTERONE
Egg travels through fallopian tube, where it may be fertilized, and into uterus
Fluid-filled space
Egg released from ovary
Bursting follicle
Egg inside follicle
Fertilized egg latches onto uterus lining
Fringe of tissue fingers called fimbriae helps guide egg into fallopian tube
1 Primary follicles form
FSH stimulates the growth of several dormant follicles in the ovaries, all of which begin to release estrogen.
2 Dominant follicle enlarges
One dominant follicle undergoes a growth spurt. The other nondominant follicles stop growing.
3 Secondary follicle develops
Fluid-filled spaces form within the dominant follicle and the egg inside continues to develop, ready for ovulation.
4 Follicle matures
The follicle grows to about 0.8–1.2 in (2–3 cm) across and may even bulge from the ovary's surface.
5 Ovulation
A surge of FSH and LH hormones from the pituitary gland causes ovulation. The follicle ruptures, releasing the egg through the ovary wall and into the fallopian tube.
6 Degeneration
The empty follicle collapses and forms a cyst called the corpus luteum. The corpus luteum produces more progesterone hormone to keep the uterus lining thick and plump.
7 Scar forms
If pregnancy does not occur, the corpus luteum stops making progesterone. The corpus luteum is then replaced by scar tissue, and the cycle restarts.
4 Further hormones
After ovulation, progesterone is produced by the dissolving corpus luteum in the ovary. This hormone prompts the growth of arteries into the endometrium. This makes the endometrium softer and spongier, ready to receive a fertilized egg.
Hormone patterns
Visualized here are the key hormones that regulate the menstrual cycle.
Follicle-stimulating hormone (FSH) and luteinising hormone (LH)
Estrogen
Progesterone

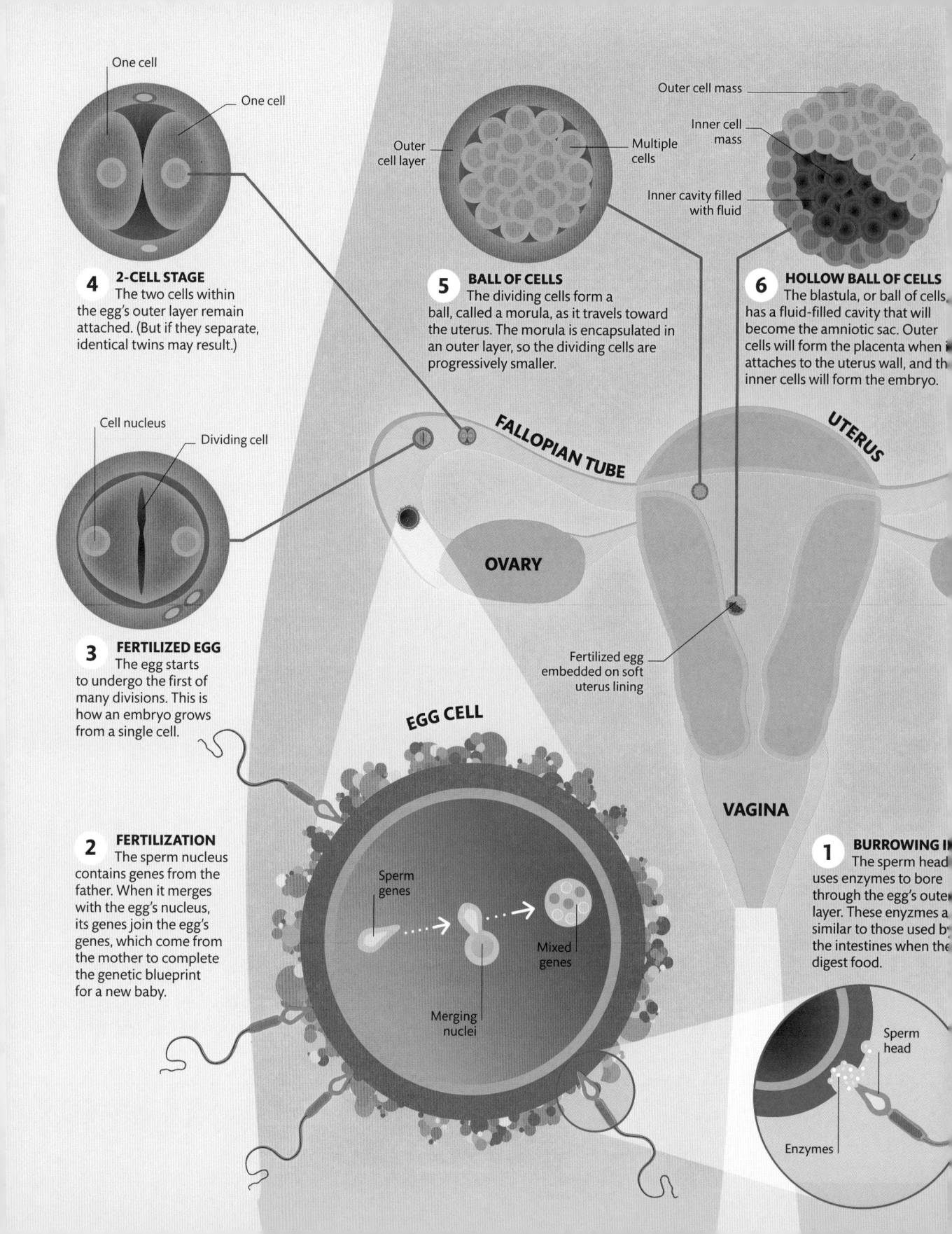

4 2-CELL STAGE
The two cells within the egg's outer layer remain attached. (But if they separate, identical twins may result.)

5 BALL OF CELLS
The dividing cells form a ball, called a morula, as it travels toward the uterus. The morula is encapsulated in an outer layer, so the dividing cells are progressively smaller.

6 HOLLOW BALL OF CELLS
The blastula, or ball of cells, has a fluid-filled cavity that will become the amniotic sac. Outer cells will form the placenta when it attaches to the uterus wall, and the inner cells will form the embryo.

3 FERTILIZED EGG
The egg starts to undergo the first of many divisions. This is how an embryo grows from a single cell.

2 FERTILIZATION
The sperm nucleus contains genes from the father. When it merges with the egg's nucleus, its genes join the egg's genes, which come from the mother to complete the genetic blueprint for a new baby.

1 BURROWING IN
The sperm head uses enzymes to bore through the egg's outer layer. These enyzmes are similar to those used by the intestines when they digest food.

Tiny beginnings

For some 48 hours after sex, around 300 million sperm race to fertilize an egg as it travels down one of the fallopian tubes. Sperm are chemically attracted to the egg, aiding them on their 6 in- (15 cm-) long journey. When a single sperm fertilizes the egg, a cascade of changes follows.

.n egg's journey
ach month, several
ggs start to mature
within the ovaries.
Normally only one
eveloped egg is
eleased at ovulation.
he released egg then
nters one of the
allopian tubes.

Fertilization

If a woman has ovulated and has had sex, there is a chance of fertilization—the joining of egg and sperm to lay the foundation for pregnancy. The moment the sperm penetrates the egg's outer layer, the egg undergoes a rapid chemical change and hardens to prevent other sperm from burrowing in. Now the combined egg and sperm is called a zygote. It begins to divide as it enters the womb (uterus). Fertilization may have been achieved, but there is a long way to go until birth.

WHEN DOES PREGNANCY BEGIN?

Pregnancy does not start until the fertilized egg successfully embeds itself in the soft lining of the uterus—at this point, new life has potentially been conceived.

THE ANSWER TO INFERTILITY

Infertility problems are common in both genders, and affect one in six couples. Some females may have problems with ovulation, their fallopian tubes may be blocked, or their eggs may be too old. Alternatively, males can suffer from a low sperm count, or their sperm may swim poorly. Nevertheless, there are a number of treatments available. One of these, in vitro fertilization, involves collecting eggs and sperm and placing them in a "test tube" for fertilization to occur. The fertilized egg is then allowed to develop before it is implanted back into the uterus to continue its development. A more advanced procedure is intracytoplasmic sperm injection, in which a sperm nucleus is injected directly into an egg.

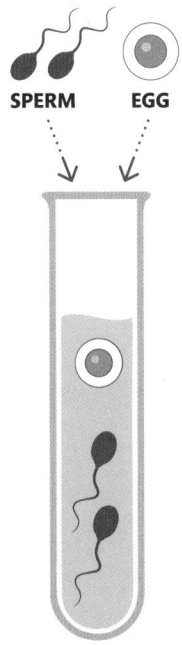

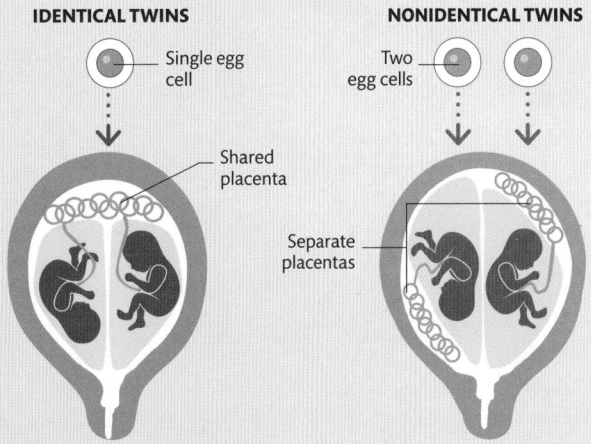

How twins form

If two eggs are released at ovulation and both are fertilized, nonidentical twins result. These can be the same or different sexes and each has its own placenta. When a single fertilized egg splits and each embryo continues to divide separately, identical twins results, each with their own placenta. If the egg divides late, identical twins share a placenta.

The generation game

Although you are a unique individual, you may have familiar features that are shared by your family. These traits are handed down from generation to generation by genes carried by the mother's eggs and father's sperm.

Hereditary traits

Genes instruct your body on how to develop (see p.23). Structures called chromosomes carry multiple genes (see p.16). Each sperm and egg cell from your father and mother contains a random selection of their genes. When these cells merge during fertilization, the sets of genes mix, forming a new and unique blueprint that makes you "you." If you have brothers or sisters, they inherit a similar selection of genes to you, so you may resemble each other in facial features or body shape and may share similar personality traits or mannerisms. Alternatively, siblings can inherit few of the same genes, and may not even look related at first glance.

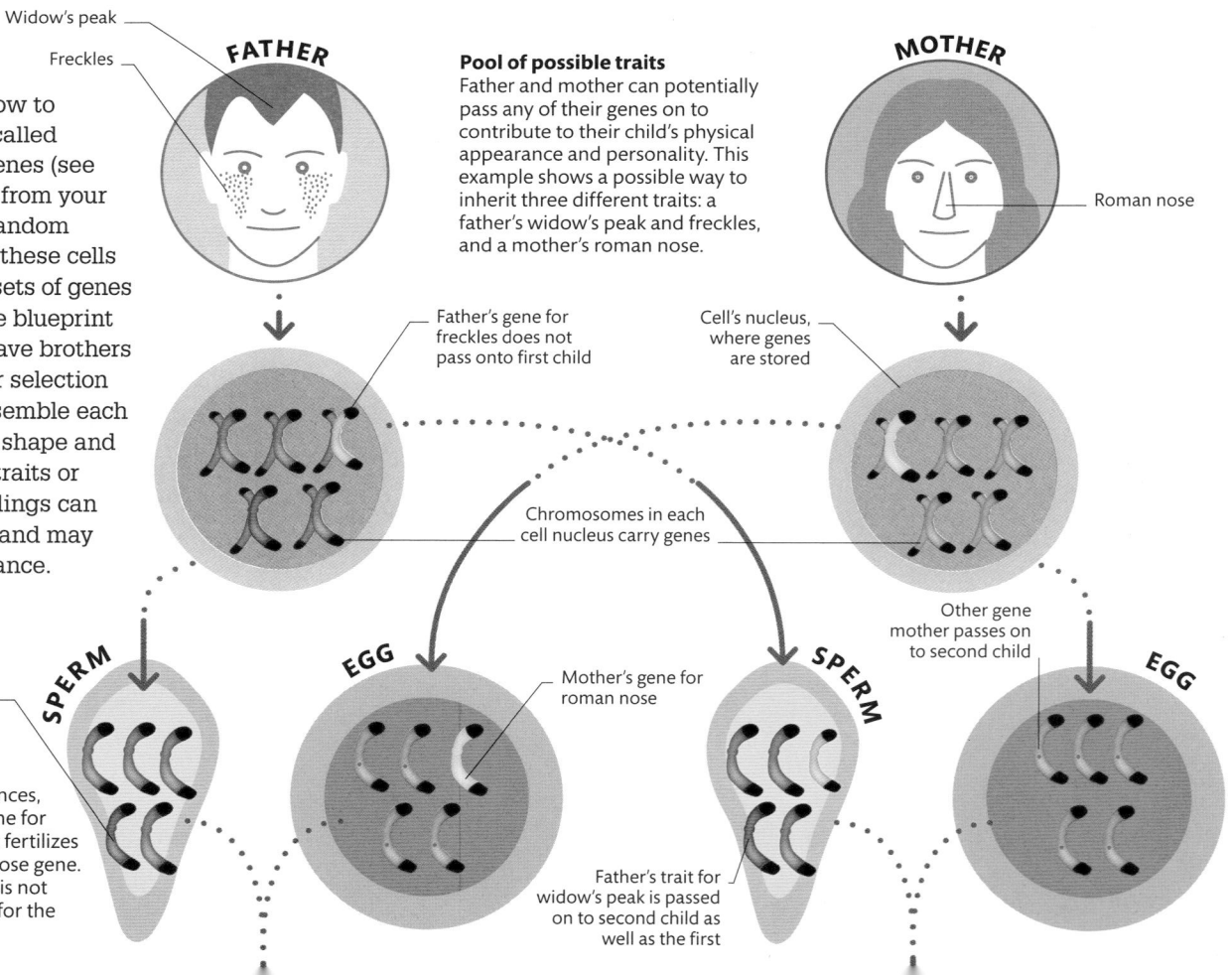

Pool of possible traits
Father and mother can potentially pass any of their genes on to contribute to their child's physical appearance and personality. This example shows a possible way to inherit three different traits: a father's widow's peak and freckles, and a mother's roman nose.

Selected traits
The combination of genes differs with each sperm and egg cell. In these instances, for the first conception, the father's gene for a widow's peak is in the sperm cell that fertilizes the mother's egg cell with the roman nose gene. However, the father's gene for freckles is not in the sperm cell that fertilizes the egg for the first child, but is for the second.

Traits from both parents
The sperm and egg that produced the first child has passed on the father's gene for a widow's peak and the mother's gene for a roman nose. As a result, this child will share traits with both parents. By chance, they have not inherited their father's freckles.

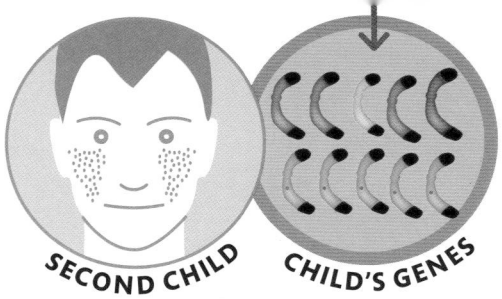

Shared traits
The second child inherits the father's genes for both a widow's peak and freckles. These siblings share at least one physical characteristic—the widow's peak.

Dominant and recessive traits

Traits can be inherited in a dominant or recessive pattern. The dominant and recessive versions of a gene are called alleles and are found at the same place on a chromosome. A dominant allele usually shows its trait whenever it is present, while a recessive gene only shows its effects if a more dominant version is absent. If you have detached earlobes, you have at least one dominant allele. Only if you have two copies of the recessive version do you show the recessive trait—the rarer attached earlobe.

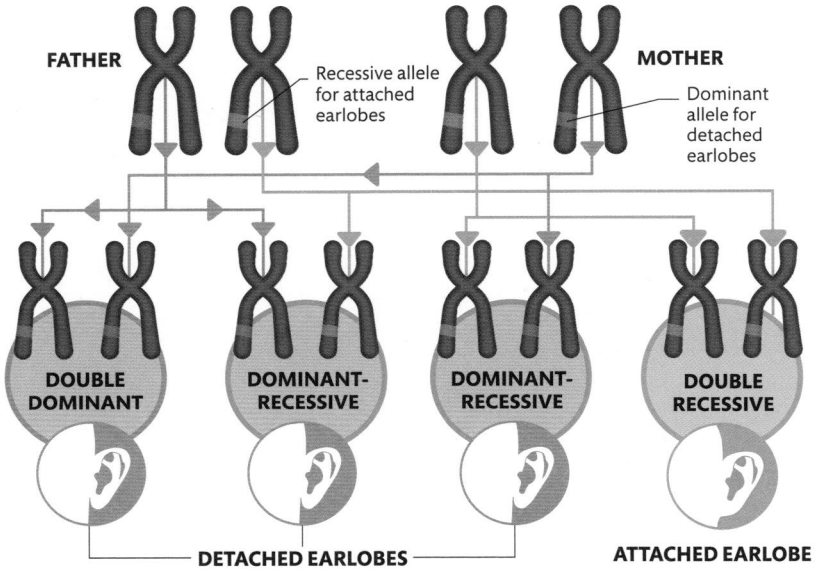

GENDER-LINKED INHERITANCE

If a mother carries a faulty recessive gene for a vision problem on one X chromosome, her body will use the fully working gene on her other X chromosome. A daughter who inherits the faulty gene will (like her mother) be a carrier and won't be affected, as the dominant gene masks its effects. However, as males have only one X chromosome, any son with the faulty gene will have deficient vision.

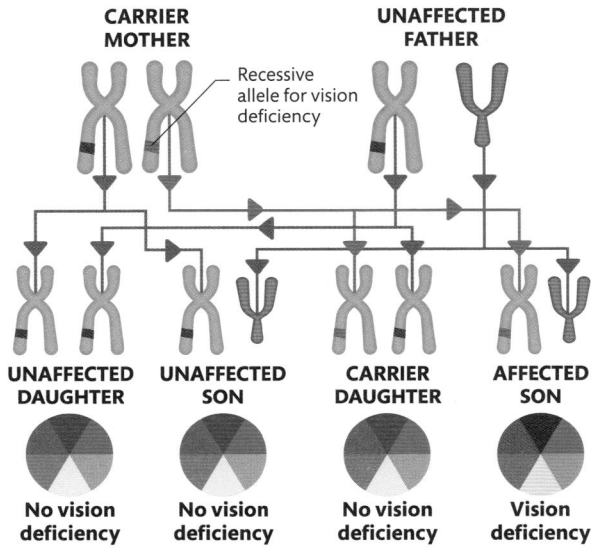

Growing life

The development of new life is a miraculous process in which a fertilized egg divides to form a full-grown baby in just 9 months. Connecting the mother and child is the placenta, a special organ that provides the growing fetus with everything it needs.

Four-week embryo
The spine, eyes, limbs, and organs have started to form. The embryo is around 3/16 in (5 mm) in length and weighs 1/32 oz (1 g).

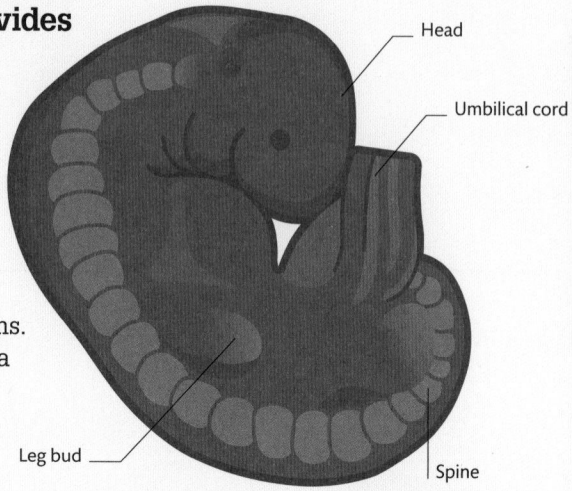

From cells to organs

During the first 8 weeks, the baby is known as an embryo. Genes switching on and off instruct cells on how to develop. Cells in the outer layer of the embryo form brain, nerve, and skin cells. The inner layer becomes the main organs, such as the intestines, while cells connecting the two layers develop into the muscles, bones, blood vessels, and reproductive organs. Once these main structures are laid down, the baby is called a fetus until birth.

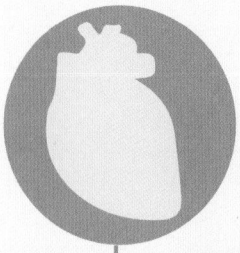

First heartbeat
Heart growth is almost complete by 6 weeks and all four chambers beat rapidly at around 144 beats per minute. This beating can be detected during an ultrasound scan.

Releases urine
Urine is released by the kidneys into the amniotic fluid every 30 minutes. It is diluted in the fluid and can be swallowed harmlessly by the fetus. Eventually, it passes via the placenta to the mother and she excretes it with her own urine.

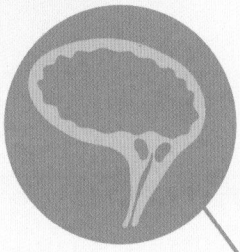

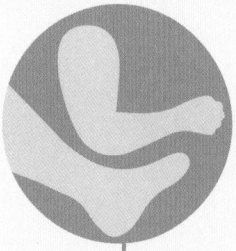

Tiny limbs
The upper limb buds will develop into the arms, while the lower limb buds will form the legs. Fingers and toes begin fused together, then separate.

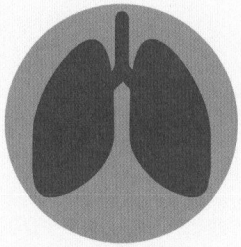

Lungs form
The two lungs begin to form around this time. They won't be ready to breathe air by themselves until the baby is almost ready to be born.

Fetal development
Every fetus develops at its own rate and the timing for key events tends to vary.

PREGNANCY TIMELINE

1 MONTH — 2 MONTHS — 3 MONTHS — 4 MONTHS

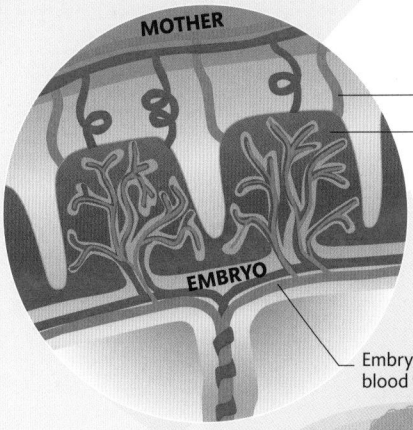

Mother's blood vessel

Mother's blood pools into space

Embryo's blood vessel

Meeting point
The baby's part of the placenta ends in a fine network of blood vessels that extends into the mother's half of the placenta—close to, but never mixing with, the mother's blood.

Support system

The baby is supported by the placenta—a unique organ that begins growing along with the embryo under the control of both the mother's and the baby's genes. In the placenta, blood vessels from both mother and fetus are intimately interwoven, but the blood never mixes. If it did, the mother's immune system would reject the fetus as "foreign." The fetus gets its oxygen and nutrients from its mother's blood, via the placenta and umbilical cord, and gets rid of wastes, such as carbon dioxide.

PLACENTA

AMNIOTIC FLUID

Umbilical cord

Sense of smell
The fetus can recognise the smell of its mother via the amniotic fluid. After birth, the baby is attracted to her smell.

Sensitive to noise
The baby is startled by loud noises. After birth, it will remember songs and voices it heard while inside the womb.

First look
The fetus's eyelids do not open until around the seventh month. When the eyes first open, they cannot form images—they can sense only light and dark.

Twitches and "kicking"
A baby's "kick" can be any number of movements the mother feels as her fetus flexes its spine and learns how to move its limbs.

5 MONTHS

6 MONTHS

7 MONTHS

8 MONTHS

9 MONTHS

Mother's new body

The growth of a baby inside a mother's body is an amazing feat—but also a demanding one. The body undergoes an incredible amount of changes and compromises during pregnancy.

Pregnancy transformation

Pregnancy is a time of great physical and emotional change. These changes prepare the mother for the extra demands of pregnancy. The body must not only supply its own needs but also provide the growing baby with all the oxygen, protein, energy, fluid, vitamins, and minerals it needs. The body also absorbs the baby's wastes, and processes them alongside its own. Organs start to support both the body and the baby, so expectant women may tire easily; however, the wonder of pregnancy is a remarkable example of the adaptability of the body.

WHAT CAUSES ODD FOOD CRAVINGS?

Food cravings are undoubtedly one of the strangest phenomena that accompany pregnancy. They may be a symptom of nutritional deficiencies. If the body or the baby is crying out for certain nutrients, this could lead to desires for odd food combinations, such as gherkins with ice cream. Cravings for non-nutritive items such as soil or coal are rarer, but do sometimes occur.

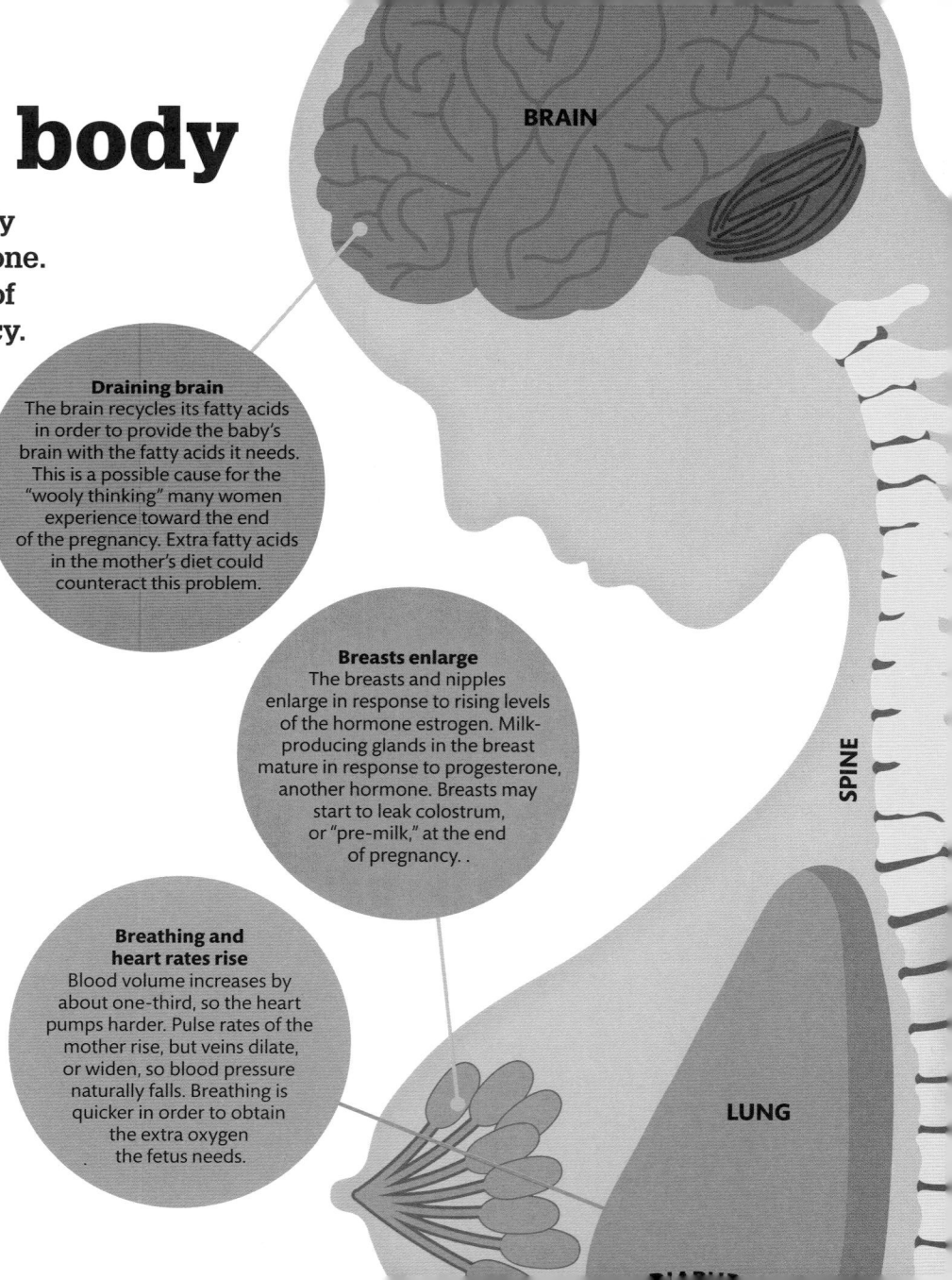

Draining brain
The brain recycles its fatty acids in order to provide the baby's brain with the fatty acids it needs. This is a possible cause for the "wooly thinking" many women experience toward the end of the pregnancy. Extra fatty acids in the mother's diet could counteract this problem.

Breasts enlarge
The breasts and nipples enlarge in response to rising levels of the hormone estrogen. Milk-producing glands in the breast mature in response to progesterone, another hormone. Breasts may start to leak colostrum, or "pre-milk," at the end of pregnancy. .

Breathing and heart rates rise
Blood volume increases by about one-third, so the heart pumps harder. Pulse rates of the mother rise, but veins dilate, or widen, so blood pressure naturally falls. Breathing is quicker in order to obtain the extra oxygen the fetus needs.

WHAT IS MORNING SICKNESS?

Early in pregnancy, hormone changes in the inner ear disrupt the balance of expectant mothers, inducing nausea and causing dizziness similar to when drunk. Morning sickness can happen at any time of day.

THE UTERUS EXPANDS TO 500 TIMES ITS NORMAL SIZE BY THE END OF PREGNANCY

STRETCH MARKS

Stretch marks are a result of rapid weight gain and stretching of the skin. Deeper in the skin, elastic fibres and collagen that normally keep skin firm and smooth wear thin over the course of pregnancy. Most women are left with stretch marks – however, some lucky women go through pregnancy unscathed.

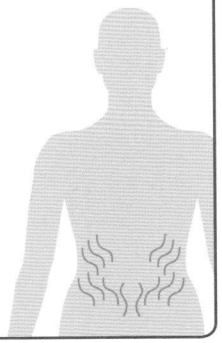

Squashed stomach
As the baby grows, so does the uterus—this pushes the mother's stomach up against the diaphragm. As a result, many expectant women experience heartburn due to acid reflux, and they may be afflicted with loud burps too!

Hormone producer
As it forms, the placenta produces a hormone, human chorionic gonadotropin (hCG), that is detected by pregnancy tests. The placenta then starts to produce estrogen and progesterone at an increasing rate, causing physical changes such as breast growth.

Abdomen growth
As the uterus expands out of the pelvis, the distance between the pubic bone and the top of the uterus (fundus) helps doctors estimate the stage of pregnancy. A fundal height of 9 in (22 cm) suggests a pregnancy is at around 22 weeks.

Squashed bladder
The bladder is squashed by the rapidly growing uterus, so it holds less urine, resulting in frequent visits to the bathroom. Late in pregnancy, the weight of the uterus stretches the muscles that support the bladder, which can lead to unfortunate leaks when coughing, laughing, or sneezing.

LIVER

STOMACH

Estrogen release

Progesterone release

Pressure on spine
As the uterus enlarges, the center of gravity of expectant women shifts forward so naturally they start to lean back. This alters their posture and puts extra strain on muscles, ligaments, and small joints in the lower spine, causing backache.

The miracle of birth

Giving birth to a new life is a daunting and exciting experience. Nine months of pregnancy have prepared mother and child for labor—which can last from 30 minutes up to a few days.

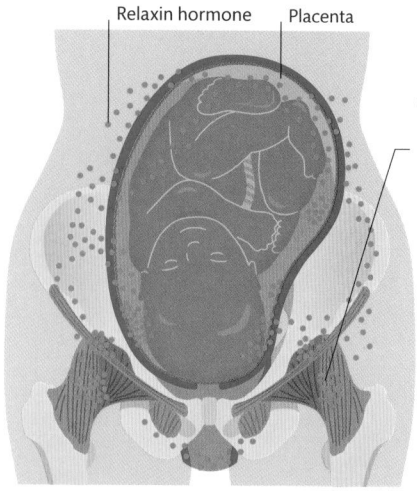

Initiating birth
Late in a pregnancy, the placenta will produce a hormone called relaxin that relaxes the pelvic ligaments to widen the pelvis, and softens and opens the cervix and vagina, in preparation for birth. The exact trigger for labor is unknown.

CONTRACTION AND DILATION

2 Cervix widens
The muscles in the uterus contract and push the baby's head against the cervix, which gradually dilates (widens) to around 4 in (10 cm) wide. Contractions are regular and painful. Typically, this stage lasts about 10 hours, but this can vary.

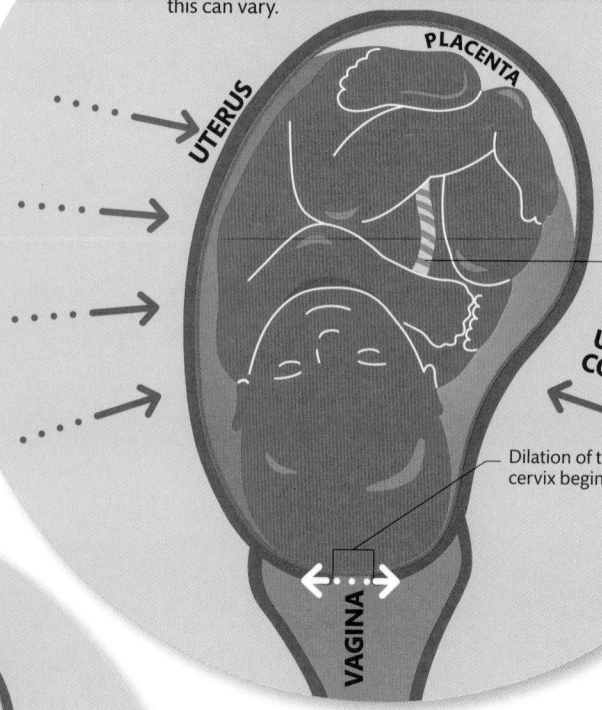

WATER BREAKS

1 Amniotic fluid
As the baby's head presses against the cervix, the amniotic sac breaks. Usually less than 10 fl oz (300 ml) of fluid leaks out. Unlike in the movies, it may not all come splashing out at once—it could be a steady trickle!

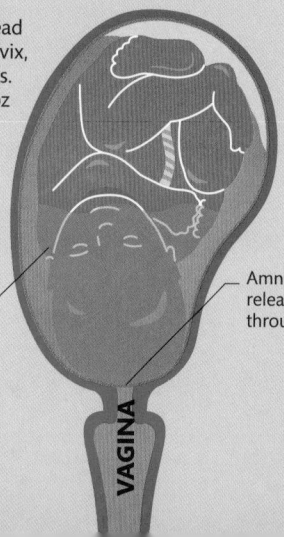

Labor differences

There are four stages to giving birth, but each stage may take a varying amount of time. Every woman has a different experience of labor, even if they have multiple births during their life. These stages can happen in quick succession, or over the course of a couple of days. During a second pregnancy, the time taken to get to the contraction stage may be shorter than it was for the first.

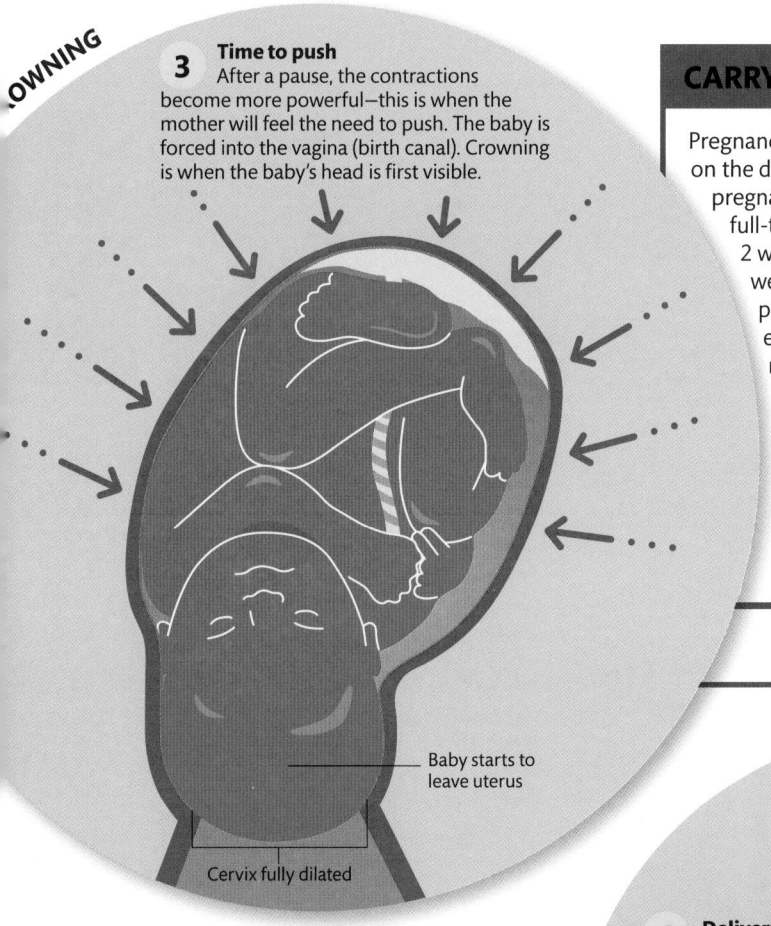

CARRYING TO FULL-TERM

Pregnancies can vary—only 1 in 20 babies are born on the due date calculated at the beginning of pregnancy. Doctors consider forty weeks to be full-term for a single pregnancy, give or take 2 weeks. For twins, doctors consider 37 weeks as full-term, and 34 weeks for a triplet pregnancy. Twins and triplets are born at a earlier stage of their development, so they require extra medical attention.

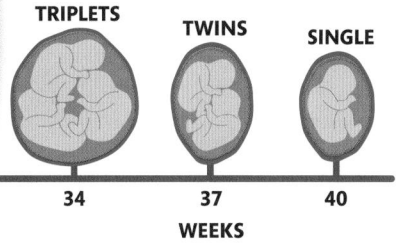

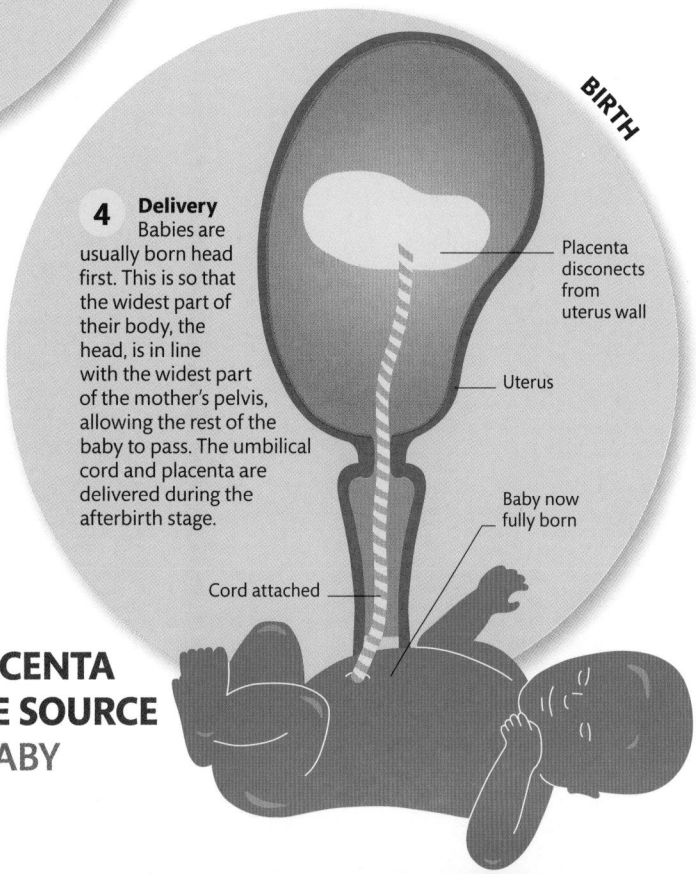

What happens after birth

After birth, the baby will take its first breath. In doing so, the baby's circulatory and respiratory systems begin to function independently from the mother for the first time. An instant rerouting of blood vessels occurs in order to obtain oxygen from the lungs. The pressure of the blood flowing back to the heart closes a hole in the heart, establishing a normal circulation.

BLOOD CAN BE COLLECTED FROM THE MOTHER'S PLACENTA AND STORED AS A FUTURE SOURCE OF STEM CELLS FOR THE BABY

Primed for life

Babies are born with features already in place that help them grow and develop. Between a newborn's skull bones are flexible, fibrous gaps that allow the head to expand as the brain gets bigger. Babies grow fast in their first year and triple their birthweight.

1 MONTH

1 Starting to smile
During the first month of life, babies listen, watch, and start to recognize people, objects, and places. They will probably smile for the first time at the age of 4–6 weeks.

3 MONTHS

2 Trying to roll
At 3 months, babies can balance their head, kick and wriggle, and try to roll over from their back onto their front.

6 MONTHS

3 Babbling begins
At this point, babies speak with babbles and coos. They imitate sounds and respond to simple commands such as "yes" and "no."

9 MONTHS

4 Sitting up
At about 9 months, babies sit up and start to shuffle or crawl. As motor functions develop, they are constantly moving.

10 MONTHS

5 Walking tall
Babies start to walk most probably between the ages of 10 and 18 months. Their first steps will occur when holding onto something.

12 MONTHS

6 Recognizing themselves
By the age of 12 months, babies know their own name, and by 18 months, they start to recognize their own image.

BABY REFLEXES

Babies are born with over 70 survival reflexes. Placing a finger next to a baby's cheek will make them turn their head and open their mouth. This is the rooting reflex, and it helps them find their mother's nipple when hungry. It fades when regular feeding is established. The grasp reflex helps stabilize them if they fall, and placing a baby on its stomach will initiate the crawling reflex Both of these are needed for a longer period of time.

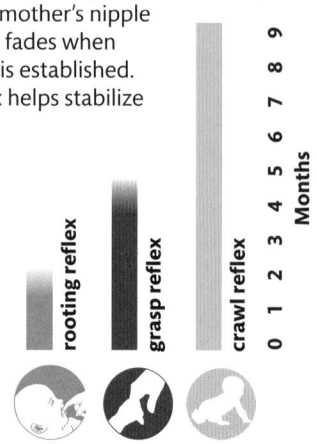

Developmental milestones

During the first year of life, babies develop skills that help them explore the world around them. Milestones of development, such as their first smile and first steps, help their caregivers monitor their progress.

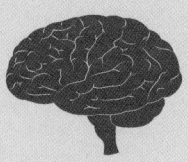

A NEWBORN'S BRAIN IS ABOUT ONE-QUARTER OF ITS ADULT SIZE

Focused senses

Newborns can focus on objects within 10 in (25 cm) and can distinguish shapes and patterns. They are familiar with their mother's voice from the womb and they are soothed by gentle, rhythmic noises similar to those of their mother's heartbeat. Babies also recognize their mother's smell.

3 days
At first, a baby can see only in black and white. A baby finds faces particularly fascinating.

1 month
Normal color and binocular vision start to develop at the age of about 1 month.

6 months
By 6 months, a baby's vision is excellent. The baby can now distinguish between faces.

Improved dental health when breastfed

Fewer respiratory problems when breastfed

Lower heart rate in breastfed infants

Occurence of food allergies decreases when breastfed for 6 months

Juvenile arthritis is less common when breastfed

Importance of breastfeeding

Breast milk is the most important source of food for a growing newborn. It is so nutritionally rich that it provides all the energy, protein, fat, vitamins, minerals, and fluids a baby needs during the first 4–6 months. Breast milk also supplies friendly bacteria, conveys antibodies and white blood cells that protect against disease, and delivers essential fatty acids that are vital for development of the brain and eyes. The benefits of breastfeeding are numerous, and influence all of a baby's bones and tissues, and most of the organs.

Understanding others

Between the ages of 1 and 5, most children develop an understanding that other people have their own minds and their own points of view. This is called the "theory of mind." Once children realize that everyone has their own thoughts and feelings, they can learn to take turns, share toys, understand emotion, and enjoy increasingly complex pretend play as they act out the roles they observe during everyday life.

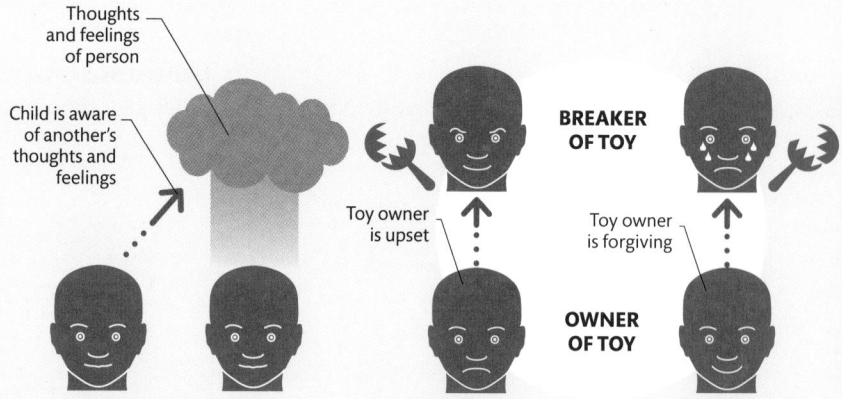

Understanding others
A child with theory of mind can predict how others might feel in a situation, can understand the intentions behind someone else's actions, and can judge how to respond.

Resentment
Realizing a friend broke a toy on purpose causes upset, as the child understands the nefarious intention.

Forgiveness
By recognizing that the break was accidental, the child understands that the friend is sorry, and the friendship is secure.

Steady growth

Childhood is a time of rapid physical and emotional growth. Social skills in adulthood are helpful, so it is imperative that children spend time with others of similar ages in order to understand themselves and each other, create boundaries, and establish social bonds. With steady physical growth comes advanced language, emotional awareness, and behavioral rules. New nerve cell connections form in the brain, laying the foundation for mental development.

Childhood development
As you grow, your body proportions develop into a more adult pattern. Growth slows between the ages of 5 and 8.

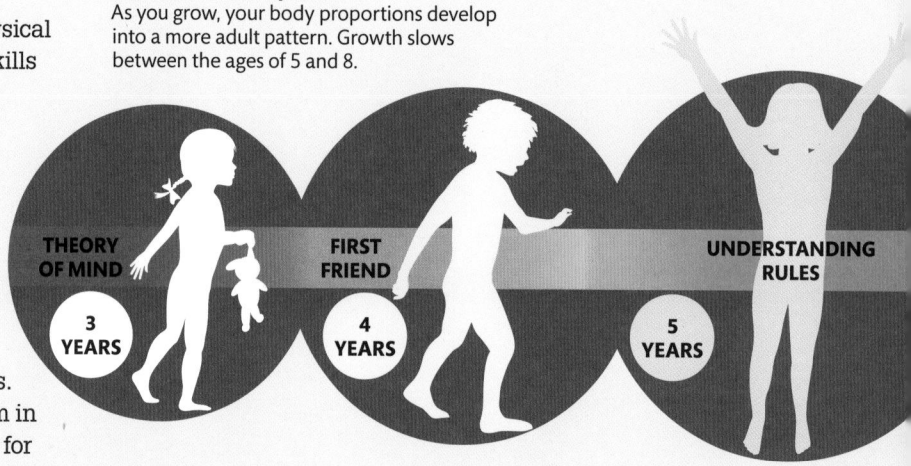

Growing up

You are full of curiosity and energy as a child. During the key stages of childhood until puberty, you gain a good grasp of language, understand that others have minds of their own, learn about the emotions of others, and actively start to explore your environment.

CHILDREN AGED BETWEEN 2 AND 10 **ASK** ABOUT **24 QUESTIONS EVERY HOUR**

Forming friendships

Many children aged 4 and above build selective friendships with others who share similar interests and activities. They have a sense of future, so they can understand the value of a friendship with someone reliable with whom secrets can be shared.

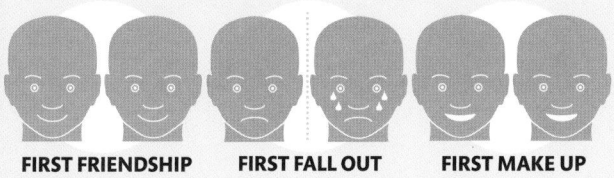

First resolution
Possessing a theory of mind helps friendships last. When they fall out, children can make up by reflecting on what made their friend upset in order to resolve the conflict.

Understanding rules

Rules-based games help children aged 5 and over to balance their desire to win with following the rules, which discourages cheating and bad behavior. This helps them recognize right from wrong and how society works later in life.

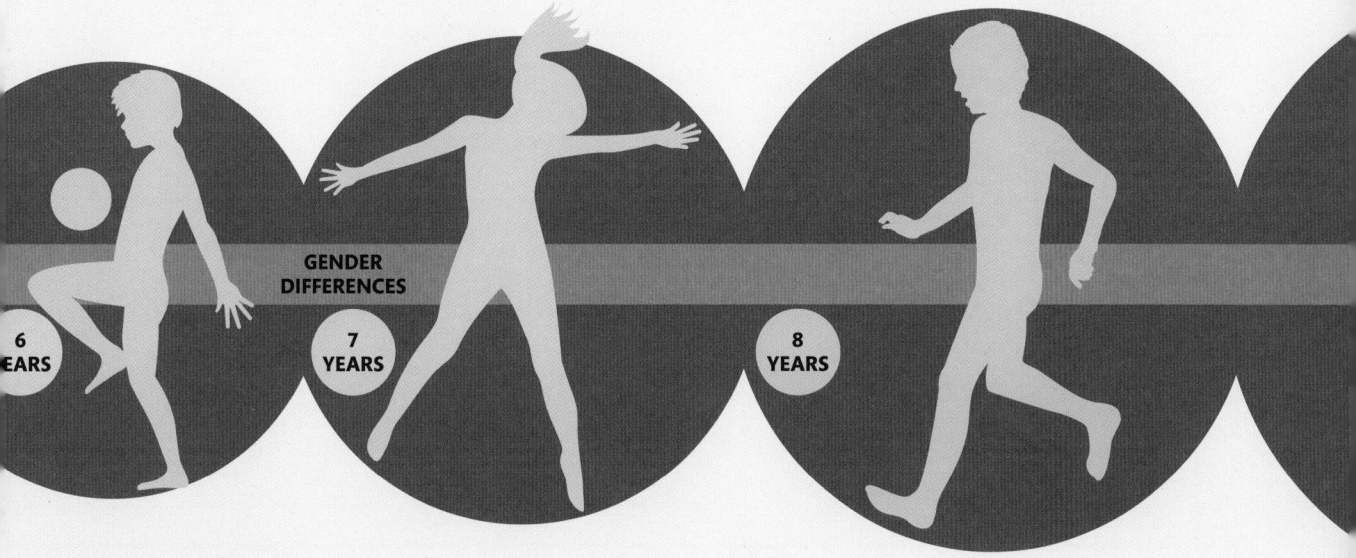

Friendship groups

Boys and girls have different types of friendship groups by the age of 7, each with their own hierarchy. Boys tend to form large groups of friends comprised of a leader, an inner ring of close friends, and peripheral followers. On the other hand, girls usually have one or two close friends with equal status. The most popular girls are highly sought after as "best" friends.

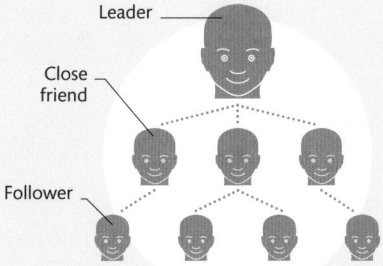

MALE FRIENDSHIPS

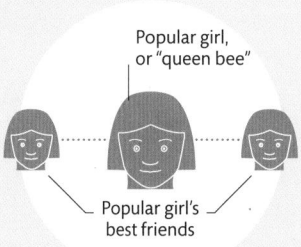

FEMALE FRIENDSHIPS

Hormonal teenagers

Puberty is the stage between childhood and adulthood, when the sex organs mature and reproduction becomes possible. Fluctuating hormone levels cause emotional and physical changes which can make teenagers feel clumsy, moody, and self-conscious.

Hypothalamus

VOICE DEEPENS

FAT CELLS

Pituitary gland

Voice breaks
Hormones cause the larynx to enlarge and the vocal cords to grow longer and thicker, deepening the voice.

Onset of puberty

When body weight and leptin (a hormone made in fat cells) reach certain levels, the hypothalamus will release pulses of gonadotrophin-releasing hormone, kickstarting changes in each gender.

Male changes

Boys usually enter puberty between the ages of 9 and 12. There is a wide variation in the rate at which it progresses, and it completes by the ages of 17 and 18.

CHEST BROADENS

The ribcage grows larger and some hair may grow, but not all males have hairy chests.

HAIR

TEENAGE BRAIN

The brain is undergoing its own changes, pruning old neural connections and forming new ones, and simply can't keep up with controlling the rapidly elongating limbs, muscles, and nerves. That's why teenagers may feel less coordinated than normal.

Female changes

Puberty generally starts a year earlier in girls than in boys, between the ages of 8 and 11. Puberty is completed by the ages of 15 and 19.

HAIR

BREAST GROWTH

Breast buds develop and may be tender. Nipples become more

DURING A **PUBERTY GROWTH SPURT,** HEIGHT MAY **INCREASE** BY AS MUCH AS **3½ IN** (9 CM) **IN A YEAR!**

Testes produce testosterone, accelerating puberty changes

PUBIC HAIR

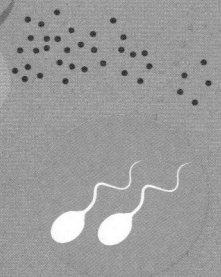

SPERM PRODUCTION IN TESTES

First ejaculation
The penis and testes grow and sperm production begins. The first ejaculation occurs, typically during sleep, as a "wet dream."

WHY DO TEENS GET ACNE?

The skin's sebaceous, or oil, glands are stimulated into action by the hormones of puberty. When newly active, they take a while to settle down to a normal rate of oil secretion, so during puberty, many teens suffer from spots.

EARLY AND LATE DEVELOPERS

Puberty starts at different ages, so some friends of the same age may be taller and seem more mature than others. Therefore, three girls at the age of 12 may differ drastically in height and weight. Girls tend to develop earlier than boys because a lower weight of around 105 lb (47 kg) seems to be the key to triggering female puberty. A higher weight average of around 120 lb (55 kg) appears to be the trigger in boys.

Less developed than her peers of same age

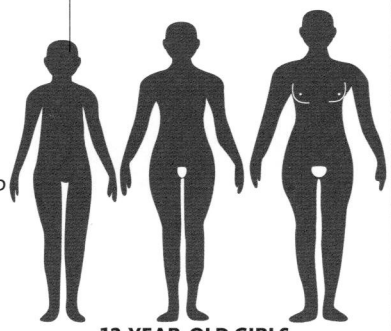

12-YEAR-OLD GIRLS

Ovaries produce estrogen, accelerating puberty changes

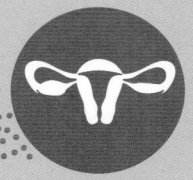

UTERUS AND OVARIES

Menstruation begins
The first period occurs between the ages of 10 and 16, at an average of 12 years. Ovulation occurs irregularly, and the uterus grows to the size of a clenched fist.

PUBIC HAIR

Vaginal secretion
The vagina lengthens and starts to secrete a clear or cream-white discharge—one of the first signs of puberty. The teenager's natural odor may also become stronger.

Getting older

Aging is a slow and inevitable process. The rate at which you age depends on interactions between your genes, diet, lifestyle, and environment.

Why do you age?

Why aging occurs is a mystery. We know that the cells in your body divide to renew themselves, but they can only do so a certain number of times. This limit is linked to the number of repeating units, called telomeres, on the end of each of their chromosomes, the X-shaped packages of DNA in every cell's nucleus. If you inherit long telomeres, your cells can undergo more divisions, and as a result you may live longer.

FREE RADICALS

Premature aging can result from genetic damage caused by free radicals. These molecular fragments are produced by sunlight, smoking, radiation, or pollution damaging your DNA. Dietary antioxidants found in fruit and vegetables help neutralize free radicals and offer an increased chance of living a longer life.

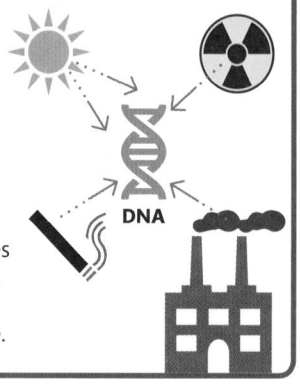

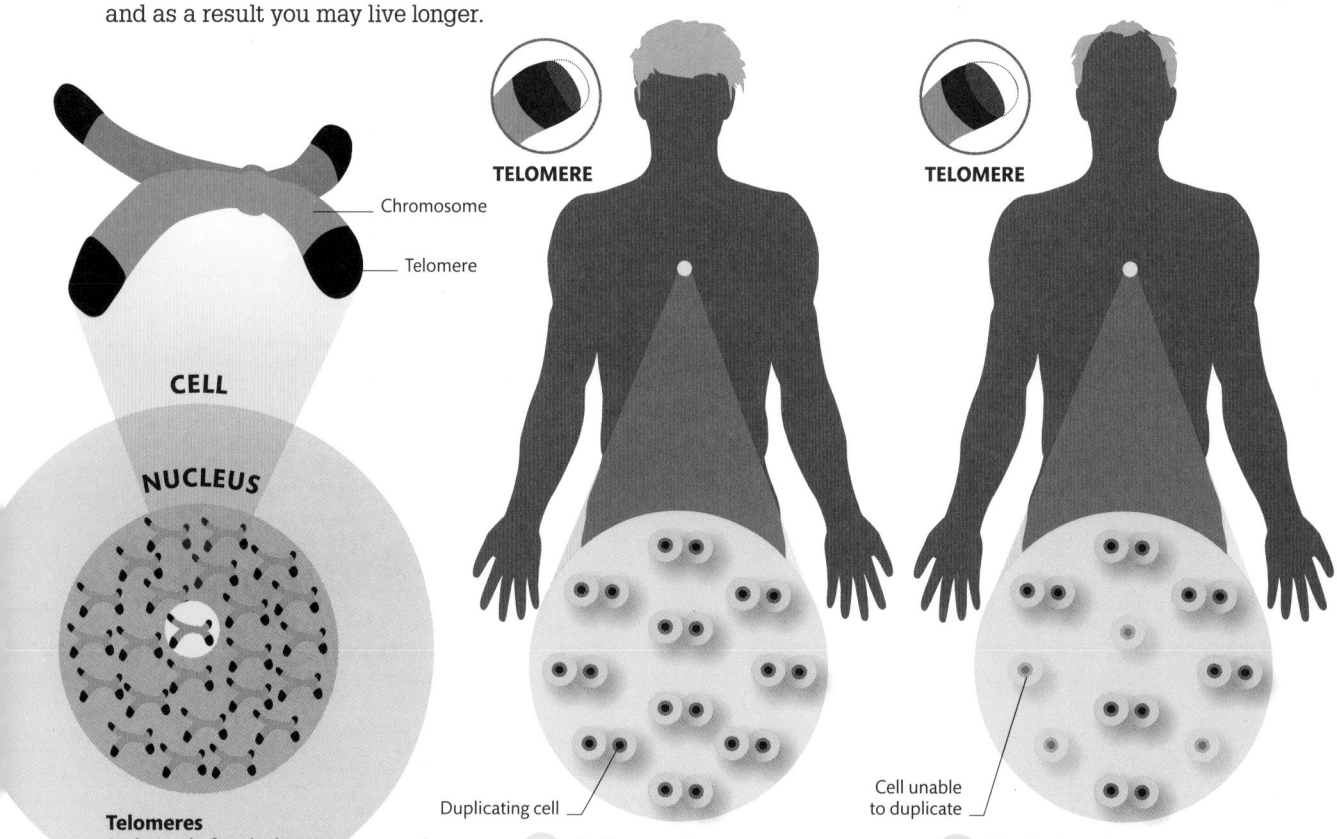

Telomeres
At the end of each chromosome arm is a telomere, a repeating section of DNA. During cell division, enzymes attach to the telomeres. These enzymes speed up the chemical reactions involved in cell division.

1 Cell renewal
The enzymes lock onto the telomeres, ready to copy each cell. When an enzyme detaches, it takes a section of telomere, so chromosomes shorten with each division.

2 Depleting telomeres
Eventually, the telomere becomes too short for enzymes to lock into place. The cells with these short telomeres can no longer duplicate or renew themselves. Cells run out of telomeres at different rates.

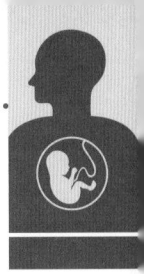

Going bald
Certain hair follicles become inactive, reaching the end of their life cycle quicker than others, resulting in different patterns of baldness in men and women. Even if you are bald, some of your hair follicles contain hairs, but they do not grow beyond the skin surface.

YOUTH
MIDDLE AGE
OLD AGE
Fine hair
Hair does not reach skin surface
Thinning hair follicle
Inactive hair follicle

OLD AGE
YOUTH
Macula
Buildup of waste products

Deteriorating sight
Light exposure can damage the macula, a key region at the back of the eye, causing a buildup of waste products that results in dark spots obscuring your vision. This is known as age-related macular degeneration.

ABSENT TELOMERE

Degeneration of hair cells in ear
OLD AGE
YOUTH

Hearing loss
Tiny hair cells within the inner ear convert sound waves into nerve signals that your brain interprets as sound. Damage or loss of these hair cells leads to increasing age-related deafness.

YOUTH
Excess melanin results in an age spot
OLD AGE
Melanocyte cell that produces the dark skin pigment melanin

Age spots
Whenever your skin is exposed to the sun, ultraviolet rays generate free radicals. These cause cells that produce pigmentation to increase their production, creating age spots.

Cell unable to duplicate

YOUTH
Healthy myelin sheath
OLD AGE
Decayed myelin sheath

Breakdown of nerves
The myelin sheaths that coat the nerve cells in the brain can deteriorate, so traveling electrical signals decelerate. This may account for slower thoughts, poor memory recall, and reduced sensation.

3 Unable to regenerate
Only a few replicating cells remain in old age. When cells can no longer renew themselves, they slowly deteriorate and the signs of aging become clear. Cells may die and be replaced with scar tissue or fat.

The end of life

Death is an inevitable part of the cycle of life. It occurs when all the biological functions that sustain living cells cease. Some deaths result from old age, while some are due to disease and injury.

Leading causes of death
Listed here are the leading causes of death worldwide in 2012, provided by WHO (World Health Organization).

High blood pressure–4%
Unchecked and uncorrected high blood pressure can be fatal late in life.

Diarrheal diseases–5%
Those suffering from chronic diarrhoea are at risk of fatal dehydration and malnutrition.

HIV–5%
Deaths caused by the Human Immunodeficiency Virus is decreasing year by year.

Traffic accidents –5%
Casualties on the road killed a large number of people in 2012.

Diabetes–5%
Those with diabetes may die due to heart disease or stroke because of their condition.

Lung infections and failures–16%
Lung cancers and lower respiratory infections together made the second-largest killer in 2012.

What can kill us

Noninfectious diseases, such as heart and lung disease, cancer, and diabetes are most commonly cited on death certificates. Many of these are related to an unhealthy diet, lack of exercise, and smoking, but some are due to nutrient deficiencies.

Heart and circulation conditions–60%
Heart attacks and strokes are the two leading causes of death worldwide.

HOW DOES WEALTH AFFECT LIFESPAN?

In high-income countries, 7 in every 10 deaths are among people aged 70 years or older, who've lived a good, long life. In the poorest countries, 1 in 10 children still die in infancy.

1 PERCENT OF THE WORLD'S POPULATION DIES EVERY YEAR

One way to determine whether a person is dead is to scan for brain activity. Brain death is diagnosed when electrical recordings (EEG) show an irreversible loss of all higher and lower brain functions, so there is no spontaneous breathing or heartbeat. Someone who is "brainstem dead" can only remain alive if artificial life support is in place.

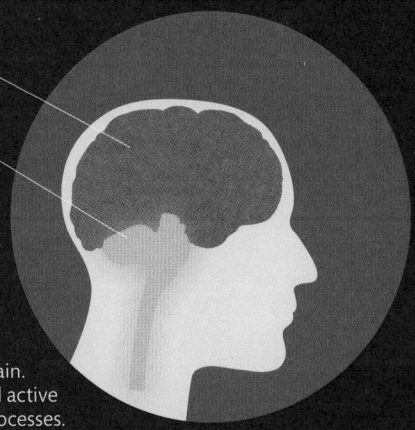

Comas
A coma is a state of unconsciousness in which someone cannot be woken, makes no movements, and does not respond to stimuli such as pain. Despite this, the brainstem is still active and can maintain some body processes.

Near-death experiences

People who almost die and are then resuscitated often report experiencing similar sensations, such as levitation, looking down on their body, and seeing a bright light at the end of a tunnel. Other common descriptions of such near-death experiences include having flashbacks, or vivid memories, of their earlier life, and being overcome by strong emotions, such as joy and serenity. The cause of these experiences may be changing oxygen levels, sudden release of brain chemicals, or surges of electrical activity—no one really knows.

YOUR BODY AFTER DEATH

When the heart stops pumping blood, the body's cells no longer receive oxygen or have toxins removed. Chemical changes in muscle cells and the overall cooling of the body cause the limbs of a corpse to stiffen after an initial period of floppiness. This stiffening is called rigor mortis and it wears off again after 2 days.

Rigor mortis
Rigor mortis starts in the eyelids and spreads to other muscles at a rate depending on surrounding temperature, age, gender, and other factors.

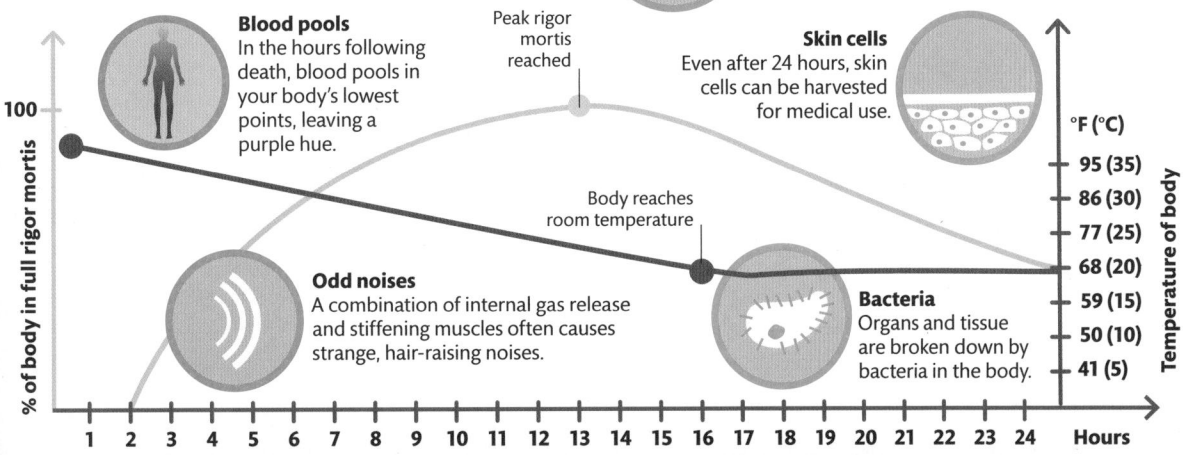

MIND MATTERS

Basis of learning

When we learn a new fact, gain a skill, or react to stimuli, connections between nerve cells form. Messages are passed from one cell to another by neurotransmitters (chemicals that are released by nerve cells). The more frequently we remember what we have learned, the more messages the cells send, and their connection becomes stronger.

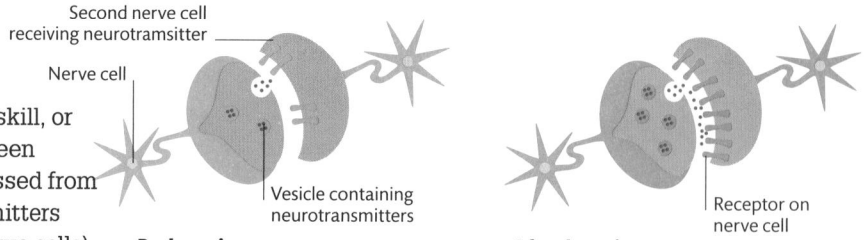

Prelearning
Initially, when the nerve cell fires, a small amount of neurotransmitter is released, and there are only a few receptors on the receiving nerve cell.

After learning
The nerve cell releases more neurotransmitter and more receptors have formed on the second nerve cell, strengthening the connection.

Types of learning

We learn information in different ways, depending on what it is and how it is presented. For some abilities, we have a "critical period" during which we can fully master the skill. Adults who have learned a new language later in life, have missed the critical period of acquiring the basic sounds of the language, and therefore may speak with an accent.

LEARNING WHAT TO IGNORE

Unimportant signals
When a stimulus is new, we automatically pay attention to it. If it doesn't signal anything important, we learn to ignore it.

Startled at sound

No response to sound

BEHAVIOR REINFORCEMEN

Rewards and reprimands
Getting rewarded for good behavior and reprimanded f or bad behavior can help reinforce our concepts of what is acceptable, and what isn't.

Rewarded behav

Behavior leading to a reprimand

LEARNING BY ASSOCIATION

Associative learning
When two events coincide on a regular basis, we learn to associate them. If you consistently eat when a bell rings, hearing the bell may stir your appetite.

Hunger caused by combined stimuli

Sound alone causes hunger

Learning skills

Connections between nerve cells in our brains allow learning to happen constantly, often with no conscious effort from us—repetition helps retain these skills.

EXPLORING A NEW CITY INCREASES YOUR BRAIN SIZE BY FORMING NEW NERVE CELL CONNECTIONS

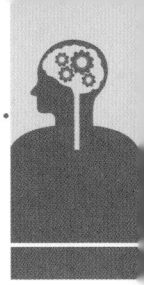

AT WHAT AGE DO WE LEARN THE MOST?

When you are a child, your cognitive, motor, and language skills advance in leaps and bounds—at the age of 2, you tend to learn 10–20 words per week!

LEARNING WHAT IS IMPORTANT

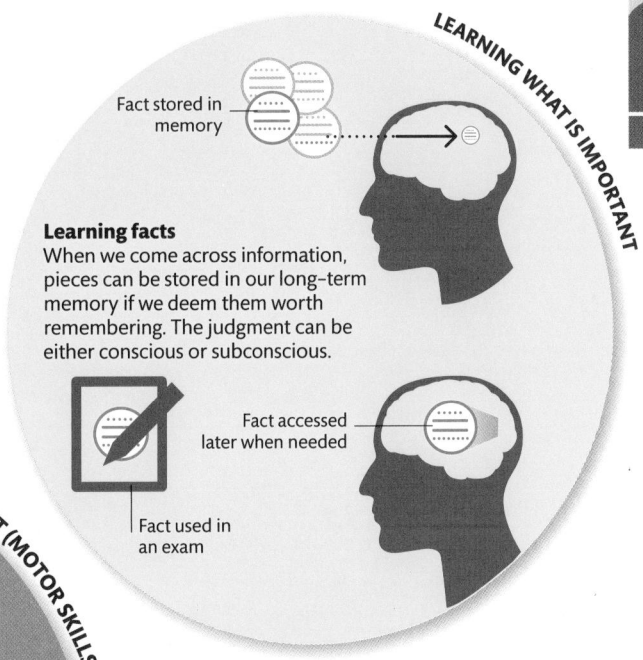

Learning facts
When we come across information, pieces can be stored in our long-term memory if we deem them worth remembering. The judgment can be either conscious or subconscious.

LEARNED MOVEMENT (MOTOR SKILLS)

Becoming automatic
When you learn to drive, you concentrate on your movements as well as the traffic. Through repetition, the driving body movements are learned and become automatic, allowing you to give attention to other things at the same time.

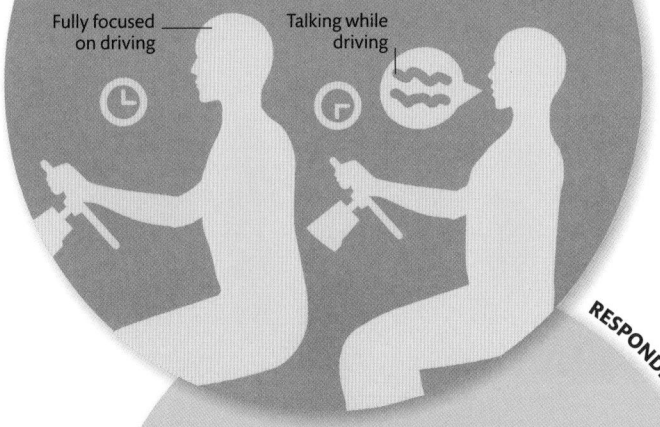

RESPONDING TO EVENTS

Episodic memory
By reviewing our experiences we learn to avoid undesirable situations, such as forgetting our umbrellas on rainy days.

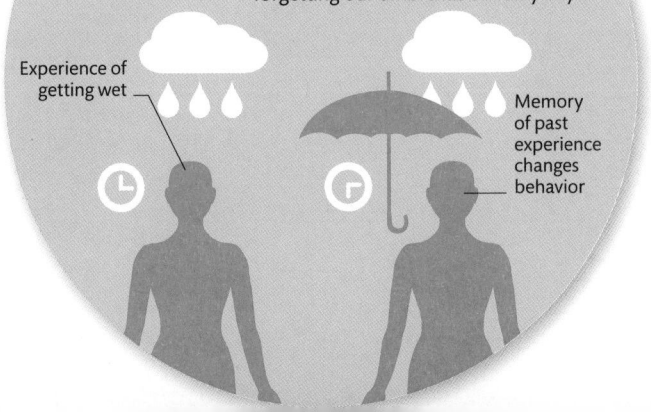

EXAM REVISION

When a memory starts to fade, revising the information increases the memory's strength with each revision session—this ensures that the learned information is stored in our long-term memory. Revisiting information little and often is best for retention. When you cram for an exam or presentation, you acquire a lot of information quickly, but it is lost without revisiting the information—this is why intense study can be useful in the short-term.

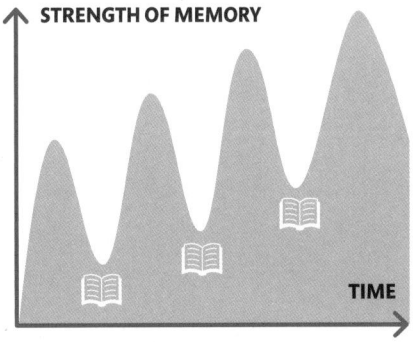

Making memories

Every time you experience something, your brain forms a memory. Inconsequential moments and life-changing events are all stored, but how often you revisit the memory determines whether it is remembered or forgotten. Memories are temporarily stored in your short-term memory, then, if important, are transferred to your long-term memory.

WHY DO WE EXPERIENCE DÉJÀ VU?

A sense of familiarity in unfamiliar situations may be because a similar memory is recalled but is confused with the present, so a sense of recognition comes without a concrete memory.

1 Sensory memory
When you sense something, you create a transitory memory, even if you are not conscious of it. It is stored in your sensory memory, and unless transferred to short-term memory, it fades in less than a second.

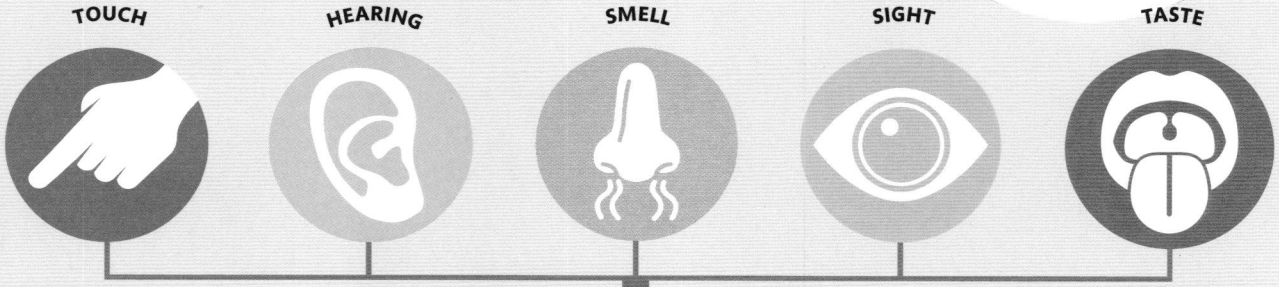

2 Nerve signals
Encoding is the process by which a sensory memory forms a true memory. When you pay attention to your sensory memory, it enters your consciousness, and the nerve cells that encode the memory fire more rapidly. Nerve cell connections strengthen temporarily to form a short-term memory.

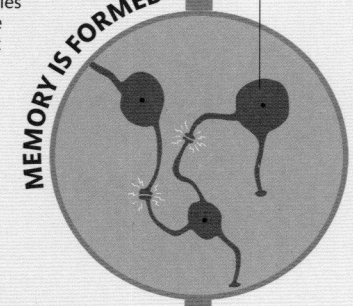

3 Consolidation
New experiences are compared against memories to provide context for new memories. Memories that have emotions and importance attached to them are stronger and less likely to be lost. Sleep is vital for consolidation to happen effectively.

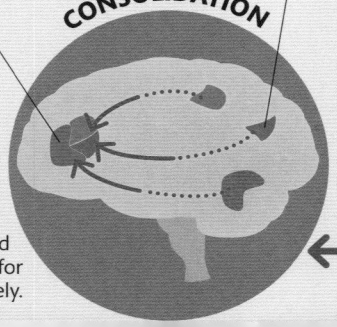

Short-term memory

Our short-term memory can retain around five to seven pieces of information. These memories, such as telephone numbers or directions, are stored only for as long as you need them. Repeating it to yourself helps prolong the memory, but if distracted, you often forget it. Short-term memory is thought to be based on temporary patterns of activity in the brain's prefrontal cortex.

Long-term memory

As far as we know, your long-term memory allows you to store unlimited amounts of information. Memories that are most likely to be stored for life include those with a high emotional impact, like a wedding, and that have a semantic value, such as your spouse's name. These memories are connected to growth in areas of the brain linked with memory, such as the hippocampus, so are more stable than short-term memories.

MEMORY CONFABULATION

When you recall a memory, the memory enters a labile, or easily altered, state. In a process called confabulation, you may unintentionally add new information to the labile memory when it is reconsolidated by your subconscious. This new information will become an inseparable part of your memory.

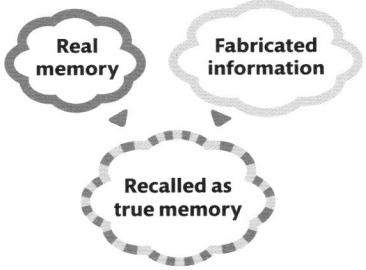

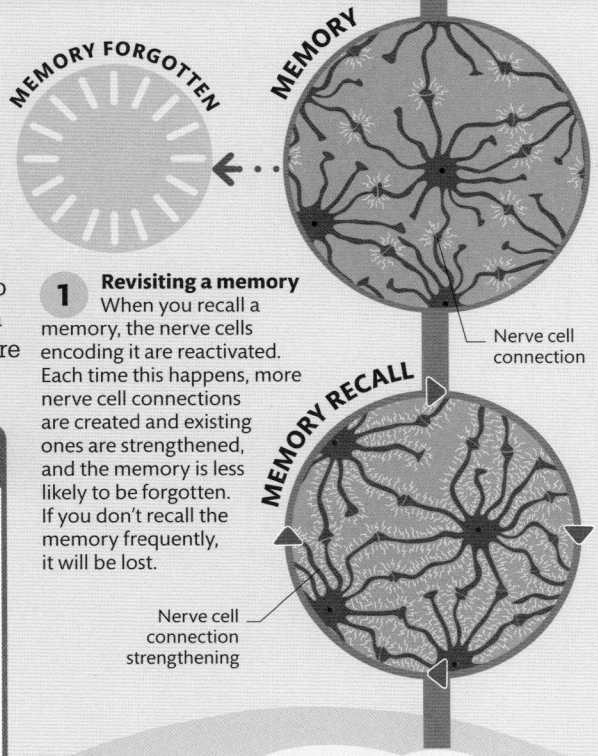

1 Revisiting a memory
When you recall a memory, the nerve cells encoding it are reactivated. Each time this happens, more nerve cell connections are created and existing ones are strengthened, and the memory is less likely to be forgotten. If you don't recall the memory frequently, it will be lost.

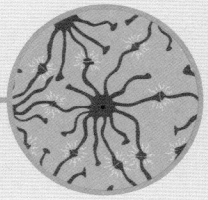

2 Storage
Months later, your nerve cell connections may become permanent. Particularly memorable experiences can jump straight into long-term storage the same day.

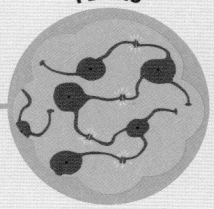

3 Memory fades
If months or years pass before you recall a memory, it is more likely to fade. Specific details about special events, such as the food you ate at your wedding, may be forgotten.

4 Losing a memory
Eventually, memories fade—even important ones! It is not known if the nerve cell connections of a memory disappear, or whether they still exist and you are unable to access them.

Falling asleep

Sleep is a curious phenomenon—we do it every day, but we don't know why. It might allow your body and brain time to repair themselves, flush out toxins that accumulate throughout the day, or strengthen memories. Depriving yourself of sleep is taxing for your body.

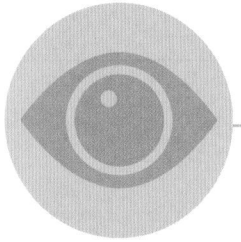

Rapid eye movement sleep (REMs)
Most dreams occur during REM sleep. If woken in this stage, you are likely to remember dreaming. Your eyes move under your lids as you dream.

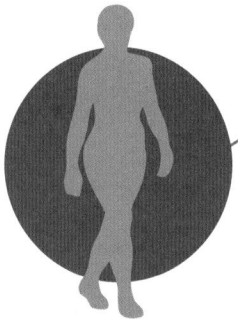

Sleep walking
Sleep walking is most likely to occur during deep sleep—but why it happens remains a mystery. You may walk around, eat, or even drive a car!

WE SPEND **ONE-THIRD OF OUR LIVES ASLEEP**, AND WE DON'T KNOW WHY

7 AM
6 AM
5 AM
4 AM
3 AM
2 AM
1 AM
12 PM

AWAKE
REM
LEVEL 1
LEVEL 2
LEVEL 3
LEVEL 4
LIGHT SLEEP
DEEP SLEEP

Body paralysis
During REM sleep, muscles are paralyzed, so you don't act out your dreams. You can wake in this stage. During this scary experience, you are semiconscious but unable to move.

After a good night's sleep, sleep pressure is low

Adenosine hormone broken down during sleep

Sleep pressure
The longer you stay awake, the larger your sleep pressure. This pressure is comprised of rising levels of chemicals such as adenosine, which causes fatigue by inhibiting neurons in the brain. More adenosine is produced if you've had an active day.

Sleep pressure peaks at bedtime

Unlikely to enter REM sleep upon falling asleep

AVOIDING SLEEP

Many of us use caffeine to help keep us awake. It makes us more alert by blocking a chemical in the brain called adenosine, which is responsible for making us sleepy. After the effect wears off, we suddenly feel very tired.

Stages of sleep

Each night you pass through different sleep levels. Level 1 is between sleep and wakefulness. In this stage, you may twitch as muscle activity slows down. As you enter true sleep, Level 2, your heart rate and breathing become even. During deep sleep, Levels 3 and 4, your brain waves slow and become regular. You tend to enter bouts of REM sleep once you have passed through other sleep levels. In REM sleep, your heart rate increases and brain waves look similar to when you are awake.

A good night's sleep
Illustrated here is a typical 8-hour night's sleep. You climb and fall between different levels of sleep in 90-minute bouts, interspersed with REM.

- Awake
- REM sleep
- Level 1 sleep
- Level 2 sleep
- Level 3 sleep
- Level 4 sleep
- Sleep pressure

Range of effects
If you don't sleep you will suffer from a range of physical and cognitive effects. Long-term sleep deprivation can even cause hallucinations.

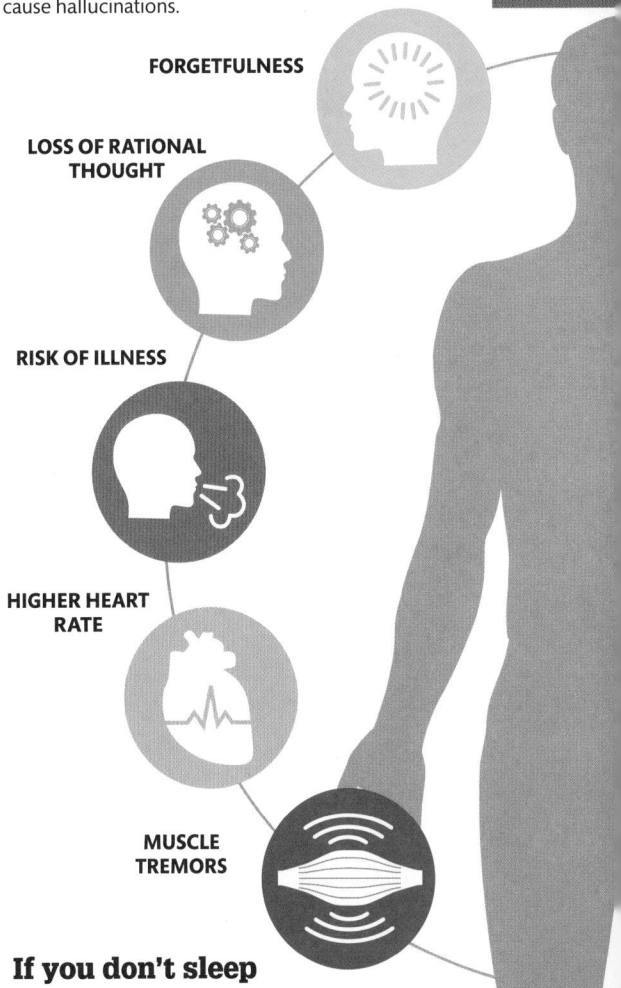

If you don't sleep

Going without sleep for a long time causes unpleasant symptoms. When you grow tired, your brain will steadily become unresponsive to neurotransmitters (chemicals) involved in regulating happiness. This is why tired people are often moody. When you sleep, your brain resets itself, and becomes sensitive to these neurotransmitters once again. The effects of sleep deprivation become progressively worse the longer you stay awake.

Entering your dreams

Your brain draws on and remixes your memories of people, places, and emotions to create sometimes complex and usually confusing virtual realities known as dreams.

Creating dreams

During REM sleep, your brain is far from asleep. It is highly active in this level of sleep, and this is when you do most of your dreaming. Areas of the brain associated with sensation and emotions are particularly active when you dream. Your heart and breathing rates are high because your brain consumes oxygen at a similar pace as when you are awake. Dreaming is thought to be linked to how the brain processes memories.

Sleepwalking and talking

Sleepwalking occurs during slow-wave, or deep, sleep. At this level of sleep, your muscles are not paralyzed, as they are during REM. The brainstem sends nervous signals to your brain's motor cortex, causing you to act out your dreams. It is more common when people are sleep deprived. Sleep talking occurs during REM sleep if nerve signals that usually paralyze your muscles are interrupted, temporarily allowing you to vocalize in your dreams. It may also happen when you are moving from one level of sleep to another.

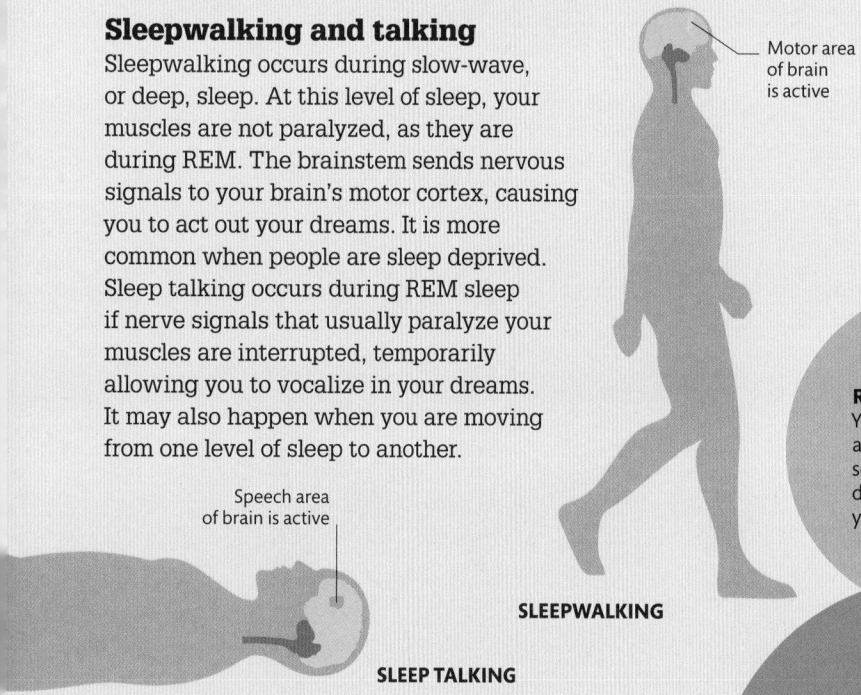

2 HOURS

THE ESTIMATED TOTAL **TIME YOU SPEND DREAMING EACH NIGHT**

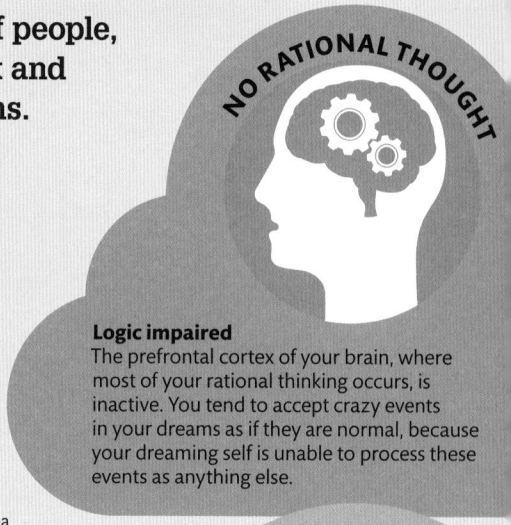

Logic impaired
The prefrontal cortex of your brain, where most of your rational thinking occurs, is inactive. You tend to accept crazy events in your dreams as if they are normal, because your dreaming self is unable to process these events as anything else.

Reliving sensations
Your brain receives little new sensory input when you are asleep, so the part of your brain that processes sensory signals is inactive. You do "sense" in your dreams, but you are reexperiencing sensations you had at some point when you were awake.

REM sleep
Nervous signals in the brainstem regulate brain activity during REM sleep. Interactions between "REM-on" and "REM-off" nerves control when and how often you pass into REM sleep. The muscles that move your eyes are the only muscles that are active in REM sleep, so your eyes move when you dream.

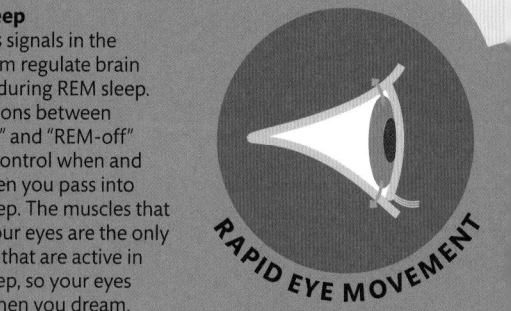

BODY PARALYZED

Inability to move
The motor cortex, which controls conscious movement, is inactive. The brainstem sends nerve signals to the spinal cord, initiating muscle paralysis, which prevents you from acting out your dreams. Production of neurotransmitters that stimulate motor nerves is completely shut down.

MEMORY CONSOLIDATION

Sleep is important for memory storage. You are more likely to retain new information after you have slept. Dreams are thought to be a by-product of your brain processing and shuffling new memories and forgetting unimportant ones.

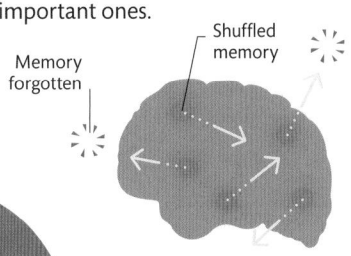

EMOTIVE RESPONSE

Emotions run wild
The emotional center in the middle of your brain is highly active, which explains the flurry of emotions you may experience when dreaming. This area encompasses the amygdala, which can be active during nightmares because it regulates your response to fear.

SPATIAL AWARENESS

Feeling of movement
Even though you don't move when you dream, you may feel as if you do. The cerebellum, which controls your spatial awareness, may become active, resulting in the feeling that you are running or falling in your dreams.

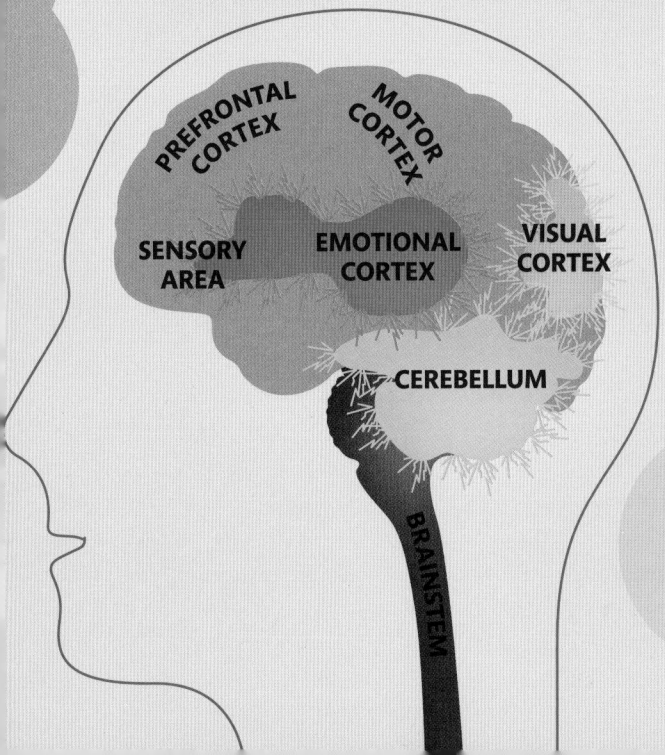

MENTAL IMAGERY

Remixed memories
The visual cortex at the rear of your brain is active, because it generates the imagery you experience in your dreams from remembered events. This can include places you've been, people you've met, and even objects you've interacted with. They can either be things you are emotionally attached to, or just completely random.

All emotional

Emotions influence our decisions and occupy much of our waking lives. Social bonds were vital for our ancestors' survival, so we have evolved to be able to read emotions in others. Understanding how emotions work has led us to believe that we can influence how we feel.

WHY DO WE CRY WHEN WE ARE SAD?

When you are feeling sad or stressed, the tears you shed secrete stress hormones such as cortisol, which is why we feel better after a good cry!

Basic emotions

A few basic emotions are universally identified. Happiness, sadness, fear, and anger seem to have facial expressions that are recognizable to people in the most widely separated cultures. Combining these gives rise to the huge number of complex emotions we experience.

Fear and anger
The body reactions for fear and anger are very similar, even though they involve different hormones. It is mainly your brain's interpretation that determines whether you feel angry or afraid.

Happiness and sadness
Your brain and large intestine produce hormones, including serotonin, dopamine, oxytocin, and endorphins, that affect happiness. Lower levels of these hormones result in sadness.

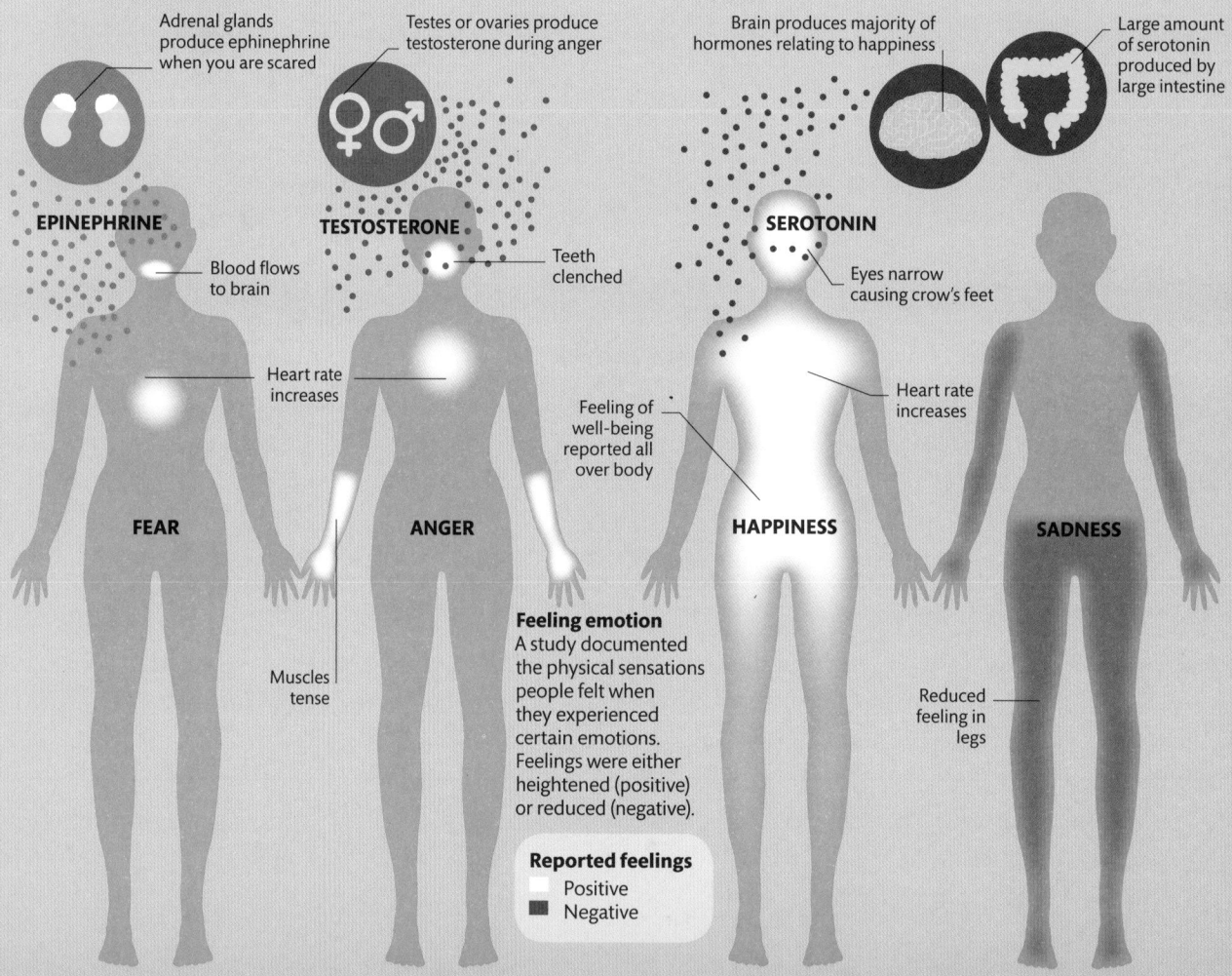

Feeling emotion
A study documented the physical sensations people felt when they experienced certain emotions. Feelings were either heightened (positive) or reduced (negative).

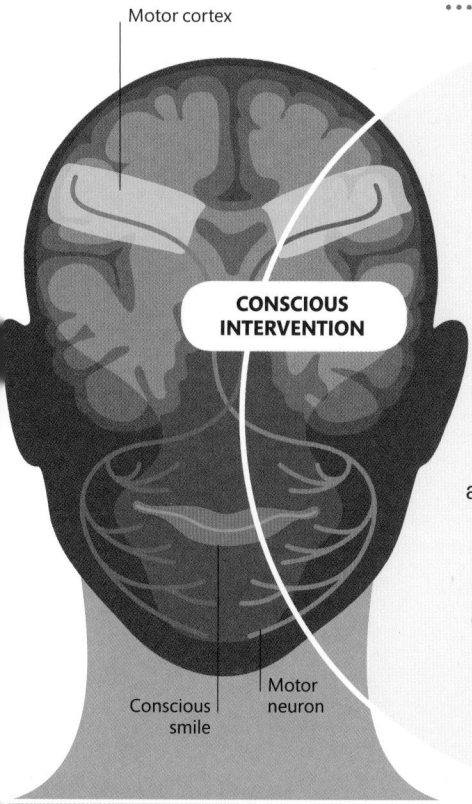

How emotions form

Emotions consist of feelings, expressions, and body symptoms. It may seem like your feelings come first, but a feedback loop allows the body to regulate your emotions and vice versa. At a certain point in this cycle you are able to reinforce, inhibit, or change emotions by altering your response. For example, if you are feeling happy, continuing to smile will make you feel even happier!

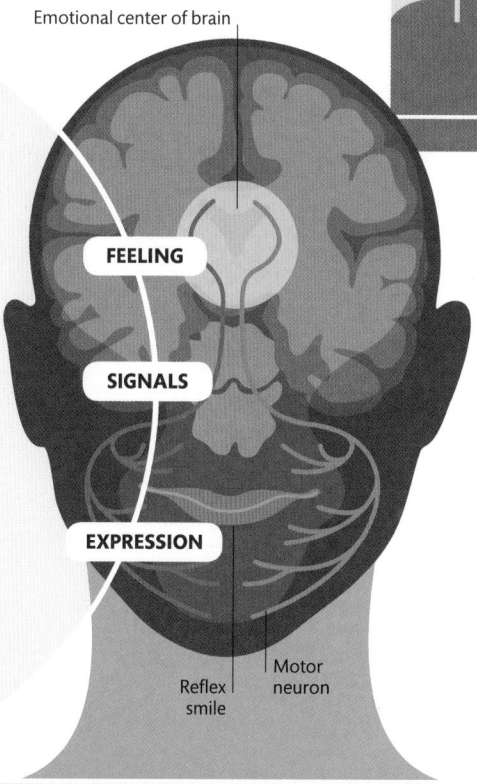

Conscious facial expressions
After you have started to experience an emotion, you are able to change your facial expression to hide or reinforce your true emotion. This action is consciously controlled by neural pathways from the motor cortex.

Reflex facial expressions
When you experience emotion, facial expressions appear without your control. For instance, when you hear good news, you cannot help but smile. These reflex actions are thought to be due to signals from the amygdala, in the brain's emotional center.

THE HAPPINESS YOU FEEL DURING A "RUNNER'S HIGH" IS CAUSED BY NATURAL CHEMICALS IN THE BRAIN CALLED OPIOIDS

WHY DO WE HAVE EMOTIONS?

Experts think that emotions evolved as a preverbal way of communicating. By understanding emotional signals, we can form stronger social bonds. Facial expressions can demonstrate that you are in need of help, are sorry for something you have done, or can warn others to stay away if you are angry. However, some scientists think there is a simpler explanation: the widened eyes of fear could help us see better, and the wrinkling of the nose in an expression of disgust could be a way of rejecting harmful chemicals in the air.

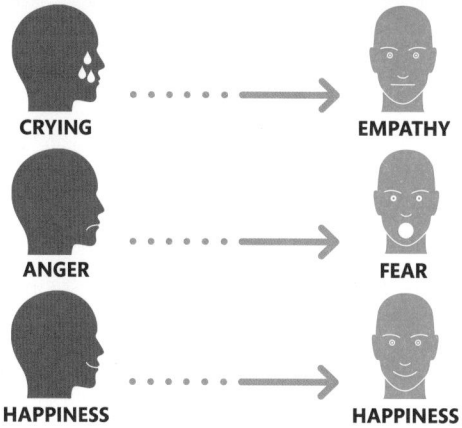

EMOTIVE RESPONSES

Fight or flight

When we are threatened, our body springs into action. Our brain sends signals to the body causing a variety of physiological changes that prepare us to face the challenge or to run away.

SNAKE

Activating a response

Have you ever been startled by a garden hose, only to realise it is not a snake and completely harmless? Before we are even consciously aware of a threat, our brain activates the nervous system, which causes the release of hormones from the adrenal glands. Meanwhile, the information also travels the longer route to our cortex where conscious brain regions can analyze whether the threat is genuine. If not, it will calm down the physical reaction.

CORTEX

Visual cortex processes the image after automatic reaction

VISUAL CORTEX

Thalamus passes sensory information as nervous signals to amygdala

THALAMUS

AMYGDALA

HIPPOCAMPUS

Amygdala activates nervous response and instructs pituitary gland to release hormones

Pituitary gland releases adrenocorticotropin (ACTH)

NERVOUS SIGNAL

HORMONE

1 Brain activity
The amygdala signals the body to take action before the fearful stimulus has even been recognized by the visual cortex—this happens when you jump at shadows. Then, the visual cortex fully analyzes the image to check if the threat is real, and your physical reactions adjust accordingly. Your cortex also consults memories stored in the hippocampus to check if the threat was faced in the past.

2 Alternative pathways
Signals from the brain are sent to the body via nerves, and also by the hormone ACTH released from the pituitary gland. The nervous signals travel faster than the hormone, so they kickstart hormone production in the adrenal glands.

IN TIMES OF **HIGH STRESS** YOU MAY EXPERIENCE **TUNNEL VISION**, IN WHICH YOU **DON'T NOTICE WHAT HAPPENS AROUND YOU**

MODERN STRESS

Modern stress tends to be very different from the type encountered by our ancestors—our stressors often overstay their welcome, and can't be dealt with by fighting or fleeing. Stress is helpful in the short term, but continued stress negatively affects your health, causing headaches and illness.

IMMEDIATE STRESS

PERSISTENT STRESS

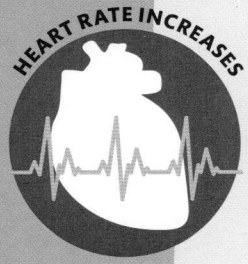

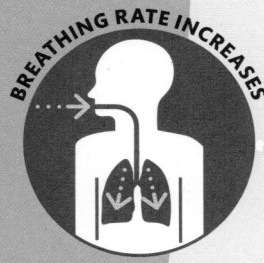

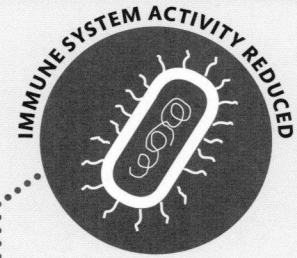

3 Hormone producer
The adrenal glands that sit on top of the kidneys produce more epinephrine and cortisol in response to the nervous signals and hormones sent by the pituitary gland. This heightens the physical effects of stress.

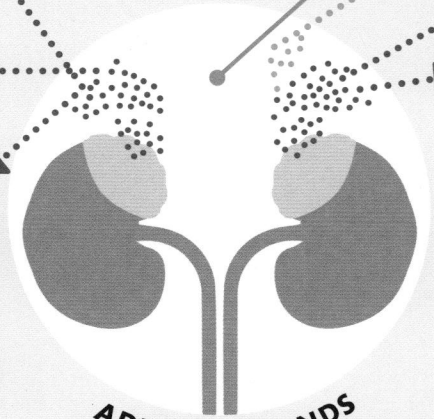

4 Short-term effects
Within seconds, heart rate and breathing increase to boost oxygen circulation. Blood vessels close to the skin constrict, leaving you pale, and your bladder muscles relax, possibly leading to embarrassing accidents!

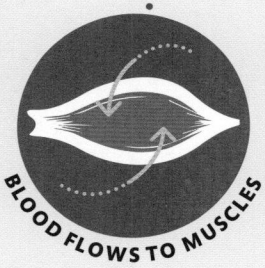

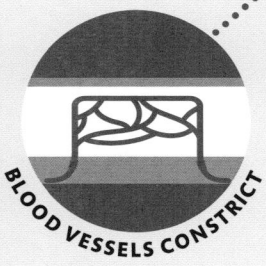

5 Long-term effects
Over minutes and hours, signals from the adrenal glands continue to cause a cascade of reactions. Blood sugar rises and fat stores are metabolized for energy so your muscles continue to work at their full potential. Nonvital processes, such as immune system activities, are shut down to conserve energy.

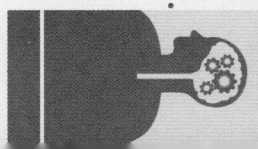

Emotional problems

Our emotions are controlled by a balance of chemicals and circuitry in the brain, so imbalances of certain chemicals can cause emotional disorders. Experts once believed they were purely psychological, but they now understand that physical changes underlie each illness.

Phobias

A fear is classed as a phobia if the fear outweighs the threat. It is logical to be wary of snakes, which can be deadly. If that fear extends to pictures or toys and begins to affect daily life, it becomes a phobia. Phobias can develop over time, be learned at an early age, or be associated with an incident involving the stimulus.

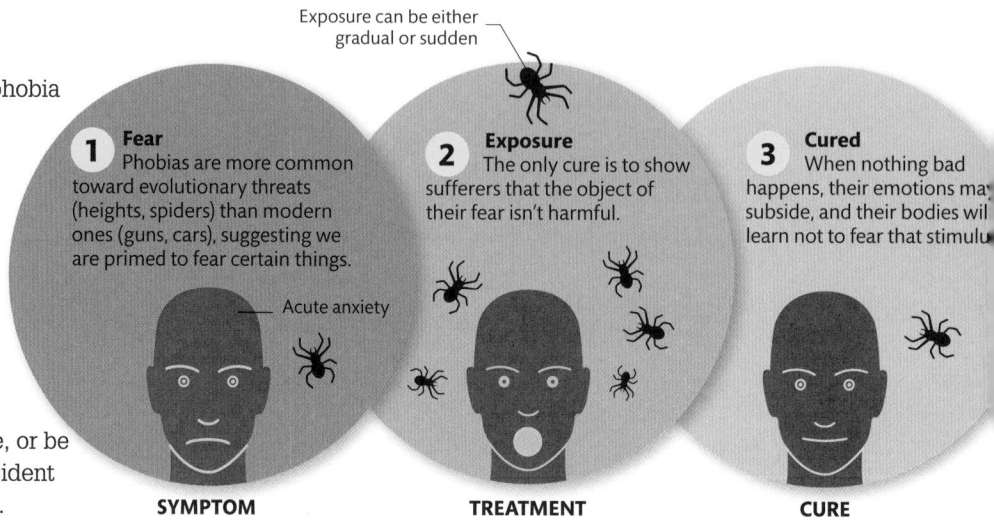

Obsessive compulsive disorder

Sufferers of obsessive compulsive disorder (OCD) experience intrusive negative thoughts leading to compulsive behaviors, which they incorrectly believe can relieve their anxiety. OCD is possibly caused by overactivity in the areas connecting the brain's frontal lobe to deeper areas. Most cases are manageable with treatment.

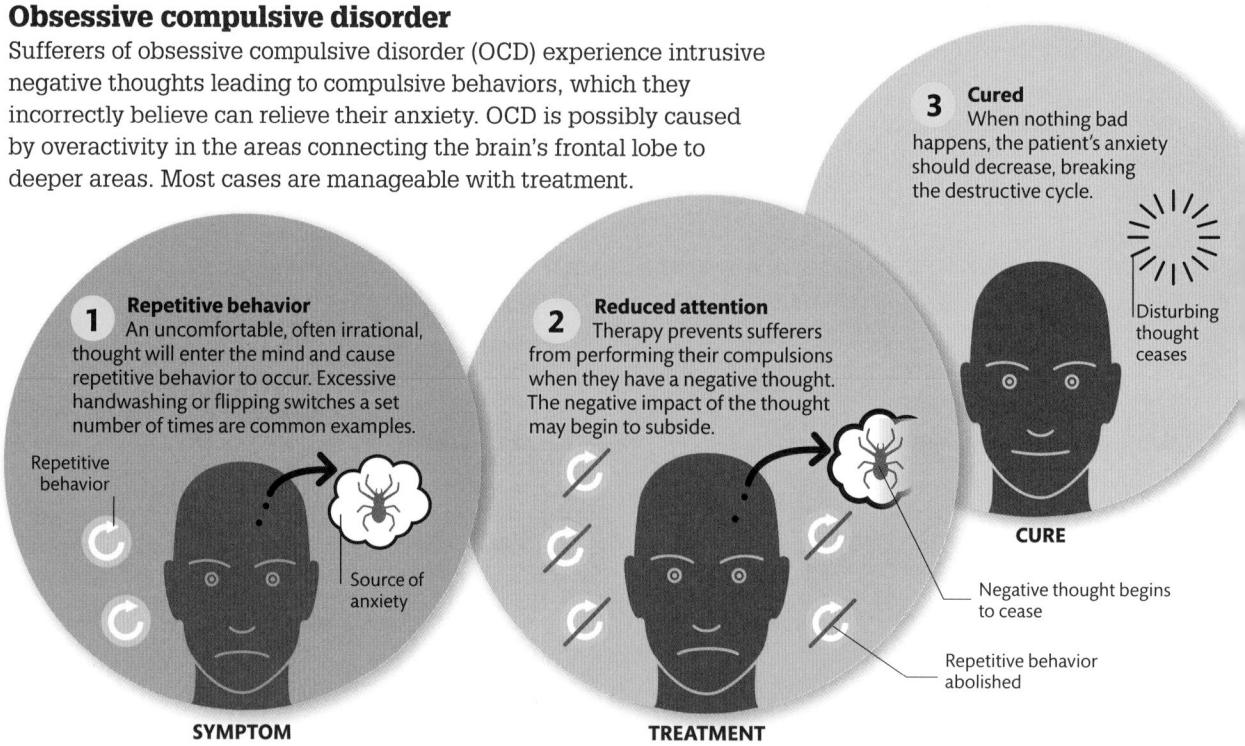

TRAUMATIC MEMORIES

After trauma, some people experience flashbacks, hypervigilance, anxiety, and depression—these are the symptoms of post traumatic stress disorder (PTSD). When you are afflicted, recalling the traumatic memory will trigger a "fight or flight" response, unlike ordinary memories. Treatments can be provided through therapy or drugs.

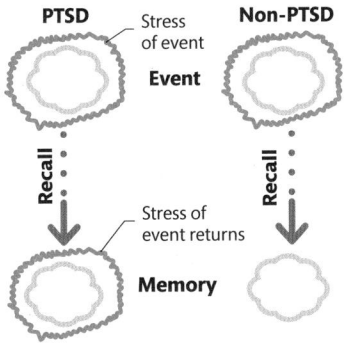

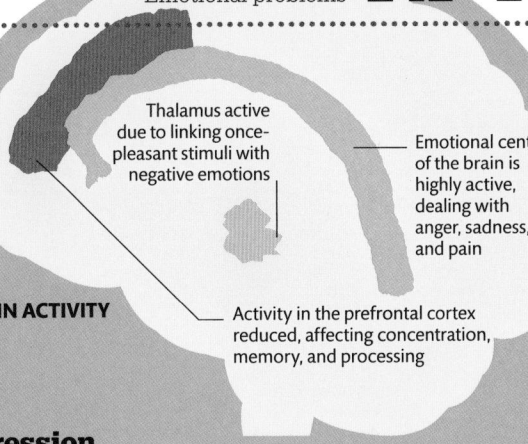

Depression

The symptoms of depression include low mood, apathy, sleeping problems, and headaches. It is thought to be caused by chemical imbalances in the brain, leading to certain areas becoming overactive or underactive. Antidepressants can help reset this balance by raising levels of chemicals, but they only address the symptoms, not the cause. Attitudes toward depression have progressed to understanding it as a condition, not a state of mind.

Bipolar disorder

Featuring changes in mood from mania to extreme depression, bipolar disorder is highly genetic—it runs in families—but it is often triggered by a stressful life event. Bipolar disorder is a type of depression. It is thought to be due to problems with the balance of certain chemicals in the brain, including norepinephrine and serotonin, and this causes the brain's synapses to become either overactive in mania or underactive in depression.

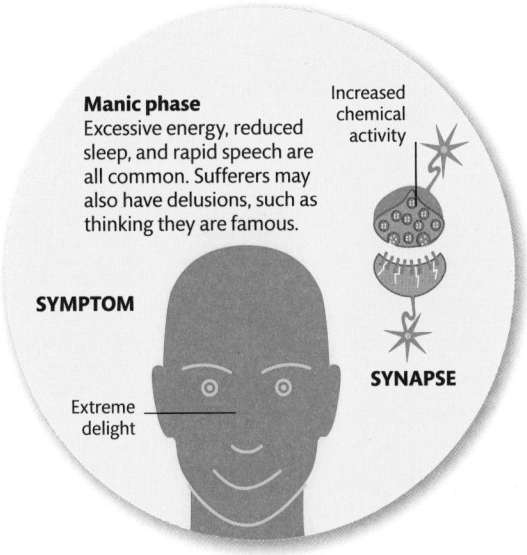

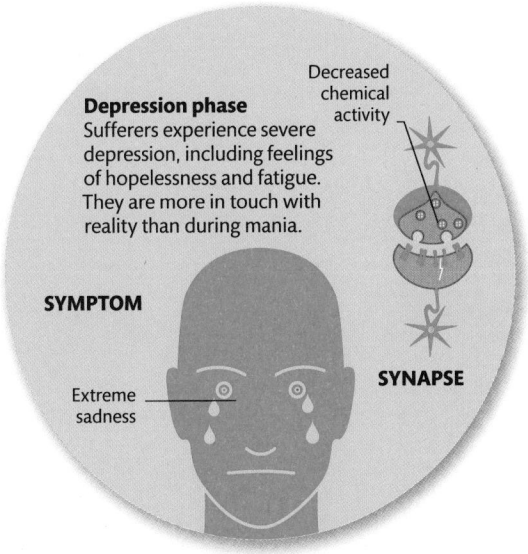

Feeling attraction

Scientists are only just beginning to understand what happens when we feel attracted to someone, why we are attracted to certain people and not others, and why we make our choices—and it is mostly to down to hormones.

DOES CULTURE AFFECT ATTRACTION?

Within a single culture, ideals of beauty change over time. In Europe, pale skin and a full figure once indicated wealth and was typically seen as attractive in a woman. Now, a thinner, more tanned figure is seen as desirable.

Chemical bond

When attraction begins, hormones play an important part in augmenting our romantic feelings. Levels of dopamine in the brain increase, providing the familiar rush of pleasure. A chemical that is converted into epinephrine is released, causing a dry mouth and sweaty palms. It also causes your pupils to widen, which signals your desire to the other person, making you increasingly attractive. Serotonin levels change and are believed to lead to obsessive, lustful thoughts.

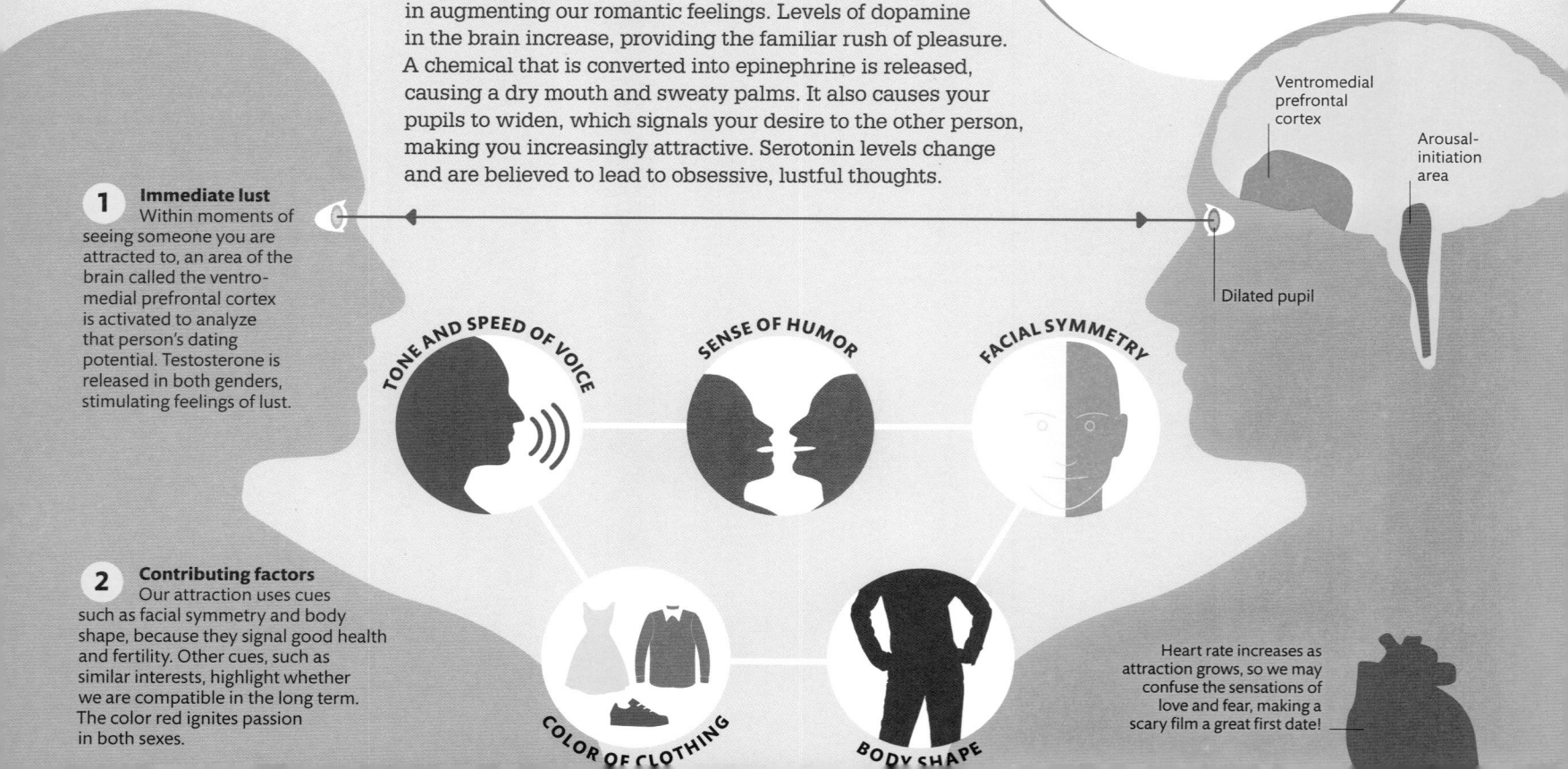

1 Immediate lust
Within moments of seeing someone you are attracted to, an area of the brain called the ventromedial prefrontal cortex is activated to analyze that person's dating potential. Testosterone is released in both genders, stimulating feelings of lust.

2 Contributing factors
Our attraction uses cues such as facial symmetry and body shape, because they signal good health and fertility. Other cues, such as similar interests, highlight whether we are compatible in the long term. The color red ignites passion in both sexes.

Heart rate increases as attraction grows, so we may confuse the sensations of love and fear, making a scary film a great first date!

3 **Long-term pair bonding**
After the initial attraction phase, relationships change, and a different set of hormones become important. Oxytocin is released after sex, and increases feelings of trust and bonding, which aids in establishing relationships. Another hormone, vasopressin, is equally important. It is released when two people spend a great deal of time together, promoting monogamy.

SEX

PROLONGED **EYE CONTACT INCREASES THE MAGNETISM** BETWEEN TWO PEOPLE

MENSTRUAL CYCLE

Subtle signals

In many animals, it is obvious when females are fertile, through bold signals such as brightly colored swellings on their body or pheromones in their urine. When it comes to humans, ovulation isn't as obvious—and it's not known why we evolved that way. Nevertheless, women do have subtle ways of advertising their fertility, such as flirting more and dressing more attractively, and men seem subconsciously to be able to pick up on these signals. One study showed that men release more testosterone when reacting to the scent of women who are ovulating than those in a less fertile phase of their cycle.

Changing signals
When women are ovulating, there are subtle changes that indicate fertility; voice pitch rises, cheeks flush, and you tend to flirt more and dress more attractively.

OVULATION

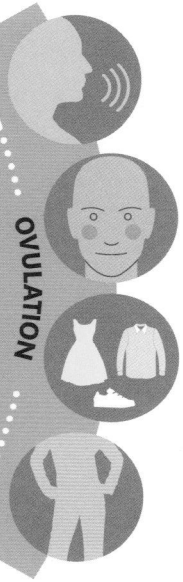

BODY ODOR

Sweat can tell us how healthy someone is, and even whether we are genetically compatible. People who have an immune system relatively different from our own tend to smell more attractive, since a mixing of these genes would encourage healthier offspring. Generally, women prefer the scent of men somewhat similar to their own over the scent of those who are either genetically identical or completely dissimilar.

Extraordinary minds

Everyone's brain is unique, but there are some people who can do amazing things that most of us can only dream of. Slight changes in the wiring of the brain, or the way we learn to use it, can give rise to these incredible abilities.

Delayed language
Children with autism (but not Asperger's) take longer to learn language, and some never become verbal. Those who do speak may have trouble using words to communicate with others as an adult.

Socializing impaired
Reduced eye contact is an early sign of autism. Autistic individuals tend to dislike socializing, finding its complex rules confusing and frightening. Nevertheless, this is not to say those with autism never form strong social bonds.

Repetitive behavior
People with autism process information differently, and this means everyday situations can be overwhelming. Self-soothing, routine behaviors are common, and can help people with autism calm themselves when anxious.

Specific interests
Those who are autistic often develop narrow, specific interests. These can be a source of comfort and enjoyment, possibly because the structure and order of familiar topics provides respite from the confusing social world.

SOMETIMES AUTISM LEADS TO

Autism spectrum

Autism spectrum disorders (including Asperger's syndrome) are probably caused by unusual patterns of connectivity in the brain. Genes are known to play a role because autism runs in the family, although why they affect some people only mildly while others need care throughout their lives isn't known.

Rare prodigious qualities
Occasionally, those with autism show incredible skills in areas such as math, music, or art. This may be due to a characteristic pattern of brain processing that focuses on details.

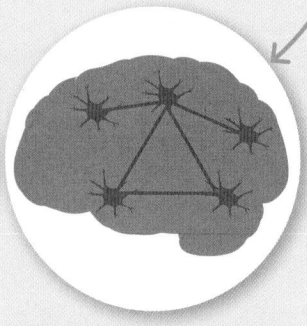

Increased connections
When any brain grows, nonessential nerve cell connections are pruned. It is thought that in autism, this process is inhibited, resulting in too many connections.

SENSORY SHORT-CIRCUITS

Some people have crossovers between their senses. Some see letters or numbers as colored while others might taste coffee when hearing a C-sharp. Their condition is called synesthesia and it happens because they do not undergo the same nerve-cell pruning process that other people do during their childhood brain development. The result is extra connections between the brain's sensory areas. Synethesia is thought to be genetic because it tends to run in families. However, since some identical twins have synesthesia while the other twin does not, genetics cannot be the entire story.

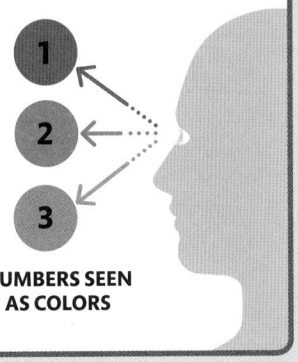

NUMBERS SEEN AS COLORS

Hallucinations

Hallucinations are surprisingly common; many recently bereaved people report seeing their spouse, and almost everyone has seen something nonexistent out of the corner of an eye. These are a normal by-product of our brains' attempts to make sense of the world.

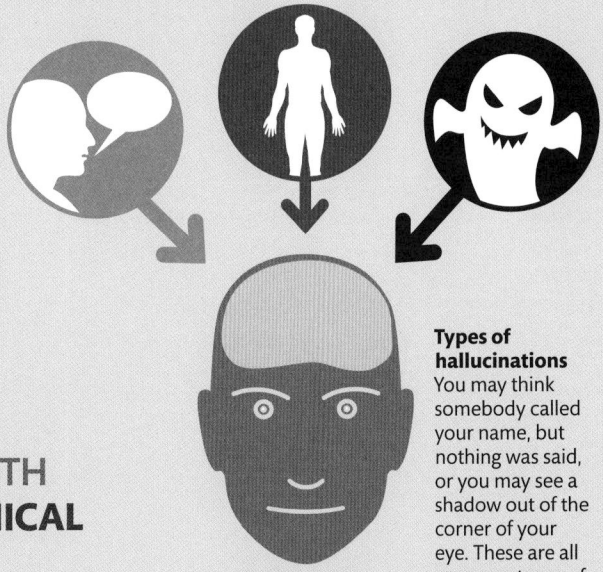

EXPERIENCING HALLUCINATIONS

Types of hallucinations
You may think somebody called your name, but nothing was said, or you may see a shadow out of the corner of your eye. These are all common types of hallucinations.

BY THE **AGE OF 5,** THOSE WITH **SUPERIOR AUTOBIOGRAPHICAL MEMORIES** START TO **REMEMBER EVERYTHING**

Memory champions

Some people have amazing memories, but they mostly use techniques such as placing the items that need to be remembered along a familiar route. A handful of people with a condition called superior autobiographical memory automatically remember every insignificant event that has happened to them for their entire lives. One individual with this condition had an enlarged temporal lobe and caudate nucleus—both areas of the brain that are linked to memory.

Memory pathway
If you need to remember a sequence of numbers, one way to do so is to associate each number with a place or object you see on your journey to work. Fitting a "3" in the window of a car or building, for instance, helps retain that number in place in the sequence.

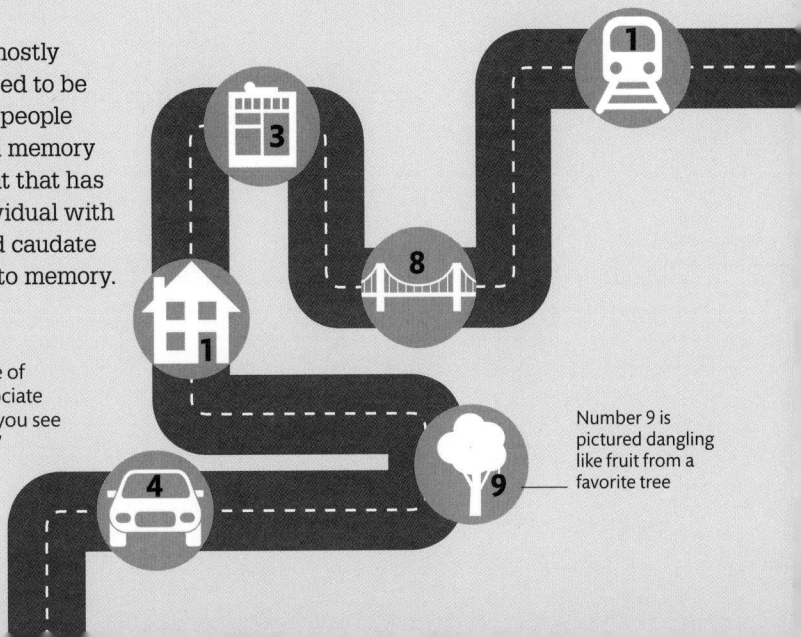

Number 9 is pictured dangling like fruit from a favorite tree

Index

A

acceleration sense 93
acetic acid 149
Achilles tendon 37, 54, 55, 71
acne 47, 223
actin 56–57, 131
adenine 16, 17
adenosine 234, 235
adenovirus 176
ADH (antidiuretic hormone) 152, 153
adipose cells 13
adrenal glands 191, 193, 238, 240, 241
aerobic respiration 130–31
afterimages 83
aging
  age spots 225
  bones 38–39, 48
  cell death 225
  DNA damage 26
  free radicals 224
  hair loss 225
  hearing loss 95, 225
  immune response 175
  memory loss 233
  nerve deterioration 225
  osteoarthritis 51
  osteoporosis 48, 50, 51
  skin 32, 225
  teeth 42
  telomere length 224
  vertebrae 71
  vision changes 86, 87, 225
  wound repair 175
airways 68, 112–13, 114, 118–19, 186; see also breathing
alcohol 61, 92, 152, 157, 196
aldosterone 199
alleles 17
allergies 186, 187
altitude effects 115
alveoli 113, 114–15
amino acids 14, 20–21, 138, 139, 144, 145, 157
amygdala 96, 97, 237, 239, 240
amylase 142, 144, 145
anaerobic respiration 131
anaphylaxis 186
angina 128
anosmia 96
antibiotics 48, 148, 173, 176
antibodies 169, 171, 172–73, 175, 178–79, 180
anticodons 20, 21
antidepressants 243
antifungal medications 177
antigens 170, 171, 178–79, 180–81
anus 147
anxiety 196
aorta 128
apoptosis 15, 27
appendix 147
armpits 172
arms
  bones 49, 54–55
  fetal development 212
  microbes 172
  muscles 54–55, 58–59
  nerve problems 70
  nonverbal communication 108
arrhythmias 129
arteries
  action/relaxation 68
  blockages 127, 128
  blood pressure 124, 125
  carrying oxygen-poor blood 114
  carrying oxygen-rich blood 124
  coronary 128
  exercise benefits 134
  hepatic 154–55
  menstrual cramps 206
  penile erection 204
  renal 150
  structure 124
arthritis 51, 186
aspergillus fungus 177
asthma 187
astigmatism 86
atherosclerosis 128
atrial fibrillation 129
auditory cortex 95
autism 107, 246
axons 62, 66, 78

B

B cells
  antibody production 175, 179, 180
  migration to lymph nodes 171
  self test 170, 171
  stimulation by T cells 178, 180
  types 178
babies
  birth 216–17
  birth weight and length 38
  bone growth 38
  breastfeeding 219
  cognitive development 218–19
  microbes 173
  senses 219
baby teeth 42
back pain 70
bacteria, friendly 146, 147, 148–49, 164, 165, 172–73
bacterial infections 43, 168, 174–75, 176, 178–79; see also infections
balance 92–93
baldness 225
belly button 172
bicarbonate 147
biceps 54, 55, 58, 59
bile 141, 144, 145, 154, 155, 157
binocular vision 84
bipolar disorder 243
birth 216–217
bite impression 42
bladder 69, 150, 151, 153, 215
blindspots 82
blisters 47
bloating 149, 164, 165
blood
  circulation through body 125
  circulation through heart 122–23
  components 120
  filtration 150–51, 154–55
  immune cells 174
  nutrient transport 145
  oxygen/carbon dioxide transport 114–17, 120–21
  production 120
  red blood cells 13, 16, 116–17, 120, 139
  supplies/waste transport 121
  volume 120
  water balance 152–53
blood clots 46, 49, 126–27
blood groups 171
blood pressure 124–25, 197
blood vessels
  action/relaxation 68–69
  at birth 217
  blockages 127
  blood flow 127, 128
  blood pressure 124–25, 197
  in bones 36
  coronary 128
  damage and repair 126–127
  fight or flight response 241
  healing 175
  immune response 174–75
  pregnancy effects 214
  water balance 152, 153
  see also arteries; capillary networks; veins
body language 108–109
body odor 245
body ownership sense 101
body position sense 100–101
bone marrow 24, 36, 120, 170, 178
bones
  aging 48, 50
  blood vessels 36
  calcium release 194
  calcium storage 195
  exercise benefits 50
  fractures 38, 48–49
  function 36
  growth 38–39, 223
  healing 49
  joints see joints
  osteoporosis 48, 50–51
  skeletal arrangement 37
  strength 36
  structure 36–37, 50

brain
- active areas 65
- alcohol, effects of 61
- autism 246
- automatic functions control 62
- baby's 218–19
- bipolar disorder 243
- body position sense 10, 101
- breathing feedback system 112
- child's 220–21
- circadian clock 198
- corpus callosum 64
- death 227
- default mode network 65
- depression 243
- development 218–21
- dreaming 236–37
- emotion formation 239
- energy supply 161
- exercise benefits 135
- fasting benefits 163
- fight or flight response 240–41
- gray matter 62
- hallucinations 247
- hemispheres 64, 76–77
- hormone production 238
- hunger/satisfaction sensing 140
- integrated senses 102–103
- itch response 79
- ketone body use 161
- learning 62, 230–31
- memory 231–33, 247
- nerve cell activity 65
- nerve cell number 65
- nerve pathways 65
- networks 65
- obsessive compulsive disorder (OCD) 242
- pain, sensing 78, 79
- phobias 242
- post-traumatic stress disorder (PTSD) 243
- pregnancy effects 214
- puberty effects 222
- reading faces 106
- sensory input/action output 60–61
- sexual attraction 244
- sight 84–85
- smell sense 97
- sound processing 90–91
- speech 104–105
- speed 60
- structure/components 62–63
- superior autobiographical memory 247
- suprachiasmatic nucleus (SCN) 198
- synesthesia 247
- thought processing 63
- touch sense 76–77
- water balance 152
- white matter 62

brainstem 61, 62, 63, 68, 79, 227, 237
breastfeeding 219
breasts 214, 222
breathing
- anaphylaxis 186
- asthma 187
- at birth 217
- during/after exercise 130–31, 132
- fight or flight response 241
- gas exchange 114–117
- hay fever 187
- mechanics 112–113
- pregnancy effects 214

bruises 126–127
burns 47
butterflies in the stomach 69
butyric acid 149

C

caffeine 152, 235
calcitonin 190, 195
calcium 36, 50, 51, 162, 190, 192, 194–95
calories 159–63
cancer 18, 27, 31, 180–81, 226
canine teeth 42–43
capillary networks
- blood pressure 124
- in broken bones 49
- diameter 117
- gas exchange 116, 117, 125
- in lungs 114, 115
- renal 150, 151
- in skin 30
- sphincters 125
- wound repair 174

carbohydrates
- diabetes 200
- digestion 144, 145
- irritable bowel syndrome (IBS) 165
- liver's use 156
- low-carbohydrate diet 161
- microbial digestion 149
- requirement 138

carbon dioxide 112, 115–17, 120–21, 125
cardiovascular exercise 132
cardiovascular system see blood vessels; heart
carpal tunnel syndrome 70
cataracts 87
cecum 146, 147
celiac disease 165
cell membrane 14, 15, 18, 139
cells
- aging 224, 225
- bacterial 148
- cell membranes 14, 15, 18, 139
- cellular respiration 116
- death 15
- diameter 15
- DNA damage, repair 26–27
- DNA molecules 16
- function 14–15
- gene expression 22–23
- hormone receptors 192, 193
- multiplication 18–19
- number 14
- self markers 170–171
- signaling 15
- structure/components 14–15
- types 13
- viral invasions 182

centrosomes 15
cerebellum 62, 63, 101, 237
cerebral cortex 62, 63, 101
cerebrum 62, 63, 76
cervix 205, 216–17
chemotherapy drugs 27
chewing 41, 44–45, 140, 142
childhood 38, 220–21, 231
chloride 146
choking 142
chromosomes
- DNA molecules 16, 17
- heredity 210–11
- meiosis 19
- mitosis 18
- telomeres 224
- X and Y 16, 23, 210–11

chyme 141, 142, 143, 144, 145
ciliary muscles 80, 81
circadian clock 198
circulatory system see blood vessels; heart
coccidiodes fungus 177
cochlea 88, 89, 91, 95
cochlear implants 95
codons 21
colds 182–183
collagen 32, 36, 41, 46, 49, 139
colon 144, 146–47, 148, 149, 165; see also intestines
color blindness 87
color consistancy 85
color vision 83
comas 227
communication
- nonverbal 108–109
- verbal 104–105

complement proteins 168
cones 82–83
connective tissue 13; see also ligaments; tendons
copper 157
cornea 25, 80, 86, 87
corpus callosum 64
corpus luteum 207
cortisol 109, 196, 198, 199, 238, 240, 241
coughing 118–119, 183
CPR (cardiopulmonary resuscitation) 129
cranium 37, 45
cryptococcus fungus 177
cupula 92
cytokines 174–75, 178, 181, 183
cytoplasm 14, 15, 20–21, 24
cytosine 16, 17

D

daily rhythms 198-199
deafness 94-95
death 226-27
deep vein thrombosis 126
defibrillation 129
dehydration 131, 152
deltoid muscle 59
dendritic cells 168
deoxyhemoglobin 117
depression 243
dermis 31, 32, 47, 74-75, 174-75
diabetes 200-201, 226
dialysis 151
diaphragm 112-13, 118-19
diarrheal diseases 226
diets 161-63, 200
digestive system 10-11
  action/relaxation 68-69
  beneficial microbes 148-49
  bile 141, 144, 145, 155, 157
  bloating 149
  chewing and swallowing 142-43
  colon 144, 146-47, 148, 149, 165
  enzymes 141, 142, 144-45
  esophagus 12, 140, 141, 142-43, 165
  flatulence 149
  food passage hours 140-41
  gas production 149
  hunger/satisfaction hormones 140, 198
  intestines see intestines
  large intestine 141, 146-47; see also colon; intestines
  liver see liver
  mouth 140
  nutrient absorption 145
  problems 141, 164-165
  small intestine 141, 144-45; see also intestines
  stomach 141, 143; see also stomach
  structure/components 142-47
  vomiting 165
direction sense 92
diseases
  death from 226
  from DNA mutations 18, 26, 27
  infectious diseases 176-77
  vaccines 184-85
  see also infections
dislocations 49, 70
dizziness 92-93
DNA (deoxyribonucleic acid)
  bases 16, 17
  chemotherapy drugs 27
  copying 20
  damage and repair 26-27
  double helix 16-17
  exchange and recombination 19
  free radicals 224
  functions 16
  gene expression 17, 22
  gene therapy 27
  mutations 26
  replication and division 18
  transcription 20
dopamine 197, 238, 244
double-jointed people 41
Down syndrome 19
dreaming 234, 236-37

E

eardrum 88, 94
ears
  aging 95, 225
  balance function 92-93, 100
  damage and disorders 94-95
  hair cells 13
  joints 41
  localizing sound 90-91
  ringing 95
  sound reception 88-89
  structure/components 36, 88-89, 92-93
eating 140, 142-43, 198; see also digestive system
eczema 186
eggs 13
  fertilization 205, 208-209
  heredity 210-11
  menstrual cycle 206-207
  production 19
  size 205
elastin 32
elbows 37, 54, 55, 71
electrocardiograms (ECGs) 122-23
embryos 212-13
emotions
  basic 238
  development 220
  in dreams 237
  facial expressions 107
  formation 239
  hormones 238-39
  and music 91
  problems/disorders 242-43
  purpose 239
  sexual attraction 244-45
  and smells 97
endocrine system 10-11
  adrenal glands 191, 193, 238, 240, 241
  hypothalamus 140, 152, 190, 193, 222
  ovaries 191, 192, 223, 238
  pancreas 140, 141, 144, 158, 159, 191, 200-201
  pituitary gland 152, 190, 193, 206, 207, 240, 241
  testes 191, 205, 223, 238
  thyroid gland 190, 195, 199
  see also hormones
energy
  action/relaxation 68
  balance 158-59, 190
  cellular respiration 116
  fatty acids 158-59
  fight or flight response 241
  glucose 156, 157, 158-59, 200
  from microbial digestion 149
  from nutrients 138, 139
  sugar trap 160-61
  use during exercise 130-31, 135
epidermis 31, 32, 47, 74-75, 174-75
epiglottis 119, 142
epinephrine
  anaphylaxis, treating 186
  circulation in blood 120
  fight or flight response 193, 238, 240, 241
  production 191
  sexual attraction 244
epithelial cells/tissue 13, 168
esophagus 12, 140, 141, 142-43, 165
estrogen
  functions 192
  menstrual cycle 206-207
  osteoporosis 50
  during pregnancy 214, 215
  production 191, 192, 223
excretory system 141, 146-47, 150-51
exercise
  aerobic respiration 130-31
  anaerobic respiration 131
  benefits 134-35
  bone production 50
  cardiovascular 132
  fat-burning 133, 163
  fitness level 134-35
  heart rate 133
  hormonal benefits 197
  hydration 131
  intensity 133
  lactic acid levels 130-31
  resistance training 133
  runner's high 239
  timing 163
  VO2 max 134-35
  warm up 58
eyelashes 81, 139
eyelids 81
eyes
  afterimages 83
  aging 86, 87, 225
  astigmatism 86
  balance function 93
  binocular vision 84
  blindspots 82
  body position sense 100
  building from nutrients 139
  cataracts 87
  correction reflex 93
  damage and disorders 86-87
  farsightedness 86
  fight or flight response 241
  focus 81

glaucoma 87
image capture 80
images, forming 82–83
light spots 82
lubrication 81
nearsightedness 86
protection 81
pupil control 68–69
rods and cones 82–83
structure/components 80
vision tests 87
see also sight

F

facial expressions 106, 107, 108, 109, 239
facial recognition 106
fascicles 56
fasting 162–63
fat
burning 133, 163
metabolism 15, 156
production 15
storage 158, 160
fat, dietary 138, 145
fat tissues 140
fatty acids 145, 158–59, 161
fear 242
feces 141, 146–47
feelings see emotions
feet 36, 37, 173
femur 37, 51
fertility treatments 209
fetuses 212–13
fever 176, 182, 183
fibula 36, 37
fight or flight response 191, 193, 240–41
fingerprints 32
fingers 55
fingertips 75, 76
fitness level 134–35
flatulence 149
flicker fusion 85
flu 182–83
follicle-stimulating hormone (FSH) 206–207
food
allergies 186
cravings 214
nutrient requirements 138–39
processing see digestive system; eating
unhealthy choices 196
fractures 38, 48–49
freckles 33
free radicals 224
friendships 221
fungal infections 168, 177
fungi, friendly 172–73
funny bone 37, 70, 71

G

galactose 164
gallbladder 144, 155, 157
gastrin 191
gastrointestinal tract 138, 140–41; see also intestines; stomach
gene swapping 19
gene therapy 27
genes
alleles 17
complement at fertilization 208, 210
dominant and recessive forms 211
expression 17, 22–23
function 20–21
gender determination 23
gender-linked 211
gene swapping 19
heredity 210–11
length 16
mutations 21
number 16, 17
transcription 22
translation 20–21
genitals, female
menstrual cycle 206–207
microbes 172
pregnancy 208–209
sperm travel 205
structure/components 205, 207
genitals, male 172, 204–205
ghrelin 140, 198
giardia 177
glaucoma 87
glucagon 159, 191, 193
glucose
balance 158–59
conversion from lactic acid 121
conversion from lactose 164
diabetes 200–201
fight or flight response 241
glycemic index (GI) 163
hormones regulating 191, 193, 196
lifestyle impacts 196–97
production in liver 68, 156, 157
sugar trap 160–161
use during exercise 130, 131
gluten 165
gluten intolerance 165
glycemic index (GI) 163
glycogen 157, 159, 193
Golgi apparatus 14
gonadotrophin-releasing hormone 222
goose bumps 30, 31
granulocytes 46, 169, 174–75
gravity sense 93
grayscale vision 83
grinding teeth 43
growth charts 39
growth hormone 38, 193, 197, 199
guanine 16, 17

H

hair, body 30, 74, 222
hair cells, ear 13, 89, 92, 93, 95
hair, scalp 34–35, 225
hallucinations 247
hands 36, 55, 75, 109, 172
hay fever 187
headaches 63
hearing
aging 95, 225
cochlear implants 95
hair cells 13, 89, 92, 93, 95
localizing sound 90–91
pitches 89
problems 94–95
sound processing 90–91
sound reception 88–89
heart
arrhythmias 129
arteries 128
beating 122–23
blood circulation 122–23
blood flow 121, 128
cardiovascular exercise 132
cycle 122–23
damage 128
defibrillation 129
ECG recording 122–23
exercise benefits 135
fetal growth and development 212
hormone production 191
problems 128–29, 226
sinoatrial node/pacemaker 122, 123
structure 122–23
tissue repair 25, 128
volume of blood 123
heart attacks 127, 128, 226
heart disease 226
heart rate
action/relaxation 68–69
cardiovascular exercise 132
exercise benefits 135
exercise intensity 133
fight or flight response 241
sexual attraction 244
heartbeat 123, 128, 129
heartburn 143
height 39
Helicobacter pylori 148
hemoglobin 116–17, 127, 139, 157
heparin 156
herd immunity 184
heredity 17, 210–211
herpesvirus 176
hiccups 119
high blood pressure 226
high-protein diet 161
higher brain 62, 63
higher cortex 79
hippocampus 233, 240
histamines 46, 174–175, 183, 186, 187

HIV (Human Immunodeficiency Virus) 176, 226
hormone therapy 193
hormones
adenosine 234, 235
ADH 152–153
aldosterone 199
balance 194–95
calcitonin 190, 195
cortisol 109, 196, 198, 199, 238, 240, 241
daily rhythms 198–99
dopamine 197, 238, 244
emotions 238–239
epinephrine see epinephrine
exercise benefits 197
fat-insoluble 193
fat-soluble 192
feedback loop 194–95
"feel good" 197
fight or flight response 240
follicle-stimulating hormone (FSH) 206–207
functions 192–193
gastrin 191
ghrelin 140
glucagon 159, 191, 193
gonadotropin-releasing hormone 222
growth hormone 38, 193, 197, 199
human chorionic gonadotropin (hCG) 215
hunger/satisfaction 140, 198
estrogen see estrogen
insulin 149, 158, 160–61, 191, 196, 197, 200–201
fatigue 234
leptin 140, 198, 222
lifestyle impacts 196–97
long-term pair bonding 245
melatonin 190, 196, 198–99
menstrual cycle 206–207
muscle growth 133
norepinephrine 243
oxytocin 192, 197, 238, 245
placental 215
production 15, 157, 190–91
progesterone 191, 206–207, 214, 215
PTH 192, 194, 195
puberty 222–223
receptors in cells 192, 193
relaxin 153, 216
serotonin 79, 135, 198–199, 238, 243, 244
sex hormones 38
sexual attraction 244
stress-management 198, 199
testosterone see testosterone
thyroid-stimulating hormone 199
triggers 192–193
vasopressin 245
housekeeping proteins 23
human chorionic gonadotropin (hCG) 215
human genome 16
humerus 36, 49, 54, 55
hunger 140, 198
hydration 131, 152–53
hypodermis 31, 32, 47
hypothalamus 140, 152, 190, 193, 222

I

immune system
adaptive immune system 169
aging 175
antibodies 172–73, 178–79
B cells see B cells
barriers to pathogens 168
calming 173
cancer, fighting 180–81
fight or flight response 241
innate immune system 169
lifestyle impacts 196–197
memory 169
problems 186–87
response to bacterial infections 174–75
response to viruses 183
Rhesus babies 179
self/nonself distinction 170–71
T cells see T cells
immunocompromised conditions 187
immunotherapy 180–81
incisors 42–43
incus (anvil) bone 88
indigestion 143
infections
bacterial 174–75, 176
barriers to pathogens 168–69
in broken bones 48
defenses against 31, 168–69
in ears 94
fungal 177
immune response 174–75, 180–81, 183
infectious diseases 176–77
necrosis of cells 15
protozoan 177
spread 177
in teeth 43
testing for antibodies 178
types 168
vaccines against 184–85
viruses 176, 182–83
infertility 209
influenza 182–83
injuries 70–71
insulin 149, 158, 160–61, 191, 196, 197, 200–201
intermittent fasting 162
intestines
action/relaxation 68–69
calcium absorption 194, 195
celiac disease 165
digestion 141, 144–47
gas production 149
hormone production 238
irritable bowel syndrome (IBS) 165
lactose intolerance 164
microbes 148, 172
nutrient absorption 149
touch sense 74
iris 80, 81
iron 116, 117, 139, 149, 157
irritable bowel syndrome (IBS) 165
itching 46, 79

J

jaws 41, 44–45, 101
jet lag 199
joints
dislocations 49, 70
double-jointed people 41
flexion and extension 54
function 40
osteoarthritis 51
position receptors 101
replacement 51
rheumatoid arthritis 186
sprains 71
structure/components 40
types 40–41

K

keratin 21, 32, 34–35, 139
ketone bodies 161
kidney stones 150
kidneys
calcium excretion 195
filtration 150–51
hormone production 191
structure/components 150–51
vitamin D activation 194
water balance 153

L

lactase 164
lactic acid 43, 57, 121, 130–31
Lactobacillus 148
lactose intolerance 164
larynx 104, 222
laughter 128
laxatives 141
learning 62, 230–31
left-handedness 64
legs 37, 41, 46, 127, 173, 212
lens 80, 81, 86, 87, 201
lentivirus 176
leptin 140, 198, 222
levers, body 55
lifestyle 196–197
ligaments 37, 40, 41, 49, 70, 71, 216
light spots 82
lipase 143, 145
liver
action/relaxation 68–69
bile 141, 144, 155, 157

blood supply 154
damage 157
fat metabolism 156
filtration speed 155
functions 156–157
gluconeogenesis 156
glycogen storage 157
heat generation 157
hormone production 157
ketone body production 161
manufacturing function 157
mineral storage 157
nutrient processing 154–55, 156, 157
protein synthesis 157
regeneration 156
storage function 157
structure/components 154–55
toxin detoxification 156
vitamin storage 157
lock and key theory 97
low-carbohydrate diet 161
lungs
action/relaxation 68–69
allergy response 187
asthma 187
breathing 112–13
cancer 226
capacity 132
coughing 118–19
disease 226
exercise benefits 134
fetal growth and development 212
gas exchange 114–15, 120
infections 226
structure/components 112–13, 114–15
surface area 113
lust 244
lying 109
lymph system 170, 171, 178
lysavirus 176

M

macrophages 126, 127, 169, 174–75, 178, 179, 180
maggot therapy 175
malleus (hammer) bone 88
mammary glands 172
mast cells 169, 175
meiosis 19
melanin 31, 33, 34, 35, 225
melatonin 190, 196, 198–199
memory 231–233, 237, 243, 247
menopause 206
menstrual cramps 206
menstrual cycle 206–207, 223, 245
messenger RNA (mRNA) 14, 20, 21
microbes
friendly 146, 148–49, 164, 172–73
harmful 148, 172–173; see also infections
microtubules 15
milk teeth 42
minerals
calcium 36, 50, 51, 162, 190, 192, 194–95
chloride 146
copper 157
iron 116, 117, 139, 149, 157
microbial digestion 149
phosphate 16, 17, 36
potassium 147
requirement 138, 139
sodium 146, 147
storage in liver 157
mirror box therapy 100
mitochondria 14, 15
mitosis 18
molars 42, 43
morning sickness 215
motor cortex 105, 236, 237, 239
mouth
eating 140, 142
microbes 172
speech 104
teeth 42–43
temporomandibular joints 41
tongue see tongue
muscles
aerobic respiration 131
anaerobic respiration 132
arm muscles 54–55
body heat generation 59
body levers and fulcrums 55
breathing 132
contraction/relaxation 57, 58–59
control by nervous system 60–61
cramps 57
energy use 158–59
exercise benefits 135
eye muscles 80, 81
facial expressions 107
flexion and extension 54–55
growth 133
hair erector muscles 30
jaw muscles 44–45
lactic acid 131
muscle fibers 56
muscle tissue 13
paralysis during sleep 234, 237
pulling and lengthening 59
pulling and shortening 58
pulling without shortening 59
reflex reactions 60–61
resistance training 133
skeletal muscle cells 13
slow-twitch/fast-twitch fibers 57
smooth muscle cells 13
sprains and strains 70, 71
stretch receptors 100
structure/components 56
teamwork 54
tendons, control by 55
tension 58–59
warm up, exercise 58
weight lifting 58–59
music 91
myosin 56–57, 131

N

nails 34, 35
natural killer (NK) cells 169
near-death experiences 227
nearsightedness 86
necrosis 15
nematodes 177
nerve bundles 62
nerve cell bodies 62, 67
nerve endings
in clitoris 204
in ear hair cells 95
in hands/fingers 75, 76
in nasal passages 118
in skin 30, 31, 74, 78
on the tongue 99
nerves/nervous system 10–11
action/relaxation 68–69
aging 225
autonomic 69
in the brain 64, 65
brain see brain
breathing feedback system 112
fight or flight response 240–41
hearing 89
injuries 70
learning 230–31
memory 231, 232, 233
message transmission 64, 66–67
nerve cells 13
nervous tissue 13
pain, sensing 78–79
pins and needles 66
sensory input/action output 60–61
sight 82–83
smell sense 97
speech 105
speed 66
spinal cord see spinal cord
structure/components 66–67
sympathetic and parasympathetic 69
touch sense 74–75
see also nerve endings
nervous tissue 13
neurons 66–67, 90, 91, 163, 197, 234
neurotransmitters 67, 89, 135, 197, 230, 235, 237
nonverbal communication 108–109
norepinephrine 243
nose
bone and cartilage 48
breaking 48
cold and flu symptoms 183
microbes 172

sense of smell 96–97
sneezing 118
snoring 118
nucleus, cell 14, 20
nutrition 138–39, 140, 149, 219; see also digestive system; eating

O

obesity 160
obsessive compulsive disorder (OCD) 242
olfactory receptors 97
opioids 79, 239
optic nerve 80, 81, 87
organs 12, 212–13; see also specific organ
orgasms 204
ossicle bones 88, 94
osteoarthritis 51
osteoporosis 48, 50–51
ovaries
hormone production 191, 192, 223, 238
menstrual cycle 206–207
oxygen
altitude effects 115
circulation 115, 116–17, 120–21, 128, 134
daily amount used 116
energy function 116, 130
gas exchange 115, 117
heart's requirement 128
oxygen intake
action/relaxation 68–69
breathing feedback system 112
during/after exercise 130–31, 132
exercise benefits 134
extraction in alveoli 114
VO2 max 135
oxyhemoglobin 117
oxytocin 192, 197, 238, 245

P

pacemaker 122–23
pain
back pain 70
nerve pinching 70
purpose 78
referred 78
reflex reactions 60–61
sensing 78–79, 99
strains and sprains 71
whiplash 71
painkillers 78, 79
paleolithic diets 162–63
pancreas 140, 141, 144, 158, 159, 191, 200–201
parasites 168, 176–177
parathyroid glands 190, 192, 194
pathogens 168, 176–77, 180–181, 184; see also infections
pelvis 37
penis 204–205
perspective 85
phagocytes 168, 169, 179
phantom limb pain 100
phobias 242
phosphate 16, 17, 36
photoreceptor cells 13
pineal gland 190
pins and needles 66
pituitary gland 152, 190, 193, 206, 207, 240, 241
placenta 179, 208, 209, 212–13, 215, 216, 217
post-traumatic stress disorder (PTSD) 243
potassium 147
prefrontal cortex 236, 243, 244
pregnancy
birth 216–17
changes to mother's body 214–15
fertilization 208–209
fetal growth and development 212–13
food cravings 214
length 217
morning sickness 215
Rhesus babies 179
stretch marks 215
premolars 42
PRICE technique 71
primitive brain 62, 63, 68–69
probiotics 149
progesterone 191, 206–207, 214, 215
propyonic acid 149
prostaglandins 78
protease 145
protein, dietary 138, 161
proteins
functions 16
housekeeping 23
packaging in vesicles 14
production in liver 157
transfer out of cell 14
translation 14, 16, 20–21
protozoan infections 177
PTH (parathyroid hormone) 192, 194, 195
puberty 222–223
pupils 68, 81, 244

R

radius 49, 54
reading faces 106–107
rectum 147
red blood cells 13, 16, 116–17, 120, 139
reflex reactions 60, 218
relaxin 153, 216
REM (rapid eye movement) sleep 234, 235, 236, 237
repressor protein 22
reproductive system
fertilization 208–209
gender determination 23
genitals, female 205, 207
genitals, male 204–205
heredity 210–11
menstrual cycle 206–207
reproductive cells 13
sexual intercourse 204–205
resistance training 133
respiration see breathing
respiratory system 10–11
lungs see lungs
trachea 112, 114
retinas 80, 81, 82, 83, 86, 87, 93
Rhesus factor (Rh) 179
rheumatoid arthritis 186
ribosomes 14, 20–21
right-handedness 64
rigor mortis 227
RNA polymerase 22
road accidents 226
rods 82–83
rough endoplasmic reticulum 14

S

saccule 93
sacrum 37
salivary glands 142
salmonella bacteria 176
sarcomeres 57
scabs 46
scars 47
scratching 79
sebum oil 31, 47
semen 204–205
senses
baby's 219
body ownership 101
body position 100–101
in dreams 236
fetal development 213
hearing see hearing
integrated 102–103
memory 231
sight see sight
smell 96–97
taste 98–99
touch 74–77, 99
sensory cortex 76–77
sensory homunculus 76
serotonin 79, 135, 198–99, 238, 243, 244
sesamoid bones 36
sex 204–205
sex determination 16
sexual attraction 244–45
shoulder blades 37, 54, 55
sight
3-D vision 84
aging 86, 87, 225
astigmatism 86
binocular vision 84
brain processing 84–85
bright and dim light 81
cataracts 86
color blindness 87

color consistancy 85
color vision 83
damage and disorders 86–87
development 219
flicker fusion 85
focus 81
glaucoma 86
grayscale vision 83
image formation 81, 82–83
long-sightedness 86
nerve signals 83
nutrition for 139
perspective 85
photoreceptor cells 13
short-sightedness 86
tunnel vision 240
vision tests 87
see also eyes
singing 105
sinoatrial node 122–23
skeletal system see bones
skin
acne 47, 223
aging 32, 225
area 31
blisters 47
burns 47
color 33
damage 46–47
as defensive barrier 31, 32–33
eczema 186
functions 30–31
healing 46–47
layers 31, 32
microbes 173
permeability 31
scabs 46
scars 47
self-renewal 32
stretch marks 215
stretch receptors 101
structure/components 30–31, 32
temperature control 30
touch sense 74–75
vitamin D synthesis 33
weight 12
wound response 174–175
skull 37
sleep
body paralysis 234, 237
circadian clock 198–99
deprivation 235
dreaming 236–37
levels 234, 235
melatonin 196, 198–99
memory consolidation 237
REM (rapid eye movement) sleep 234, 235, 236, 237
sleep pressure 234
sleep talking 236
sleepwalking 234, 236
sleep deprivation 235
sleep talking 236
sleepwalking 234, 236
slipped discs 70
smell, sense of 96–97
smoking 196
smooth endoplasmic reticulum 15
smooth muscle 12, 13, 124, 125
sneezing 118, 176
snoring 118
sodium 146, 147
spatial awareness 237
speech 101, 104–105
sperm 13
ejaculation 204, 205, 223
fertilization of egg 208–209
heredity 210–211
journey to fallopian tubes 205
production 19, 223
size 205
spinal cord
message transmission 62, 68, 69, 78, 79, 237
reflex reactions 60, 61
sensory input/action output 60, 61
spine 36, 50, 70, 71, 212, 215
sprains 71
stapes (stirrup) bone 36, 88
stem cells 23, 24–25, 170, 217
stomach
action/relaxation 68–69
butterflies 69
digestion 141, 143
gastric juices 143
hormone production 140, 191
microbes 148
pregnancy effects 215
structure/components 12, 143
ulcers 148
strains 71
streptococcus bacteria 176
stress 198, 199, 241
stretch marks 215
strokes 226
sugar trap 160–161
sugars 57, 68, 116, 138, 139, 144–45, 147; see also glucose
superman pose 109
suprachiasmatic nucleus (SCN) 198
swallowing 142
sweat glands 30
synaesthesia 247
synovial fluid 40
systems 10–11

T

T cells
B cell stimulation 178, 180
bactericidal chemicals 175
cancer, fighting 180–181
count 181
maturation/release 171
migration to lymph nodes 171
production 170, 190
self test 171
types 181
taste 98–99
taste buds 98
tattoos 32
tears 81
teenagers 38, 222–223
teeth 42–43
telomeres 224, 225
temperature, control 30
temperature, sensing 77, 99
temporomandibular joints 41, 44–45
tendons
Achilles tendon 37, 54, 55, 71
in fingers 55
function 54
length 54, 55
sesamoid bones 36
sprains and strains 71
temporalis tendon 44
tension receptors 100
testes 191, 205, 223, 238
testosterone
emotions 238
exercise benefits 197
nightly surge 198
production 191, 223
sexual attraction 244, 245
superman pose 109
thalamus 62, 63, 79, 240, 243
theory of mind 220
thymine 16, 17
thymus 171, 190
thyroid gland 190, 195, 199
thyroid-stimulating hormone 199
tibia 37
tickle response 77
tiredness 234–235
tissues
stem cell engineering 25
transplants 25
types 13
tongue
body position sensors 101
eating 142
other receptors 99
sensitivity 76
snoring 118
speech 104
taste receptors 98–99
touch 74–77, 99
toxins
damage from 15, 26, 168
detoxification in liver 156
released as defense 148
trachea 112, 114
transfer RNA (tRNA) 20, 21
transplants 25, 170
treponema bacteria 176
triceps 54, 55
trichomonas 177
turning sense 92
twins 170, 208, 209

U

ulcers 148
ulna 49, 55
ulnar nerve 37, 70, 71
ultraviolet light 26, 31, 33
urea 150, 151
ureters 150, 151, 153
urinary system 10–11, 150–53
  bladder 69, 150, 151, 153, 215
  kidneys see kidneys
  ureters 150, 151, 153
urine 150–51, 153, 195, 212
uterus
  birth 216–17
  embedding of fertilized egg 208, 209
  expansion during pregnancy 215
  menstrual cycle 206, 207
utricle 93

V

vaccines 176, 180–81, 184–85
vagina 204–205, 223
varicose veins 127
vasopressin 245
veins
  blood pressure 124
  carrying oxygen-poor blood 116, 125
  carrying oxygen-rich blood 114
  deep vein thrombosis 126
  hepatic 154–55
  penile erection 204
  renal 150, 151
  varicose veins 127
vertebrae 50, 70
vesicles 14, 15
vibrio bacteria 176
viruses
  cell invasion 168, 182
  cold and flu 182–83
  friendly 172–73
  gene therapy 27
  symptoms 183
  types 176
  vaccines 176, 184–85
vision see sight
vision tests 87
visual cortex 85, 237, 240
vitamins
  microbial production/secretion 149
  requirement 138
  storage in liver 157
  vitamin A 139, 155, 157
  vitamin B 146, 149, 157
  vitamin C 126, 138
  vitamin D 30, 31, 33, 36, 50, 157, 194, 195
  vitamin E 157
  vitamin K 126, 146, 149, 157
VO2 max 135
vocal chords 104
vomiting 165

W

water
  absorption 146
  balance 152–53
  elimination 151
  requirement 138, 139
weight 158, 159, 160–63, 196
whiplash 71
wisdom teeth 42, 43
wounds 46–47, 174–75

Y

yawning 119

Z

zoonotic diseases 177

Acknowledgments

DK would like to thank the following people for help in preparing this book: Amy Child, Michael Duffy, Jon Durbin, Phil Gamble, Alex Lloyd, and Katherine Raj for design assistance; Nadine King, Dragana Puvacic, and Gillian Reid for pre-production assistance; Caroline Jones for indexing; and Angeles Gavira Guerrero and Andy Szudek for proofreading.

The publisher would like to thank the following for their kind permission to reproduce their images:

p.85: Edward H Adelson
p.87: Photolibrary: Steve Allen

For further information see:
www.dkimages.com